TRANSPORT AND CLIMATE CHANGE

TRANSPORT AND SUSTAINABILITY

Series Editors: Stephen Ison and Jon Shaw

Recent Volume:

Volume 1: Cycling and Sustainability – Edited by John Parkin

TRANSPORT AND SUSTAINABILITY VOLUME 2

TRANSPORT AND CLIMATE CHANGE

EDITED BY

TIM RYLEY
Loughborough University, Loughborough, UK

LEE CHAPMAN
University of Birmingham, Birmingham, UK

Emerald

United Kingdom – North America – Japan
India – Malaysia – China

Emerald Group Publishing Limited
Howard House, Wagon Lane, Bingley BD16 1WA, UK

First edition 2012

Reprints and permission service
Contact: booksandseries@emeraldinsight.com

British Library Cataloguing in Publication Data
A catalogue record for this book is available from the British Library

ISBN: 978-1-78052-440-5
ISSN: 2044-9941 (Series)

ISOQAR REGISTERED
UKAS MANAGEMENT SYSTEMS
0026
Certificate Number 1985
ISO 9001
ISO 14001

ISOQAR certified Management Systems, awarded to Emerald for adherence to Quality and Environmental standards ISO 9001:2008 and 14001:2004, respectively

INVESTOR IN PEOPLE

CONTENTS

LIST OF CONTRIBUTORS

Erel Avineri	Centre for Transport and Society, University of the West of England (UWE), Bristol, UK
Ian Bache	Department of Politics, University of Sheffield, Sheffield, UK
David Banister	Transport Studies Unit, Oxford University Centre for the Environment, Oxford, UK
Abigail L. Bristow	Transport Studies Group, School of Civil and Building Engineering, Loughborough University, Loughborough, UK
Lucy Budd	Transport Studies Group, School of Civil and Building Engineering, Loughborough University, Loughborough, UK
Lee Chapman	School of Geography, Earth and Environmental Sciences, University of Birmingham, Birmingham, UK
Alison Chisholm	Department of Planning, Oxford Brookes University, Oxford, UK
Kevin Cullinane	Transport Research Institute, Edinburgh Napier University, Edinburgh, UK
Robert Dingwall	Dingwall Enterprises, Nottingham, UK
Murray Goulden	Horizon Digital Economy Research Institute, Nottingham University, Nottingham, UK
Helen Harwatt	Sustainability Research Institute, University of Leeds, Leeds, UK
Robin Hickman	Bartlett School of Planning, University College London, London, UK

Elizabeth Hooper School of Geography, Earth and Environmental Sciences, University of Birmingham, Birmingham, UK

Dave Horton Lancaster Environment Centre, Lancaster University, Lancaster, UK

Candice Howarth Global Sustainability Institute, Anglia Ruskin University, Cambridge, UK

David Jaroszweski School of Civil Engineering, University of Birmingham, Birmingham, UK

Tim Jones Faculty of Technology Design and Environment, Oxford Brookes University, Oxford, UK

Ann Jopson Institute for Transport Studies, University of Leeds, Leeds, UK

Charlotte Kelly Institute for Transport Studies, University of Leeds, Leeds, UK

Karen Lucas Transport Studies Unit, University of Oxford, Oxford, UK

Glenn Lyons Centre for Transport and Society, University of the West of England (UWE), Bristol, UK

Greg Marsden Institute for Transport Studies, University of Leeds, Leeds, UK

Caroline Mullen Institute for Transport Studies, University of Leeds, Leeds, UK

Kate Pangbourne Geography and Environment, University of Aberdeen, Aberdeen, UK

Colin G. Pooley Lancaster Environment Centre, Lancaster University, Lancaster, UK

Tim Ryley Transport Studies Group, School of Civil and Building Engineering, Loughborough University, Loughborough, UK

Griet Scheldeman Lancaster Environment Centre, Lancaster University, Lancaster, UK

Miles Tight	School of Civil Engineering, University of Birmingham, Birmingham, UK
Owen Waygood	Centre for Transport and Society, University of the West of England (UWE), Bristol, UK
Alberto M. Zanni	Transport Studies Group, School of Civil and Building Engineering, Loughborough University, Loughborough, UK

ACKNOWLEDGEMENTS

The editors, Dr Tim Ryley and Dr Lee Chapman, would like to thank the chapter authors for their contributions to this book, together with the two anonymous reviewers for each chapter. They would also like to acknowledge the valuable guidance of the *Transport and Sustainability* series editors, Professor Stephen Ison and Professor Jon Shaw. Many of the contacts developed for this book project came initially from the EPSRC FUTURENET (Future Resilient Transport Networks) project (both editors are co-investigators of this project) and the 'Transport and Climate Change' session within the RGS-IBG Conference in London, September 2010. Individuals involved in the FUTURENET project and the RGS-IBG 'Transport and Climate Change' sessions are also gratefully acknowledged by the editor.

Finally, both editors would like thank their families for their support during the development of the book project.

CHAPTER 1

INTRODUCTION

Tim Ryley and Lee Chapman

BACKGROUND

Climate change is recognised as one of the greatest challenges that contemporary global society faces. The IPCC (Intergovernmental Panel on Climate Change) Fourth Assessment Report (2007) states that it is 'very likely' that anthropogenic global warming will result in a temperature rise of between 1.8°C and 4°C by the end of the 21st century. Temperatures at the upper end of this range are considered 'dangerous', and the international community is focused on attempting to limit the increase to within 2°C (Meinshausen et al., 2009). Increasing global temperatures are just one consequence. The world will face an increasing level of unpredictable and extreme weather patterns, each with different, but in many cases, serious consequences for life on earth (IPCC, 2007).

Greenhouse gas emissions, particularly carbon dioxide (CO_2), are the underlying root cause of climate change and there is a desperate and urgent need to decarbonise the global economy in order to avoid catastrophic climate change (Fox & Chapman, 2011). Transport accounts for 23% of global CO_2 emissions (IEA, 2011) with the majority of emissions originating from industrialised countries. Transport is one of the few sectors where emissions have continued to grow, fuelled by increasing demand, particularly for high polluting forms of travel such as by car and by aeroplane; even though emissions are set to fall for transport in the United Kingdom, concerns remain that they will not reduce by as much as other

Transport and Climate Change
Transport and Sustainability, Volume 2, 1–8

ISSN: 2044-9941/doi:10.1108/S2044-9941(2012)0000002004

sectors (see AEA Group, 2011; Department for Transport, 2007 for further information).

There is a pressing requirement for transport to begin to reduce emissions, as with other sectors, to help *mitigate* the scale of climate change (Chapman, 2007). However, this is only half of the problem. Due to historic emissions and the long atmospheric residence time of CO_2 (Penner, Lister, Griggs, Dokken, & McFarland, 1999), the world is already committed to some degree of climate change (at best 2°C), and the future climate will impact upon transport networks which will need to be *adapted* to cope (Fig. 1). Using UK climate change scenarios as a central theme, this book analyses both mitigation and adaptation in the transport sector from a range of perspectives. Indeed, the UK was the first country to make a legally binding commitment to cut greenhouse gases via the UK Climate Change Act (DEFRA, 2008). This makes the UK a logical and unique place to study and this book documents the early progress and challenges made in implementing this act in the transport sector.

Overall, given the twin challenges associated with the increased importance of dealing with issues associated with climate change and the increased focus

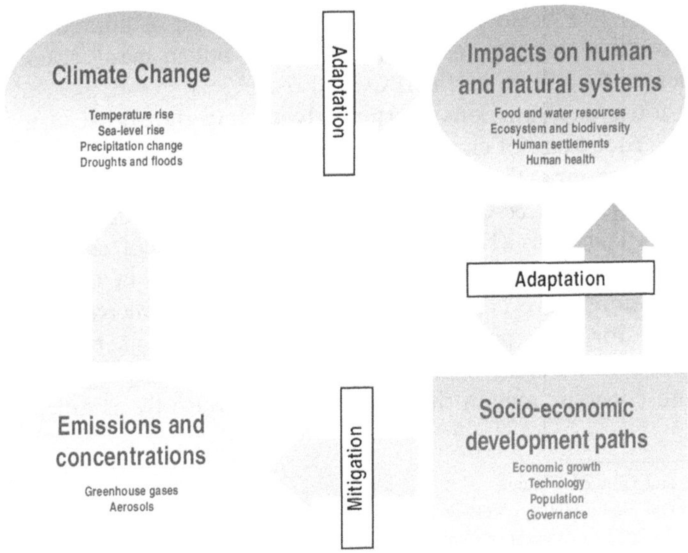

Fig. 1. The Climate Change Integrated Framework. *Source*: Reproduced with permission from: Climate change 2001: Synthesis report (IPCC, 2001).

within this on transport mitigation and adaptation, this book aims to provide a timely and original contribution to the issues surrounding the relationship between transport and climate change.

SCENARIOS

Predicting the exact nature of the impacts of climate change is a complicated and inexact science. Current predictions indicate that the 2°C target will be breached by 2050, but realistically, overall warming will be far greater depending on the speed that society reacts to reduce emissions (Anderson & Bows, 2008). This uncertainty presents a challenge in determining the exact impacts of climate change. As a result, meteorologists need to consider several different scenarios, essentially storylines of future emissions, when setting the boundary conditions of the global climate models used to predict future climates (Fig. 2). The end result is the large range of potential values (and associated impacts) cited in the scientific literature.

Global climate models in themselves are far from perfect. They are essentially an oversimplified model of a chaotic system and as Fig. 2 shows,

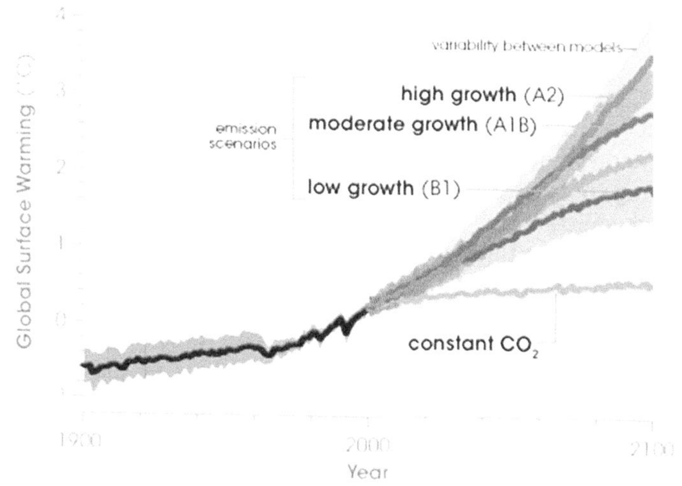

Fig. 2. Temperature Projections to the Year 2100, Based on a Range of Emission Scenarios and Global Climate Models. *Source*: Reproduced with permission from Climate Change 2007: Synthesis report (IPCC, 2007).

the unknowns associated with future emissions are a large potential source of error. As a result, all aspects of the transport and climate change debate are focused on improving or using scenarios. For example, the cumulative impact of mitigation in the transport sector is a key aspect in determining the future level of emissions. Detailed emission scenarios, however, can only be produced by investigating the individual components of each sector, and in the case of transport, each mode. Ultimately, a detailed knowledge of current emissions and a clear appraisal of the various available mitigation strategies are required to inform global climate models. Any uncertainties will then be propagated into the downscaled output from the model which is subsequently used in climate change impact assessments to inform adaptation. In summary, getting the future socio-economic storyline correct is crucial, as the accuracy of the climate science depends upon these boundary conditions. Each chapter in this book adds weight to inform this bigger picture, yet all are ultimately underpinned by climate change scenarios. Consequently, the difficulties in forecasting the future is the key starting point for this book and *Chapter 2* examines the complexities involved in producing socio-economic scenarios for inclusion in global climate models. Indeed, the first half of this book deals with the mitigation issues surrounding individual modes of transport which when considered collectively can be used to inform scenario development.

MITIGATION

With respect to climate change and transport, much of the focus within the academic literature (see Chapman, 2007 for a detailed review), and indeed, climate change policies, has been on the introduction of mitigation measures to reduce greenhouse gas emissions. Policy has largely been driven by the 1997 Kyoto Protocol (later Kyoto Treaty), which aimed to commit industrialised countries to a 5.2% reduction in greenhouse gases emissions between 1990 and 2012. A major criticism of the Kyoto Protocol was its inherent failure to consider international emissions from transport, in particular aviation. Indeed, as Kyoto and subsequent climate summits have shown, multilateral agreement on climate change cannot be taken for granted and presents the major obstacle which needs to be overcome (Fox & Chapman, 2011). These areas remain perhaps some of the most challenging in the discipline and are covered in detail early in this book (*Chapters 3 and 4*). As a result, policy is quicker to implement at a national scale where it is easier to quantify current emissions and to implement decarbonisation measures. For this reason, the focus of this book is predominately set in a national context, using UK case studies to illustrate the various perspectives on the subject (note that 39% of all energy in the UK is used for transport purposes – Department for Transport, 2010). Despite this focus, in many cases, the examples and material are relevant to policy across the world. The mitigation measures on surface transport networks (road, rail, freight, walking and cycling) are covered in *Chapters 5, 6 and 7*, but the finer nuances involved in their implementation are discussed throughout the book.

It could be argued that Kyoto has encouraged countries to largely work independently as opposed to collectively to respond to the challenges of climate change. There is certainly a sense of each country needing to get its own house in order before tackling the problem internationally. The UK Climate Change Act has a commitment for a UK reduction of 80% in CO_2 emissions by 2050 based on 1990 levels, and an interim target of at least 34% by 2020. *Chapter 8* provides the background to this policy, whereas the economic perspective is covered in *Chapter 9* which argues the case for intervention and mitigation. These chapters set the scene for the second half of this book which provides a series of commentaries offering various perspectives on the climate change and transport debate.

Linking with policy are the various mitigation measures, or tools, available to transport stakeholders. Most of these measures are devised to promote behavioural change via modal shift. For example, *Chapter 10* discusses the stick and carrot approach (Stradling, Meadow, & Beatty, 2000). Sticks are effective and are used to fiscally constrain demand (e.g. tax) whereas carrots are softer measures (e.g. improved or subsidised public transport). In reality, both approaches are required, although as *Chapter 11* argues, carrots or soft measures are far more acceptable. These can be viewed as socially progressive policies and ultimately 'vote-winners' for government. Overall, despite the numerous behavioural approaches outlined in this book, there is a consensus that policy is too slow and it is argued that technological change will spearhead mitigation in the sector (Chapman, 2007). Indeed, technological improvements are evident across the transport sector and are discussed throughout the book. Even incremental technological changes can make a difference, particularly considering the speed of potential implementation. *Chapter 12* provides an example of this, specifying the potential future role of information delivery in terms of reducing transport emissions. For larger scale decarbonisation of the various transport modes, fuel diversification (e.g. biofuels, natural gas, hydrogen and electric vehicles) provide the most feasible potential solution to the problem (Maclean, 2004). The implementation of alternative fuels, however, is offset by the cost of the new infrastructure required to deliver the fuel, therefore requiring a strong policy implementation (Hart, 2003).

ADAPTATION

No single mitigation measure (whether behavioural or technological) provides a solution to stabilise emissions from transport. Reductions to date fall a long way short of the stabilisation targets that will limit global climate change to the 2°C target (Anderson & Bows, 2008). The effects of climate change will vary geographically, with some regions being exposed to greater extremes than others. As a result, all transport systems will need to adapt to continue to function. The nature of the effects, impacts and required adaptations are dealt with in a climate change impact assessment (Jaroszweski, Chapman, & Petts, 2010). Again, scenarios are central to the analysis, but this time *climate* scenarios are delivered by global climate models in contrast to the *socio-economic* scenarios used to inform them. The global climate model is then downscaled to a local level where the impacts are quantified (e.g. railway buckling in Dobney, Baker, Chapman, & Quinn, 2010; road melting in DEFRA, 2011) and adaptation measures devised (e.g. dual use infrastructure in RAE, 2011). Hence, probabilistic mitigation scenarios are indirectly used to drive adaptation policy and thus a holistic approach is required based on a range of detailed scenarios (Jaroszweski et al., 2010). Adaptation is covered in the first half of this book where detail is provided for the various modes. In particular, *Chapter 4* provides a detailed case study of the potential impacts of climate change on the UK surface transport network, which highlights the fact that whilst climate change is a global problem, the impacts will be local. Hence, whilst climate change mitigation policy is ultimately driven by national policy, climate change adaptation is driven by national importance and localised impacts.

SUMMARY

As is true for all sectors, there is a strong case for mitigation and adaptation in the transport sector: 'The scientific evidence is now overwhelming: climate change presents very serious global risks, and it demands an urgent global response … the benefits of strong, early action considerably outweigh the costs' (Stern et al., 2006, pp. 1–2). In effect, both mitigation and adaptation are linked and a holistic framework for research and investment is required to improve the relationship between transport and climate change. For future transport networks to cope with predicted climate change, it is essential to implement mitigation and adaptation measures for both existing and new transport infrastructure to cope with the predicted changes. There is growing pressure on policy makers to provide sustainable transport (Chapman, 2007) not only in the volume of emissions produced, but also in its ability to adapt to and continue to operate under predicted climate change.

Overall, the climate change issue is complex with international and national dimensions. However, the complexities surrounding the international debate forces the need for each country to forge strong national policy to deliver the needed reductions in global greenhouse gas emissions. Many impacts will be local and therefore adaptation measures will be the responsibility of local governments – not national assemblies. For this reason, it is also sensible to think about mitigation locally as discussed in the regional case study presented in *Chapter 13* before subsequently scaling up. Hence, whilst this book uses the UK as a basis for case studies and examples, the content is clearly equally applicable whether working at a local or global scale – the solution is the sum of many small actions, discussed in the concluding chapter, *Chapter 14*.

REFERENCES

AEA Group. (2011). *Pathways to 2050 – Key results. MARKAL model review and scenarios for DECC's 4th carbon budget evidence base*. Report for the Department of Energy and Climate Change. Retrieved from http://www.decc.gov.uk/assets/decc/11/cutting-emissions/carbon-budgets/2290-pathways-to-2050-key-results.pdf

Anderson, K., & Bows, A. (2008). Reframing the climate change challenge in light of post-2000 emission trends. *Philosophical Transactions of the Royal Society A*, *366*, 3863–3882.

Chapman, L. (2007). Transport and climate change: A review. *Journal of Transport Geography*, *15*, 354–367.

Department for Environment, Food and Rural Affairs (DEFRA). (2008). *Climate change bill*. London: HMSO.

Department for Environment, Food and Rural Affairs (DEFRA). (2011). *Climate resilient infrastructure: Preparing for a changing climate*. London: HMSO.

Department for Transport (DfT). (2007). *Towards a sustainable transport system. Supporting economic growth in a low carbon world* (Retrieved from http://webarchive.nationalarchives.gov.uk/+/http://www.dft.gov.uk/about/strategy/transportstrategy/pdfsustaintranssystem.pdf). London: HMSO.

Department for Transport (DfT). (2010). *Transport statistics Great Britain 2010* (36th ed.). London: HMSO.

Dobney, K., Baker, C. J., Chapman, L., & Quinn, A. D. (2010). The future cost to the UK's railway network of heat related delays and buckles caused by the predicted increase in high summer temperatures due to climate change. *Proceedings of the Institution of Mechanical Engineers – Part F: Journal of Rail and Rapid Transit*, *224*, 25–34.

Fox, T., & Chapman, L. (2011). Engineering geo-engineering. *Meteorological Applications*, *18*, 1–8.

Hart, D. (2003). Hydrogen – A truly sustainable transport fuel? *Frontiers in Ecology and the Environment*, *1*, 138–145.

Intergovernmental Panel on Climate Change (IPCC). (2001). *Climate change 2001: Synthesis report*. A contribution of Working Groups I, II and III to the Third Assessment Report of the IPCC. Cambridge, UK: Cambridge University Press (Fig. SPM-1).

Intergovernmental Panel on Climate Change (IPCC). (2007). *Climate change 2007: Synthesis Report*. A contribution of Working Groups I, II and III to the Fourth Assessment Report of the IPCC. Geneva: IPCC (Fig. 3.2.).

International Energy Agency (IEA). (2011). *CO_2 emissions from fuel combustion*. Retrieved from http://www.iea.org/co2highlights/co2highlights.pdf

Jaroszweski, D., Chapman, L., & Petts, J. (2010). Assessing the potential impact of climate change on transportation: The need for an interdisciplinary approach. *Journal of Transport Geography*, *18*, 331–335.

Maclean, H. L. (2004). Alternative transport fuels for the future. *International Journal of Vehicle Design*, *35*, 27–49.

Meinshausen, M., Meinshausen, N., Hare, W., Raper, S. C. B., Frieler, K., Knutti, R., … Allen, M. R. (2009). Greenhouse-gas emission targets for limiting global warming to 2°C. *Nature*, *458*, 1158–1162.

Penner, J. E., Lister, D. H., Griggs, D. J., Dokken, D. J., & McFarland, M. (1999). *IPCC special report: Aviation and the global atmosphere*. Cambridge, UK: Cambridge University Press.

Royal Academy of Engineering (RAE). (2011). *Infrastructure, engineering and climate change adaptation-ensuring services in an uncertain future*. Retrieved from http://www.raeng.org.uk/adaptation

Stern, N., Peters, S., Bakhshi, V., Bowen, A., Cameron, C., Catovsky, S., … Zenghelis, D. (2006). *Stern review: The economics of climate change*. London: HM Treasury.

Stradling, S. G., Meadows, M. L., & Beatty, S. (2000). Helping drivers out of their cars. Integrating transport policy and social psychology for sustainable change. *Transport Policy*, *7*, 207–215.

CHAPTER 2

MANAGING THE FUTURE: MODELS, SCENARIOS AND THE CONTROL OF UNCERTAINTY

Murray Goulden and Robert Dingwall

ABSTRACT

Purpose – *To consider the emerging qualitative approaches that consider the future, exploring ways in which these may be better used to complement and extend established practice in global climate models.*

Methodology/approach – *A review of contemporary modelling approaches that consider the future for transport applications, with a focus on complexity and uncertainty issues, the search for foresight, the role of engagement and the credibility of qualitative foresight.*

Findings – *The importance of the need to incorporate socio-economic scenarios in climate change models and the possibilities offered by the tool of systematic qualitative foresight are demonstrated.*

Originality/value – *Explores limitations of quantitative modelling approaches to foresight and introduces a potentially complementary, innovative approach based on the systematic use of qualitative methodologies.*

Keywords: Transport; future; modelling; scenarios; foresight

Transport and Climate Change
Transport and Sustainability, Volume 2, 9–37

ISSN: 2044-9941/doi:10.1108/S2044-9941(2012)0000002005

INTRODUCTION

As Norman (1993) pithily states, 'attempting to predict the future is both foolhardy and essential'. All human societies face the problem that actions in the present have implications for the future which, by its very nature, is uncertain. However, those actions can be highly consequential. At best, they consume resources that could be put to uses that would turn out to be more efficient, effective, equitable or humane. At worst, they can lead to catastrophe. The questions generated by acting on assumptions about the nature of events yet to occur, such as those attributable to climate change, are critically important in the field of transport because of the length of the investment cycle in key areas of infrastructure. The layout of the UK trunk road network was substantially determined until the 1960s by the route decisions of Roman military engineers nearly two thousand years earlier. The UK rail network still experiences important constraints from decisions made by Victorian engineers working to the limits of their technology: building to small gauges that were also compatible with lower construction costs in smaller tunnels, tighter curves and narrower clearances. The UK HST125 train sets introduced to celebrate Queen Elizabeth's II's Silver Jubilee are still in operation as the monarch celebrates her Diamond Jubilee, thirty-five years later. Transport planners necessarily need to look into futures that may be quite distant.

The wide impact of transport on society also means that planning cannot be a purely technocratic matter (Hall, 1980). The scale of investment in large public goods means that citizens may be forced to pay the costs, either directly through taxation or indirectly through guarantees to private investors. The choices made have both immediate local impacts and longer term implications for the configuration of society in, for example, allocations of time between work, family and leisure, and travel between the spaces in which the former activities are carried out. Transport planning, then, necessarily has a social and political dimension that generates challenges in securing the legitimacy of investment decisions through the mobilisation of public engagement and consent.

Investment decision-making normally rests on quantitative calculations that reduce the various factors involved to the common metric of money. These calculations may take the form of models that allow for various options to be considered and evaluated for their rate of return. Such approaches have, however, tended to prove inadequate in addressing the problem of legitimising their outcomes since the process of selecting and generating the calculable factors involves a set of judgements and normative assumptions that may not be widely acceptable, but which may be obscured by the complexity of the modelling.

Modelling climate change may produce a 'rational' outcome – but only if it is generally agreed to have included the right elements in an appropriate manner. As a result, it co-exists with scenario approaches that are more qualitative in nature. There are, though, few principles for the rigorous development of scenarios.[1] At one end of a continuum, there are acknowledged works of social science fiction, which may have an important role in stimulating the imagination of publics, policymakers and scientists alike, thereby framing the reception of innovations. At the other, there are ostensibly neutral constructions that conceal the prior normative positions of their authors. This chapter reviews current approaches to thinking about the future and explores ways in which qualitative methods may be used to complement and extend the established practice of modelling.

THE COUNTING IMPULSE

When considering issues surrounding transport and climate change, the following question emerges: What must be done today in order to live with uncertainty about what will happen tomorrow? In many traditional societies, solutions are found through communication with deities, ancestors or others who live in a different temporality that is thought to allow them to look into, or even shape, the future, if correctly requested or incentivised by gifts, sacrifices or offerings. Specialised classes of intermediary, like priests, shamans, diviners and oracles, emerge to transmit human requests for information and to interpret the responses of the entities that inhabit that other world. In the transition to modernity, such solutions lost much of their power to legitimate action. They were neither rationally explicable nor calculable. Other strategies developed, particularly that of insurance (Bernstein, 1996). A seventeenth-century merchant could not forecast with certainty the outcome of investing in a specific long-distance trading voyage, given the problems of navigation and the frailty of ships in the face of natural hazards. They could, though, use general experience of such voyages to estimate the probability of a safe return and to place a bet on this that would contribute to covering his losses if a particular enterprise should fail. The management of risk and uncertainty drove the creation of both a great industry and the mathematical tools to sustain it, in the form of statistics.

Insurance is not, however, a practical solution to all problems associated with uncertain futures. Some risks are potentially so catastrophic that individuals would prefer to prevent them rather than to be compensated for their consequences: they may insure against the explosion of a nuclear plant but it is preferable that it not happen in the first place. Some risks approach certainty to the degree that insurance can become impracticable: Lloyds of London, providers of the first aviation insurance in 1911, withdrew their policies just a year later as the difficulties of mastering this new socio-technical system of flight were realised in numerous crashes during bad weather (Wells & Chadbourne, 2007). This does not, though, mean that stakeholders would not like reasonable warning of a catastrophic event so that they can take such measures as are possible to minimise its impact. Weather gods give way to meteorology. The mathematics of risk that was created in the seventeenth century has enrolled new sources of quantitative data and harnessed the growing power of information technology to create increasingly detailed representations of the future. These models assemble such available evidence as can be expressed in numerical forms and linked by means of equations or algorithms to project possible future states of the dataset. If, for example, the UK population in 2011 is x, the fertility rate is y and the mortality rate is z, then a range of population estimates can be made well into the future that reflect different assumptions about the movement of these underlying rates and the relationship between them. Such estimates are important for policy purposes. Predicting a UK population of 71.6 million in 2035, or at least a population that can with reasonable certainty be expected to fall within a range of ± 4 million either side of that estimate, is essential for calculations of aspects such as social security and pensions (Office of National Statistics, 2011).

Modelling becomes more problematic, however, when it seeks to incorporate elements that are not intrinsically countable. As Bowker and Star (2000) state, all metrics are classifications that embed a set of prior social decisions about how to impose an order on the world: there is necessarily a politics of metrics. If there is a politics of metrics, there is also a potential for conflict about them and a need to provide for means of resolving those conflicts that will be regarded as legitimate by a sufficiency of interested parties. In the physical sciences, this is not a severe problem, since phenomena can be defined in terms of the measurement practices used to describe them, which can, in turn, be standardised by agreement among all those for whom this is a practical concern and codified by international organisations like the *Bureau International des Poids et Mesures*.

In social contexts, such as transport planning, however, the politics of metrics cannot be so simply resolved. This has been well recognised since the Report of the *Commission on the Third London Airport* (1971), chaired by Mr Justice Roskill (later Lord Roskill). Although Cost-Benefit Analysis (CBA) was implemented in UK transport planning as early as 1960 (Coburn, Beesley, & Reynolds, 1960), the Roskill Commission was the first attempt in a major UK public investment decision to capture all planning uncertainties through CBA methods (see the review in Adams, 1971). It was intended to bring closure to a public debate about the location of a major transport investment and to create a sufficient degree of certainty about the outcome to justify the scale of social disruption and public expenditure that would be associated with any choice. Every identifiable cost and benefit associated with each potential site was reduced to a uniform metric of pounds sterling and entered onto a balance sheet, which produced the recommendation that the airport should be constructed at Cublington in North Buckinghamshire. It was clearly believed by many at the time that this was an objective process which would produce a socially and technically optimal solution, freed from stakeholder politics. This would be accepted as a legitimate closure of the debate and allow investment and construction to proceed. However, as the resulting debate showed, the process had merely served to mystify a set of value assumptions and re-present them in a technical form: in effect, the uncertainty and its resolution were 'black-boxed' (Latour, 1987, p. 2). Parry Lewis (1971) raised technical criticisms of the way in which travellers' time was costed and incorporated into the black box. Undermining just this one component of the box was enough radically to change its outputs: removing this calculation from the cost-benefit algorithm meant that the proposed site at Foulness moved from the bottom of the rankings to the top. The absurdity of other outputs – for example the calculation that a life lost in an air crash cost the victim's community £4,360 if male, but *gained* the community £1,120 if female (Adams, 1971, p. 477) – further undermined efforts to keep it closed.

More fundamentally, Mishan (1970) introduced a question that has become increasingly debated in more recent years, namely whether unlimited mobility should be assumed to be an unquestioned social good. This assumption, held the potential for generating social conflict because of the way in which it glossed over different choices about societal futures and the balance between environment and growth. The crux of the problem was, and is, the two major assumptions upon which CBA relies to deal with the uncertainties of future action. The first is that all relevant variables can be both identified and quantified. Complex systems thinking suggests that identifying all relevant variables is a highly problematic notion, and the difficulties of quantifying are shown above in Adams' account (1971). Perhaps even more intractable, however, is the second assumption: that there exists such a thing as a benefit (*singular*) to society. This relies on a notion of society as essentially homogenous, sharing both circumstances and aspirations. Modern societies are in fact extremely heterogeneous assemblages of actors and institutions, and what may benefit one group may well cost another.

Roskill's CBA failed to address the normative assumptions embedded in its own process and choice of metrics. The report's recommendation was eventually abandoned in favour of the second choice, Foulness, to the east of London, which was, in turn, dropped in favour of expanding existing provision at Stansted in Essex and, without an explicit policy decision, at Luton in Bedfordshire. Despite the Commission's prodigious efforts to enrol the future of aviation demand, and the social, economic and cultural environment in which it would unfurl, into its scheme of metrication, the phenomenon escaped and reverted to the messy wilderness of values and politics. Roskill's CBA had failed to engage sufficient, or sufficiently influential, public support, despite its exhaustive claims to scientific method and objectivity. As a result it failed to be accepted as a legitimate closure of debate prior to action.

As Adams (1981) noted, CBA was not dealt a mortal blow by the Roskill fiasco, although this did blight its use in public policy for twenty years or so (Pearce, 1998, p. 89). Its subsequent resurgence invites sociological analysis. What is the appeal of what Adams calls a *Solomon Machine ... that embodies in quantified form the principles of profit maximization, and distributive justice, a machine into which can be fed accurately measured valuations of all the relevant facts, and out of which will flow wealth and even-handed justice for all*? (1981, p. 177). In parallel with the revival of the 1950s aspiration to reduce all social phenomena to countable descriptions, there has been a new wave of critique, mainly from the United States, exploring the politics of quantification and its role in the supply of legitimacy to contested societal decisions (e.g. Espeland & Vannebo, 2011; Martin & Lynch, 2009). The Solomon Machine feeds the illusion that there is an impersonal, objective and neutral resolution to all conflicts in a world that is marked by fungible differences or variations in interest rather than by fundamental structures of social inclusion and exclusion.

It is important to stress that this is a specific critique of the over-extended and uncritical application of quantitative techniques to human affairs. If a suspension bridge is built, modelling is a vital exercise in seeking to predict whether it is likely to fail under a range of stresses and environmental conditions. Even here, though, the range considered is necessarily a product of normative judgements: a structure built to meet every theoretically possible eventuality is likely to be so costly as to be impracticable. The construction decisions behind the Fukushima nuclear reactors are a good example. They incorporated a measure of earthquake resilience, reflecting the range of recent historical experience in Japan, but were overwhelmed in the 2011 earthquake by the stress of peak ground acceleration that was double that for which they had been designed. The same company's Kashiwazaki-Kariwa plant escaped comparable damage in 2007 when it survived a 6.8-magnitude earthquake, but construction costs would have been massively increased in designing for the 9.0 magnitude experienced in 2011. The selection of technology rests on a social judgement about risk and the appropriate level of investment that should be made in managing this. At one price, the construction of the power stations is competitive with other investments in securing future energy supplies: at another, the price of certainty is too high.

In other contexts, social judgements may lead to apparently paradoxical willingness to pay a premium for risk reduction. Carlsson, Johansson-Stenman, and Martinsson (2004), examining travellers' responses to risk, note that, if individuals took a purely rational approach, they would compare the relative danger of different modes and invest accordingly. However, even when informed of the safety record of specific forms of transport, travellers are, on average, twice as willing to pay for risk reduction when travelling by air than by (considerably more dangerous) taxi. Carlsson et al. explain this finding by an *ad hoc* argument that air travellers tend to experience greater *mental suffering* (*ibid.*, p. 149), and are willing to pay more to counter this. There are a number of other possible psychological explanations, such as level of worry (Moen & Rundmo, 2006) or perceived control (Horswill & McKenna, 1999) over the risk, but the key point is that the 'rational' calculus does not, on its own, generate an output that the traveller accepts as a legitimate basis for their actions.

COMPLEXITY AND UNCERTAINTY

Bounding the range of future conditions to be considered is a major task facing transport planners dealing with large infrastructure systems. The structural integrity of a suspension bridge consists of a relatively 'closed' system, which is fairly straightforward to model. There are a finite number of factors relevant to the bridge's ongoing structural integrity, which stand in a limited number of relationships to each other: the material properties of the steel and concrete used, and the configuration in which the components are assembled; the physical environmental factors – the local topography, the bedrock, the local climate profile; and the human environmental factors – the loads placed on the bridge and the skills and resources available for routine maintenance. These factors are mostly quantifiable and – allowing for 'wildcard' events such as that which struck Fukushima – modelling them yields a sufficient degree of certainty to cover most eventualities. There may still be debate about, for example, the trade-off between resilience and cost in actual construction decisions but that debate is constrained by the bounds of the system.

Transport planning, however, also has to deal with much more 'open' systems. Building on the work of Friend and Jessop (1969), Hall (1980) identifies three dimensions to the uncertainty that bedevils planning activities. These are *uncertainty about the planning environment* (UE), which, for Hall, typically concerns future demand and the delivery costs of the planned scheme; *uncertainty about decisions in related areas* (UR), concerning the potential for decision-makers outside the immediate planning process to change the environment within which future actions occur; and *uncertainty about value judgements* (UV). Hall notes that UE can often be determined by UR, which can, in turn, be considered a product of UV. Indeed UV ultimately determines the other axes (p. 10). For example, predictions of future demand for a new airport rely in part on population estimates, which are themselves influenced by factors such as national immigration policies (UR), which in turn are determined by contemporary ideas about what is socially and culturally desirable (UV). Given current concerns about global climatic stability, physical-environment factors should be added to Hall's sources of UE. However these factors can also be argued to be ultimately determined by UV, as the climate change likely to be experienced will overwhelmingly be the product of how societies respond to the challenges of reducing greenhouse gas emissions and the impact of climate change on future networks.

Open systems are exposed to shocks stemming from one, or a combination, of Hall's dimensions of uncertainty: natural disasters; energy shortages; international agreements on emissions reductions; locally organised protest movements. The scale of such systems may mean their fundamental assumptions about social organisation are more transparent and, hence, debatable: the supposed role, for example, of mobility per se and its implications for the orientation of citizen actions towards local, national, or even global relationships. Such issues arise even in the example of the bridge discussed above, once it is considered as part of a transport *system*. The investment decision involves assumptions about travel demand, which include a normative valuation of mobility within a future society. Such open questions, with multiple and indeterminate answers, serve to make these systems 'complex' in a technical sense. In what can be thought of as an extension of Hall's 'nesting' of dimensions of uncertainty within one another, change is generated both from changes in individual elements, and from the interactions *between* them, with their outputs non-linear and emergent. The system is not then reducible to the sum of its components. The most famous example of this is the 'butterfly effect' described computationally by Lorenz (1963). A tiny change in starting conditions (the flap of a butterfly's wings) can have a series of knock-on effects which result in major system-level changes (the formation of a hurricane hundreds of miles away). This process is observed in the formation of traffic jams: once a certain density of vehicles is reached, their interactions can result in cascading effects that bring the system to a standstill even in the absence of observable bottlenecks like road works (Daijian & Xiao, 2009; Sugiyama et al., 2008). Practical certainty about the future is difficult to achieve in such systems, since equilibrium is not their 'normal' state. They are never entirely stable, but in varying degrees of flux. They are dynamic and contradictory, being both patterned and unpredictable. This is known as 'orderly disorder'.

Two notions are important to understanding complex open systems: 'feedback' and 'path dependency'. The principles of negative feedback underpin the humble home thermostat. In this simple, closed system the thermostat switches off the boiler once the house temperature reaches a defined level. When the temperature falls low enough, the thermostat switches the boiler back on, maintaining room temperatures within a narrow equilibrium band of a few degrees. In a complex open system like the global climate or human societies, there are also positive feedbacks. These have the opposite effect, pushing a system away from homeostasis. These loops are self-reinforcing, and can quickly multiply a small effect into a large one. Glacial ice sheets reflect sunlight back into space, so reducing temperature. However, if temperatures rise high enough to melt an ice sheet, the result is that less sunlight is reflected, so the temperature may rise even higher. Positive feedback can result in a system becoming 'locked-in' to a particular pattern of change rather than equilibrium. In the social sciences this is often referred to as 'path dependency'. One small event reinforced by relevant conditions – which may well be local and transitory – can quickly be cemented by positive feedback.

Dennis and Urry (2009) describe the way in which twentieth-century Western societies became locked-in to the petroleum-based car, leaving behind the previous equilibrium of steam-powered and horse-powered mobility. In the late nineteenth century, there was little to separate the petrol engine from its rival technology, the electric battery. However, certain initial conditions favoured the former. With this head start, positive feedbacks such as investment in the technology and the supply infrastructure, and institutional inertia, ensured that the petrol engine quickly became unchallengeable, and the twentieth century was locked-in to a pattern of ever greater petroleum car use. An understanding of this process, for example, might lead to some scepticism about the probability of electricity displacing liquid biofuels as the energy source of choice for many transport applications, given modern societies' lock-in to liquid fuels. Another classic transport example of positive feedback is that generated by new motorways, which reduce journey time for travellers by providing high-speed travel between populous areas. Over time this very possibility results in travellers restructuring their lives, perhaps utilising the potential for greater mobility to live further from their place of work. This leads more users onto the road, which becomes increasingly clogged with traffic, to the point where demand for further transport infrastructure develops. This is the contradiction at the heart of 'predict and provide' transport policy that led to its formal abandonment by the Labour government in the 1998 *New Deal for Transport* White Paper.

Although complexity science has developed mathematical methods for dealing with such systems and exploring their implications through powerful and sophisticated models, these confront in more substantial ways the problems of incorporating relevant variables and relationships that are not readily subject to metrication. Their complexity also has implications for their role in legitimating decision-making. At their heart is an apparent contradiction: while '*disordered* order' means that it is difficult to generate predictions with high degrees of certainty, *order* reflects the fact that systems do follow patterns. There is structure to the process of change, and so it is possible to generate useful scenarios that describe more and less plausible future system states. At the same time, users may be frustrated both by the heterogeneity of possible futures and by the suspicion that the very complexity of the models mystifies their underlying assumptions and processes in ways that make it hard for them to generate legitimacy. Models can be thought of as 'inscription devices', *transform[ing] a material substance into a figure or diagram* whose construction is opaque but rendered authoritative by the automatic nature of the process involved (Latour &

Woolgar, 1986, p. 51). The role of the inscription device is less discovering the truth, than convincing others of the truth. Complex models may capture more dimensions than traditional cost-benefit analyses but they do not necessarily function any more successfully as inscription devices. In fields like transport, with distant time horizons, the accuracy or otherwise of predictions can often only be determined long after the planning exercise has ceased to concern policymakers. The crucial goal is the legitimisation of decision-making in the present that can justify the investment decisions that need to be made if the future is to happen at all. The important criterion for evaluating approaches to modelling is, then, their ability to generate a *perception* of accuracy: the futures created must appear plausible, both within their own logic and within that of a contemporary audience. There is no value in Cassandra's vision of the future, a true prophecy that no-one heeds because it is so discrepant with present expectations. Orwell famously wrote *Who controls the past controls the future. Who controls the present controls the past* (1998). In the case of foresight it can be reversed: who controls the (imagined) future controls the present.

THE SEARCH FOR FORESIGHT

The problems of achieving sufficient practical certainty in planning choices about future-oriented actions, and of having those choices accepted as legitimate, solely on the basis of statistical or mathematical techniques – what is generically termed 'modelling' – has led to a variety of attempts to develop other tools that may be more effective in enrolling contemporary social actors in an agreement to follow a particular pathway into the future. This shall be described collectively as foresight.

Beginning as a niche application for American military strategists in the 1950s, and popularised in corporate circles by Royal Dutch Shell's use during the 1970s, foresight exercises have become an integral feature of the policy landscape. The *European Foresight Monitoring Network* lists over 1,900 initiatives, most of which were conducted since 2000 (EFMN, 2010). This growth has prompted its own social science sub-discipline (Brown, Rappert, & Webster, 2000) and attempts to reorient the social sciences themselves to studying the future (Adam, 2007). No single approach dominates, partly reflecting the elusiveness of the task. However, it is possible to classify these exercises in terms of their primary dimensions.

One axis is expertise. High level policy initiatives often favour Delphi exercises. Here, panels of experts repeatedly review and revise their anonymised collective judgements, in the belief that a consensus will gradually emerge. This approach underpinned the *UK Technology Foresight Programme*, announced in the (1993) White Paper *Realising Our Potential*, with the aim of both creating a network of scientists and industrialists who together might better exploit opportunities, and guiding future public funding of science and technology (Georghiou, 1996). Delphi expert panels lend themselves to generating networks, and its systematic approach to generating agreement from groups of respected agents creates considerable legitimacy for the outputs. They may, though, be vulnerable to counter-charges that only certain kinds of expert have been selected to participate. Some participants may be asked to make judgements on issues outside their legitimate expertise, while other experts may be excluded because their knowledge is not valued, recognised or considered capable of contributing to a consensual outcome.

These closed tools may be contrasted with projects like *Sciencehorizons*, a public engagement tool created for the then Department for Innovation, Universities and Skills (DIUS) that sought to elicit public attitudes towards emerging technologies. Far from aiming to generate credible knowledge upon which to base future funding decisions, the predictive element is largely generated a priori, as a starting point for semi-structured discussion. Credibility is generated through the (apparent) absence of recognised expertise, in a process that owes more to the principles of democratic than technocratic legitimacy, an outcome based on citizen judgements rather than specialised knowledge. The problem here, of course, as so-called 'Third Wave' writers in science studies have noted, is that some problems do require specialised input to make correct judgements (Collins & Evans, 2002). The pre-exercise framing also creates the potential for the unacknowledged shaping of subsequent discussion.

A second axis is between forecasting and backcasting. Forecasting extrapolates current trends into the future. By this means, it expects to generate objective, 'realistic' visions, which produce certainty in the future through the assumed certainty of our knowledge of the present and recent past. As Nelson, Geltzer, and Hilgartner (2008) point out, society often seems more sensitive to a lack of certainty about the future than about uncertainty in the present, although the latter is not necessarily more securely knowable. Backcasting, in contrast, is overtly normative, starting from one or more imagined futures, and working backwards by asking how to get to such a point from our current position. These futures are chosen to express particular desirable or undesirable states. Backcasting constructs the future not as a predetermined object, but as one to be consciously created in the present. In practice, however, it can be argued that the forecasting/backcasting distinction is greatly overstated. Backcasting's imagined futures inevitably reflect current trends, if only in subverting them, whilst forecasting invariably embeds normative assumptions about what is or is not 'likely' further down the line.

The final axis lies between quantitative and qualitative approaches. As noted previously, quantitative modelling has often underpinned forecasting as an inscription device. Though often prefaced by a disclaimer that they do not aim at prediction, such forecasts do seek to generate visions of the future whose utility lies in their empirically derived legitimacy, that is to say their numerical extrapolation of current trends with all detail reduced to a limited set of numbers. However, aside from the issues related to the compression of complexity, discussed above, these visions may struggle to achieve legitimacy because of their inability to represent the future to citizens in accessible terms or to characterise the pathways that take them from the here and now to the there and then. Even specialists may find it hard to visualise a future expressed purely in numerical terms: what would it be like to travel in the United Kingdom in 2033 alongside nearly 72 million other people? Can the experience of mobility in such an environment be judged from the numbers alone? Without successfully engaging stakeholders in the process there is a danger of repeating the errors of the Roskill Commission: creating a narrow vision[2] of the future that is alien for so many that it fails to gain traction in the present, and so fails outright.

The two inter-related ideals in play here – multi-disciplinarity and public engagement – converge around the issue of expertise. To a degree, the latter can be thought of as an extension of the former, going beyond the inclusion of experts from multiple scientific disciplines to include laity as well. Both demand the opening up of methods of imagining the future, by permitting diverse perspectives to shape the process. Qualitative methods are not a panacea for achieving such inclusion, but they do aid the process by providing a more universal language with which to work. Their lower precision can be counted as a cost in certain circumstances, but is beneficial when combining multiple perspectives that might not survive the compression demanded by quantitative approaches. In such multi-disciplinary environments the elasticity allowed by qualitative definitions allows for the deployment of 'boundary objects', being:

> objects which are both plastic enough to adapt to local needs and constraints of the several parties employing them, yet robust enough to maintain a common identity across

> sites [...] They have different meanings in different social worlds but their structure is common enough to more than one world to make them recognizable, a means of translation. The creation and management of boundary objects is key in developing and maintaining coherence across intersecting social worlds. (Star & Griesemer, 1989)

Of course the nature of planning means that ultimately a decision or decisions must be taken; the value of boundary objects lies in allowing a multi-party decision-making process to proceed without universal consensus at every step along the way (Bechky, 2003). Boundary objects are not a technique in themselves, but are readily observable phenomena found in decision-making processes where multiple groups collaborate. The nature of CBA, specifically its *non-dialogical quality [...] in which different moral claims are reduced to preferences measured along a single (monetary) metric* (Owens & Cowell, 2011, p. 68), denies a role to such entities, and so the processes they enable.

Some recognition of these problems can be seen in the Intergovernmental Panel on Climate Change's [IPCC] evolving approach to scenario development. The IPCC's Special Report on Emissions Scenarios (SRES) (2000) was the organisation's attempt at plotting future emissions scenarios. This took a series of qualitative narratives about future socio-economic development, and then proceeded to quantify them through economic and climate computer modelling. This was not only a laborious process that struggled to keep up with actual events on the ground (Inman, 2011), but also fared poorly against observed emissions: actual emissions between 2000 and 2007 outstripped even the most emissions-intensive SRES scenario (Raupach et al., 2007, Fig. 1). Wynne's (2010) argument, bolstered by the observed failure of the SRES pathways to predict the scale of recorded emissions, is that – contrary to the claims of 'climate sceptics' – socio-political considerations mean that the IPCC's work may consistently *understate* the threat of climate change. This is at least in part an outcome of the inherent ambiguity of predicting future interactions between climate systems and human systems, ambiguity that no amount of modelling can overcome. Instead assumptions are required to close down the profusion of possible future states,[3] and Wynne finds that these assumptions are shaped by expectations of what policymakers and the public could countenance in regard to the speed and severity of climate change (pp. 295–298).

The SRES process faced further criticism on several fronts relating to expertise and openness: for being centred on developed nations, and so failing to effectively model developing nations (van Ruijven et al., 2008); for the IPCC's monopoly over the generation of such information for governments (Henderson, 2005); and for failing to make clear assumptions made regarding emission reductions (Girod, Wiek, Mieg, & Hulme, 2009, pp. 114–115).

Presumably in response to such criticisms, Girod et al.'s (*ibid.*) history of IPCC scenarios shows a gradual move away from closed, deterministic 'predictions' to increasingly open, uncertain 'projections'. This process continues, with the IPCC now pursuing an approach using 'representative carbon pathways' (RCPs). This reverses the SRES approach: beginning with a series of possible modelled emissions trajectories, it backcasts from these to create qualitative narratives about how society might come to follow particular pathways. The assumptions that were hidden by SRES are now brought into the foreground. In the words of physicist Angela Wilkinson:

> If we don't put our assumptions about the future on the table, then we have the same situation we had in the run-up to the financial crisis, and that is, we blindly follow the assumptions about the future that are built into our disciplines and models. (Inman, 2011, p. 9)

This opening up of the future to innovative rethinking is given prominence in Banister and Hickman's (2005) report 'Visioning and Back-casting for UK Transport'. The authors echo Wilkinson in noting the value of the approach for challenging expert assumptions:

> The opening up of the debate for alternative challenging futures has generated considerable interest ... [and] very usefully brought together experts from many different backgrounds. This is a critical part of the work-a lower carbon future requires innovative thinking across many disciplines – multi-disciplinary analysis is required. (p. 7)

Although Banister and Hickman's use of backcasting is predicated here on the inclusion of formally recognised expertise, backcasting is also a powerful tool for addressing the challenge of securing democratic legitimacy through means of public engagement.

ENGAGEMENT

A substantial literature has developed in the last decade on the challenge of securing public support for scientific and technological developments (Irwin, 1995; Wilsdon & Willis, 2004). From a position where differences between lay people and experts were characterised mainly a result of a deficit in understanding among the former, to be filled by rational education pouring knowledge into otherwise empty vessels, most leading scholars and policymakers have now come to recognise that these differences reflect different starting points for approaching, and criteria for judging, the consequences of innovations (the same variations that make CBA's attempts at determining a social benefit so problematic).

The value of engagement can be thought about on numerous levels. One is improving the quality of science and technology by recognising the value of informal expertise and approaching engagement as a process of mutual education (Collins & Evans, 2002; Wynne, 1992). Another is meeting the obligations of a democratic system in which all members of society are effectively stakeholders in major scientific and technological developments (which may take the form of major infrastructure projects) (Bohman, 1996; European Commission, 2007). From a pragmatic perspective, there is the need to avoid the kind of negative public reaction that followed the introduction of genetically modified crops to the United Kingdom in the early 2000s or that blighted the Roskill Commission. The process is, in essence, one of securing legitimation for some kind of change in social arrangements resulting from a change in technology. How do people come to accept a change in their relationship with their cars that might involve leaving them half a mile from their home in order to connect to an electric charging point? This may seem rational to a transport planner but perverse to a car owner who is more worried about the risk of vandalism than the possible environmental benefits. The benefits of a smart grid using electric cars' batteries for decentralised storage may be clear to policymakers tasked with maximising the worth of fluctuating wind turbine output, but for the car owner, who finds their car unavailable for unplanned or emergency journeys due to a flat battery, the costs may be more readily apparent.

The process of engagement may itself achieve some or all of the effect targeted in planned-for infrastructure. King's review of electric cars (2007) makes clear that in the near term, whilst the technology remains immature, carbon savings of 10–15% can still be made by drivers adopting more efficient examples of current ICE (internal combustion engine) technology, and adjusting their usage of them (e.g. by reducing braking and aggressive acceleration). The likely success of approaches directed at changing individual attitudes, and so behaviour, as envisaged by King, may be debatable (cf. Shove, 2010), but it seems indisputable that an engaged public is a necessary precondition.

While policy-focused foresight has favoured quantitative methods, which are treated as legitimate by other technocrats, engagement-focused foresight has been drawn to qualitative approaches. In part this is predicated on the notion that lay people have a deficit in relation to dealing with quantitative data. However, it is also clear that effective engagement requires the restoration of much of the detail that quantification suppresses and the exposure of the modellers' assumptions so that lay people can assess both their common sense plausibility and their moral or ideological loading. It is notable, for instance, that Adams (1971) concludes his detailed critique of the Roskill Commission's quantification practices and regressions by inducing a picture of the future from the report:

> It is a vision of a completely polarized society. The rich and powerful are highly mobile, flitting overhead in helicopters and in instant electronic touch with anywhere and everywhere. The poor inhabit a dismal, dirty, pedestrian world. They lead alienated, depersonalized and apathetic lives, quite overwhelmed by the power and scale of the State with governs them. They feel, and are, incapable of taking any action that can improve the circumstances of their existence. It is an extreme and profoundly pessimistic vision. It is George Orwell's view of England in 1984. Certainly in many of its details it can be argued that it is improbable. But one does not need to accept the whole of it in order to appreciate the appropriateness of the name he chose for is. He called it 'Airstrip One'. (p. 492)

In effect, to engage non-experts qualitative scenarios require more complexity, or at least a different take on complexity, rather than being the simplifications they are often dismissed as.

QUALITATIVE FORESIGHT

The credibility of qualitative foresight is often thought to be tarnished by its ill-defined frontier with science fiction, which can be the subject of considerable boundary work (Miller & Bennett, 2008, p. 597). Nevertheless, science fiction exemplifies the capacity of qualitative work to engage wide audiences, being one of the dominant genres of contemporary entertainment (Sterling, 1999). The potential for its tropes to frame public perceptions of new technologies in highly influential ways has been demonstrated by the 'Frankenfoods' metaphor that characterised UK public resistance to GM crop technology in the late 1990s (Hellsten, 2000; Nerlich, Clarke, & Dingwall, 2001). Popular tropes and narratives are valuable in engaging the wider public in new science and technology (Nerlich, 2008), and can be influential on scientists themselves (Goulden & Balmer, 2009). A number of science fiction writers have explicitly positioned themselves as writers of 'social science fiction'. Frederik Pohl, one of the leading exponents of this approach, commented in a 2006 interview:

> I've done a lot of reading on various political thinkers, all of the Utopians, and the kind of societies that I thought would be something to write about ... one of the advantages of writing SF is that you can think about how other societies might work if the rules were a little different, and then write a story and see how people would be affected by that.

Millar and Bennett (2008) make a strong case for the value of science fiction in a foresight context: *Even in its current literary forms, science fiction is, de facto, technology assessment for the rest of us* (p. 598). They argue that it offers an escape from determinist discourses of technical rationality, focusing instead on fundamentally human-centred accounts of technology in which the interplay with identity, meaning and community is brought to the fore. Consider the current debate around the merits of the High Speed 2 rail line (HS2) in the UK. The highly techno-rationalist argument put forward by Government treats mobility as an intrinsic good,[4] much as the Roskill Commission did forty years previously. Just as before, opposition from residents along the proposed route, particularly as it crosses the same areas of North Buckinghamshire, focuses on the impact on their community and way of life. Moreover, the exercise treats the highest goal of people living in the North of the country as travelling to London as quickly as possible rather than, for example, examining the possibility that the key to developing an alternative pole of economic and political development might be to improve links between Northern cities (cf. Hall, 2005a, 2005b). The silence on links to Scotland is even more compelling: will citizens of an increasingly autonomous nation within the United Kingdom actually generate the supposed demand for travel to the South-East of England when economic and political authority is ever more firmly centred in Edinburgh? If Scotland needs new rail investment, is this not more likely to focus on journey times to its own capital city? The value of Millar and Bennett's approach is grounded not in the accuracy or otherwise of the visions that it generates, but in its exploration of fundamental aspects of human social organisation, which shape, and are shaped by, science and technology.

Science fiction can also be a useful source of boundary objects in complex organisational contexts:

> A large technology company or lab might employ hundreds or thousands of persons, each of whom can address only a thin slice of the overall problem. Communication among them can become a mare's nest of email threads and Powerpoints. The fondness that many such people have for SF [science fiction] reflects, in part, the usefulness of an over-arching narrative that supplies them and their colleagues with a shared vision. (Stephenson, 2011)

In terms of their stylistic features, some qualitative foresight outputs are indeed difficult to distinguish from science fiction. Consider these two extracts:

> It's 2 August, 2055. A raft of developments in ICT and intelligent transport throughout the 2010s and 2020s have helped to fuel the 'always on' society of the past 50 years.

> Intelligent positioning systems, encryption technology, real-time telepresencing and a shift towards a low-carbon economy have all played their part in driving the rampant consumerism that shows few signs of abating. If energy issues have been addressed, other sustainability problems have not. Europe's waste footprint is still far larger than Europe is; we are still a 'three-planet' society, sustained only because, globally, the poorest are poorer than ever.

> My model of a long term high volume data storage medium is a synthetic diamond [...] We can speculate that if molecular nanotechnology as described by, among others, Eric Drexler, is possible, we can build a device that will create a diamond, one layer at a time, atom by atom, by stacking individual atoms [...] The capacity of memory diamond storage is of the order of Avogadro's number of bits per two molar weights. For diamond, that works out at 6.022×10^{23} bits per 25 grams [...] by the best estimate I can track down, in 2003 we as a species recorded 2500 petabytes — 2.5×10^{18} bytes — of data. That's almost ten milligrams. [...] Six hundred grams of this material would be enough to store lifelogs for everyone on the planet.

The first comes from the UK Office of Science and Technology's (OST) report *Intelligent Infrastructure Futures* (Office of Science and Technology, 2006) which outlines four scenarios for the period up to 2055. These scenarios are given titles that might be those of science fiction short stories, such as *Tribal Trading* and *Perpetual Motion*, from which the extract is taken. The second is from a lecture by Charlie Stross, a science fiction writer, entitled *Shaping the Future* (2007). Stross's language is markedly more 'scientific' than that of the OST!

The important difference, however, is the way in which each of these works generates its legitimacy. As with any literary author, Stross's work is ultimately the product of his individual creativity and/or judgement. The authority of the work derives from the authority of the author's imagination. In foresight this is known as 'genius forecasting', the method that underpinned the work of Herman Kahn, the 'grandfather of foresight' at the RAND Corporation in the 1950s. Genius forecasters marshal 'facts' in support of their argument – the molar weight of carbon atoms, or the Gross Domestic Product of the United States during 1960 – but the legitimacy of their narrative about the future relies on their personal credibility in the eyes of the reader. This is not problematic in science *fiction*; a successful work is an *exploration* of possible worlds, not a *projection* of a world that might be expected to come into being. Foresight work, though, is expressly created for performative purposes – to bring a future into being and to enrol readers as travellers along the pathway towards it. As such it is forced to justify its claim to authority.

At the time Kahn wrote *The Year 2000*, Western societies were already undergoing the metamorphosis that spawned countless new social movements (Kriesi, 1995) emerging into the public realm and demanding the right to be heard. The old elites – political, economic or intellectual – no longer spoke for 'everybody'. Trust in elite judgements can no longer be assumed, but must be earned by allowing the public access to their inner workings. In this context, genius forecasting is no longer viable. The OST scenarios respond by deploying three legitimising techniques. The first is an emphasis on the scenarios being the product of *groups* of experts rather than just individuals:

> Experts from the research community, business and the public sector took part in workshops that identified the key drivers and trends, and explored possible futures based on future uncertainties (Office of Science and Technology, 2006, p. 1).

This is a safety-in-numbers approach to creating trust, analogous to the stuffing of academic writings with references to other authors (Latour, 1987, p. 33). The second technique is that of ensuring public (i.e. non-elite) participation (Levi, 1998, p. 92). The OST report deploys public participation – *ranging from local and regional government to sixth-form students and representatives from businesses* (Office of Science and Technology, 2006, p. 6) – in a rather limited manner: not in the formation of the scenarios, but as a test bed for the finished product. Despite this, public involvement receives prominent billing in the Executive Summary. Finally, there is the attention given to an open display of the systematic methods used to develop the scenarios. The two-by-two matrix of variables used to frame the four scenarios is detailed, and each scenario is concluded with an analysis of the 'core and major causal loops' which together create a 'system' whose internal 'logic' (Office of Science and Technology, 2006, p. 26) generates the scenario. This is the same means by which science – which despite the travails of recent years remains the dominant form of knowledge generation – works: *the public display of one's methods ... is the essence of science* (Strong, 1983, pp. 69–70). This 'literary technology' (Shapin & Schaffer, 1989) acts to remove fallible, untrustworthy actors from the process of knowledge creation, leaving an impersonal mechanistic process that looks remarkably like a qualitative inscription device.

SYSTEMATIC QUALITATIVE FORESIGHT

Thus far, some common methods for dealing with uncertainty when planning large-scale infrastructure projects have been discussed, which neglect their fundamental properties as complex open systems. Two primary lines of attack on these attempts to leverage the future to act in the present are set out. First, that uncertainty is an unavoidable property of such systems, so that approaches which seek to achieve legitimacy by claiming to produce certainty are guilty of obfuscation. Second, that the attempt to proclaim simple certainties seeks to engineer away the political dimensions of present choices about the future. As such, current approaches are likely to struggle to achieve democratic legitimacy and to produce practical closure on major decisions about infrastructure investment.

To conclude, an alternative approach to dealing with uncertainty in such systems is sketched. This approach, labelled as Systematic Qualitative Foresight (SQF), is not intended to be a one-size-fits-all solution, or – at this stage – even to be comprehensive in its own terms. Rather, it is an attempt to think further through the problems identified, and to consider possible solutions. The challenge is to consider how qualitative methods can be used in ways that command confidence beyond that of the creative genius or the institutional authority involved in developing scenarios. SQF acknowledges the importance of sustaining its own legitimacy through the rigour, transparency and reflexivity of its methods rather than the power of its creative imagination, its use of language or its prior normative commitments. These allow it to seek legitimacy from contemporary democratic principles of inclusivity and openness. SQF dismisses quantitative analysis only where it claims unwarrantable and spurious precision, particularly through the complexity of the underlying calculations. Complexity alone does not guarantee the correctness of a model: it may simply close off engagement to all but a technocratic elite. Traditional technology assessments are often criticised for failing to embed technology in a social context, with the result that the anticipatory knowledge generated is not fit for purpose. SQF addresses this by bringing the social and the cultural into the process.

This view does not leave the future unchallengeable so much as uncertain. The future may be shaped by performative actions in the present but desired outcomes cannot be guaranteed. SQF recognises that any foresight work, no matter how much its visions are abstracted to numerical data, or book-ended by qualifiers about multiple options and nondeterminism, remains a political act, opening some possibilities while closing others. The IEA's *Energy Technology Perspectives* (2010) is a good example. This 700-page report, with 17 chapters spanning everything from energy networks to China and consumer behaviour to global finance, presents itself as a comprehensive assessment of energy *strategy and scenarios to 2050* (report subheading). However despite the forty-year time span and the stated aim of *demonstrate[ing] the many opportunities to create a more secure and sustainable energy future* (*ibid.*, p. 3), no discussion is given to the underlying socio-economic system that drives our resource demands. The word 'capitalism', for example, does not appear on any page. At numerous points, the report recognises the link between economic growth and rising energy demand and carbon emissions – the very problem the report seeks to solve – yet the value of unceasing growth is never questioned. By contrast the Sustainable Development Commission produced an entire report on this issue: *Prosperity without Growth?* (2009).

SQF does not rule out the use of quantitative data, but insists that it is only used where it adds, rather than removes, clarity. The 'systematic' element lies in the process by which variable data are collected and possible/likely interactions are identified. Policy elites may see the results as insufficiently quantitative to command technical legitimacy. However, it may be more valuable to them in its capacity to generate democratic legitimacy by presenting visions of the future that are accessible to, and engaging for, wider publics. As such, there is potential for integrating the output of SQF, the narratives constructed by public participants, into more traditional technology assessment, anchoring an overwhelmingly rational-technical framework in a real-world setting populated by non-rational actors.

It can be proposed that SQF should be based on the following, empirically derived, principles:

1. Traditional top-down technical-rational policymaking fails to incorporate many of the meaningful ways in which people experience technology. This neglect of the socio-cultural dimension of technology policy generates resistance, hostility and obstruction, which may block innovation.
2. Foresight exists to influence decisions in the present. Its influence is derived from the legitimacy that is achieved by its processes rather than the (unknowable) accuracy of its predictions.
3. Legitimacy is created by a diversity of expert and lay inputs, systematic display of the processes by which variables are identified and enrolled into the scenarios, openness about the values and assumptions embedded in the process of scenario generation, and 'fair dealing' with positions other than those held by the scenario authors.
4. Scenarios, whether qualitative or quantitative, are of limited value in telling what the future *will* bring.

5. Scenarios are of great value in engaging people in a discussion about what the future *might* bring, and what that *does* mean for the present.
6. Through guided scenario creation, lay groups can think through technology in terms meaningful to themselves, with a focus on linkages, interactions and choices.

The first stages of SQF are shared with other foresight methods and will not be detailed here. The initial, planning, stage involves the enrolment of relevant experts, stakeholders and bodies of literature into a network with the technology. At the scoping stage, the network relations are mapped and key features identified. While it is tempting to reduce network complexity by limiting membership to expert workshops and literature, this presents a danger that the project becomes irreversibly framed in a manner than blocks public engagement: what MacLean and Burgess call 'stakeholder capture' (2010, p. 487). Their method for avoiding this is to keep experts and lay participants largely apart. Experts were asked to present their evidence and to answer questions at the start of the process, but were then excluded from the subsequent deliberations. Another defence against stakeholder capture may be to include as diverse a body of experts as possible, to lessen the chance that any one perspective dominates. The aim is to navigate a course between being so informed that the participants feel constrained by the framing provided by experts and stakeholders, and spontaneously generated deliberation that is of limited value because it is unfocused and uninformed (*ibid.*, p. 494). The former risks outcomes that lack democratic legitimacy and the latter risks a lack of technical legitimacy.

Expert advice in SQF is structured through a matrix of key variables. These variables must be as generic as possible – for example 'socio-economic', 'technology' – as their purpose is simply to provide a basic framework from which scenario narratives can be created. There is a careful balancing act at this stage between reducing complexity to manageable levels, and not losing it completely, as it must be brought back during the engagement process. If necessary for the subject in question, a three-by-three 'cube' can be used. This may be the case if, for example, the exercise is concerned with how a particular technology (A) might interact with socio-economic factors (X) and environmental-physical factors (Y). However complexity rises exponentially with the number of cells in the matrix, making the process of engagement more difficult, so this should only be employed where essential.

SQF becomes most distinctive at the engagement stage. The process can be thought of by analogy to building with Lego. The systematically collected variable data is presented to participants in individual 'blocks'. These are used to 'build' the scenarios, around a theme set by the variable matrix. Each key variable is associated with a set of these blocks, each of which concerns a macro-level development. The metaphor of blocks to be assembled into coherent shapes focuses attention on the interaction between elements. Having assembled a structure, participants are tasked with inhabiting it through a narrative about themselves or a character of their choosing. Participants are encouraged to link macro-level developments with the micro-level of their everyday existence, to consider the choices made, and to identify the winners and losers. This approach echoes that of Miller and Bennett (2008), specifically their approach of encouraging participants to write science-fiction narratives to explore the human dimensions of technology change. However, it is recognised that the genre associations of science fiction would constitute a barrier to some participants. It can also be considered that this approach does not reflect our insistence on a comprehensive, systematic and disinterested assembly of information as the foundation of the engagement process. As a result, it is preferable to describe the participants' task as 'scenario development'. The information blocks could take a variety of forms, from bullet points on pieces of card to touch screen displays that could be assembled to create a scaffold for the scenario. The scenario itself could take a variety of forms, depending on audience and resources: a short story, a press or online feature, a short film for a regional or online news channel, a graphic novel, a playlet, etc. In a larger exercise, different groups of participants can be given different blocks and asked to present the outputs back to each other. In a smaller group, they may work on a single task but be encouraged by facilitators to reflect on and debate the choices that are being made as the output is developed.

This vision is a long way from that of CBA and other techno-rationalist attempts at planning, but it can be argued that it is a necessary one for decision-making regarding large public works in a twenty-first century democracy. Elite institutions no longer command sufficient inherent legitimacy to dominate such processes as they might once have done, and even in 'simpler times', overly determinist approaches tended to fail, as the Roskill Commission discovered. The SQF approach is not intended as to replace quantitative modelling, but rather to monitor its assumptions and processes and to address those normative policy dimensions that it inherently struggles to accommodate.

CONCLUSIONS

Foresight has become a popular tool in policymaking, such as in areas concerned with transport and climate change for a number of reasons. One is as a means of generating legitimacy for political decisions at a time when trust in authority and expertise are thought to have declined. The governance of new technologies has been a particular object of public suspicion. The failure to gain legitimacy for decisions about their future has in some cases, like GM crops, led to outright public rejection. SQF seeks to address this by generating democratic legitimacy – not through the projection of greater certainty about future events, but through the engagement of public groups with the uncertainties and potential of the innovation, both positive and negative, in ways that are designed to connect with their everyday knowledge and activities.

SQF promotes public engagement with socio-technical policymaking in ways that reduce the risk of 'stakeholder capture' (MacLean & Burgess, 2010) while still facilitating expert-informed discussion. Its approach is based on the belief that thinking about the future allows a creative space in which participants can escape their everyday constraints: the 'grindingly ordinary' elements whose very power lies in their unseenness (Shove, 2003, p. 1). A further benefit may be the response to the 'paradox of participation' (Collingridge, 1980), namely that technologies generally only become well known to the public at a point where they are already fixed. As Rogers-Hayden, Mohr, and Pidgeon (2007) point out in their discussion around upstream engagement with nanotechnology, 'many of the more radical impacts of nanotechnologies have yet to be envisioned, remain hypothetical, or will depend upon its convergence with other technologies' (p. 126). SQF can create fora in which both lay and expert participants are encouraged to imagine, and react to, future conditions in both technology and society.

A remaining challenge is the meaningful incorporation of such work into policymaking. The difficulties of ensuring that public engagement is not a mere exercise in public relations is well recognised (see, e.g. Gavelin, Wilson, & Doubleday, 2007). SQF cannot necessarily solve this problem, although it can limit it by the voluntary constraints adopted in its systematic approach. If, however, it were to be more extensively incorporated into technology assessment exercises, then one could at least ensure that the anticipatory knowledge informing policy was better fitted for purpose and that the outcomes of the exercise could fairly claim both democratic and technical legitimacy.

NOTES

1. In the current chapter, the following terminology is used: *Foresight* is used a collective term for efforts to enrol the future in current planning; *scenarios* is applied to qualitative foresight approaches; *modelling* to quantitative approaches.

2. A powerful example of danger of narrow visions comes from Adams (1971) analysis of the Roskill Commission's hyper-rationalist approach: in the early stages of the CBA, the value of a historic Norman church, threatened by the construction of a new airport, was based upon the amount for which it was insured against fire (though as Pearce (1998) points out, this valuation did not appear in the final report). It is difficult to imagine a more overtly symbolic artefact than a site of worship, and so it should not be surprising that many stakeholders felt that the fire insurance metric failed to speak to their understanding of the 'value' of the building.

3. Given the contested nature of this subject, it is worth underlining that this is not a statement of nihilism: that useful knowledge about future climate change is impossible to generate. What it is, is a recognition of the limits of accuracy in such a complex process and of the inevitable presence of normative elements in closing off ambiguity.

4. The Transport Secretary Phillip Hammond launched the Government's case for HS2 in the following terms: 'I believe that a national high speed rail network from London to Birmingham, with onward legs to Leeds and Manchester, could transform Britain's competitiveness as profoundly as the coming of the railways in the 19th century. It would reshape Britain's economic geography, helping bridge the north-south divide though massive improvements in journey times and better connections between cities – slashing almost an hour off the trip from London to Manchester' (Hammond, 2011).

REFERENCES

Adam, B. (2007). *Future matters: Action, knowledge, ethics*. Netherlands: Brill.

Adams, J. (1971). London's third airport. *Geographical Journal*, *137*, 468–504.

Adams, J. (1981). *Transport planning*. London: Routledge & Kegan Paul Ltd.

Anonymous. (1993). *Realising our potential: A strategy for science, engineering and technology*. White Paper, Select Committee on Science and Technology.

Banister, D., & Hickman, R. (2005). *Looking over the horizon: Visioning and backcasting for UK transport policy*. Retrieved from http://www.vibat.org/vibat_uk/pdf/vibatuk_method_issues.pdf

Bechky, B. A. (2003). Sharing meaning across occupational communities: The transformation of understanding on a production floor. *Organization Science*, *14*(3), 312–330.

Bernstein, P. L. (1996). *Against the gods: The remarkable story of risk*. New York, NY: Wiley.

Bohman, J. (1996). *Public deliberation: Pluralism, complexity, and democracy*. MIT Press.

Bowker, G. C., & Star, S. L. (2000). *Sorting things out*. Cambridge, MA: The MIT Press.

Brown, N., Rappert, B., & Webster, A. (Eds.). (2000). *Contested futures: A sociology of prospective techno-science*. Aldershot, UK: Ashgate Publishing Limited.

Carlsson, F., Johansson-Stenman, O., & Martinsson, P. (2004). Is transport safety more valuable in the air? *Journal of Risk and Uncertainty*, *28*, 147–163.

Coburn, T. M., Beesley, M. E., & Reynolds, D. J. (1960). *The London-Birmingham motorway: Traffic and economics*. Road Research Technical Paper, Pt I. Issue 46. London: HMSO.

Collingridge, D. (1980). *The social control of technology*. Milton Keynes: Open University Press.

Collins, H. M., & Evans, R. (2002). The third wave of science studies: Studies of expertise and experience. *Social Studies of Science*, *32*(2), 235–296.

Dennis, K., & Urry, J. (2009). *After the car*. Cambridge, UK: Polity Press.

EFMN. (2010). *EFMN – European foresight monitoring network*. Retrieved from http://www.efmn.info/

Espeland, W. N., & Vannebo, B. I. (2011). Accountability, quantification, and law. *Annual Review of Law and Social Science*, *3*(1), 21–43. doi:10.1146/annurev.lawsocsci.2.081805.105908

European Commission. (2007). *Public engagement in science*. Brussels: EC.

Friend, J. K., & Jessop, S. N. (1969). *Local government and strategic choice*. London: Tavistock Publications.

Gavelin, K., Wilson, R., & Doubleday, R. (2007). *Democratic technologies? The final report of the Nanotechnology Engagement Group (NEG)*. London: Involve.

Georghiou, L. (1996). The UK technology foresight programme. *Futures*, *28*(4), 359–377.

Girod, B., Wiek, A., Mieg, H., & Hulme, M. (2009). The evolution of the IPCC's emissions scenarios. *Environmental Science & Policy*, *12*(2), 103–118. doi:10.1016/j.envsci.2008.12.006

Goulden, M., & Balmer, A. S. (2009). Almost human: Scientific and popular strategies for making sense of "missing links". In S. Bauer & A. Wahlberg (Eds.), *Contested categories: Life sciences in society*. Farnham, Great Britain: Ashgate.

Hall, P. (1980). *Great planning disasters*. London: George Weidenfeld and Nicholson Limited.

Hall, P. (2005a). Divides that grow by rail. *Town and Country Planning* (June), 186–187.

Hall, P. (2005b). A low-cost high-speed rail line for the North. *Town and Country Planning* (February), 46–48.

Hammond, P. (2011, February 28). *High speed rail*. Retrieved from http://www.dft.gov.uk/press/speechesstatements/statements/hammond20110228

Hellsten, I. (2000). Dolly: Scientific breakthrough or Frankenstein's monster? Journalistic and scientific metaphors of cloning. *Metaphor and Symbol*, *15*(4), 213–221.

Henderson, D. (2005). SRES, IPCC and the treatment of economic issues: What has emerged?. *Energy & Environment*, *16*(3–4), 549–578.

Horswill, M. S., & McKenna, F. P. (1999). The effect of perceived control on risk taking. *Journal of Applied Social Psychology*, *29*, 377–391.

IEA. (2010). *Energy technology perspectives 2010 executive summary*. Paris, France: IEA Publications. Retrieved from http://www.iea.org/techno/etp/etp10/English.pdf

Inman, M. (2011). Opening the future. *Nature Climate Change*, *1*(1), 7–9. doi:10.1038/nclimate1058

IPCC. (2000). *Special report on emissions scenarios: A special report of working group III of the intergovenmental panel on climate change*. New York, NY: Cambridge University Press.

Irwin, A. (1995). *Citizen science: A study of people, expertise and sustainable development*. Oxon: Routledge.

King, J. (2007). *The king review of low carbon-cars. Part I: The potential for CO_2 reduction*. London: HM Treasury.

Kriesi, H. (1995). *New social movements in Western Europe: A comparative analysis*. Minneapolis, MN: University of Minnesota Press.

Latour, B. (1987). *Science in action*. Cambridge: Harvard University Press.

Latour, B., & Woolgar, S. (1986). *Laboratory life: The social construction of scientific facts*. Princeton, NJ: Princeton University Press.

Levi, M. (1998). A state of trust. In M. Levi & V. Braithwaite (Eds.), *Trust and governance*. New York, NY: Russell Sage Foundation.

Ling, D., & Xiao, P. (2009). Simulation research on chaos characteristics of a class of macroscopic traffic models. *International conference on engineering computation, ICEC '09* (pp. 135–138), IEEE Computer Society, Washington, DC.

Lorenz, E. N. (1963). Deterministic nonperiodic flow. *Journal of Atmospheric Sciences*, *20*, 130–141.

MacLean, S., & Burgess, M. M. (2010). In the public interest: Assessing expert and stakeholder influence in public deliberation about biobanks. *Public Understanding of Science*, *19*(4), 486–496.

Martin, A., & Lynch, M. (2009). Counting things and people: The practices and politics of counting. *Social Problems*, *56*(2), 243–266.

Miller, C. A., & Bennett, I. (2008). Thinking longer term about technology: Is there value in science fiction-inspired approaches to constructing futures? *Science and Public Policy*, *35*(8), 597–606.

Mishan, E. J. (1970). What is wrong with Roskill? *Journal of Transport Economic Policy*, *4*, 221–234.

Moen, B.-E., & Rundmo, T. (2006). Perception of transport risk in the Norwegian public. *Risk Management*, *8*(1), 43–60.

Nelson, N., Geltzer, A., & Hilgartner, S. (2008). The anticipatory state: Making policy-relevant knowledge about the future. *Science and Public Policy*, *35*(8), 546–550.

Nerlich, B. (2008). Powered by imagination: Nanobots at the science photo library. *Science as Culture*, *17*(3), 269–292.

Nerlich, B., Clarke, D. D., & Dingwall, R. (2001). Fictions, fantasies, and fears: The literary foundations of the cloning debate. *Journal of Literary Semantics*, *30*(1), 37–52.

Norman, D. (1993). *Things that make us smart*. New York, NY: Addison-Wesley Publishing Co.

Office for National Statistics. (2011). *National population projections: 2010-based statistical bulletin*. Office for National Statistics, Newport. Retrieved from http://www.ons.gov.uk/ons/rel/npp/national-population-projections/2010-based-projections/stb-2010-based-npp-principal-and-key-variants.html. Accessed on May 29, 2012.

Office of Science and Technology. (2006). *Intelligent infrastructure futures. The scenarios – towards 2055*. London: Crown.

Orwell, G. (1998). *Nineteen eighty-four*. London: Penguin.

Owens, S., & Cowell, R. (2011). *Land and limits: Interpreting sustainability in the planning process* (2nd ed.). Cornwall: Routledge.

Parry Lewis, J. (1971). *The forecasts of Roskill*. Occasional Paper No. 11. Manchester: Centre for Urban and Regional Research, University of Manchester.

Pearce, D. (1998). Cost-benefit analysis and environmental policy. *Oxford Review of Economic Policy*, *14*(4), 84–100.

Raupach, M. R., Marland, G., Ciais, P., Le Quéré, C., Canadell, J. G., Klepper, G., & Field, C. B. (2007). Global and regional drivers of accelerating CO_2 emissions. *Proceedings of the National Academy of Sciences*, *104*(24), 10288–10293.

Rogers-Hayden, T., Mohr, A., & Pidgeon, N. (2007). Engaging with nanotechnologies: Engaging differently? *NanoEthics*, *1*, 123–130.

Roskill Commission (1971). *Commission on the third London airport*. London: HMSO.

Shapin, S., & Schaffer, S. (1989). *Leviathan and the air-pump: Hobbes, Boyle, and the experimental life*. Princeton, NJ: Princeton University Press.

Shove, E. (2003). *Comfort, cleanliness and convenience: The social organization of normality*. Milton Keynes: Berg.

Shove, E. (2010). Beyond the ABC: Climate change policy and theories of social change. *Environment and Planning A*, *42*(6), 1273–1285.

Star, S. L., & Griesemer, J. R. (1989). Institutional ecology, 'translations' and boundary objects: Amateurs and professionals in Berkeley's museum of vertebrate zoology, 1907–39. *Social Studies of Science*, *19*(3), 387–420.

Stephenson, N. (2011). Innovation starvation. *World Policy Journal*, *28*, 8–9.

Sterling, B. (1999). A century of science fiction. *Time*. Retrieved from http://www.time.com/time/magazine/article/0,9171,990631-1,00.html

Strong, P. M. (1983). The rivals: An essay on the sociological trades. In R. Dingwall & P. Lewis (Eds.), *The sociology of the professions: Lawyers, doctors and others*. London: Macmillan.

Stross, C. (2007). *Shaping the future – Charlie's diary*. Retrieved from http://www.antipope.org/charlie/blog-static/2007/05/shaping-the-future.html

Sugiyama, Y., Fukui, M., Kikuchi, M., Hasebe, K., Nakayama, A., Nishinari, K., … Yukawa, S. (2008). Traffic jams without bottlenecks: Experimental evidence for the physical mechanism of the formation of a jam. *New Journal of Physics*, *10*, 1–7.

Sustainable Development Commission. (2009). *Prosperity without growth? The transition to a sustainable economy*. London: Sustainable Development Commission.

van Ruijven, B., Urban, F., Benders, R. M. J., Moll, H. C., van der Sluijs, J. P., de Vries, B., & van Vuuren, D. P. (2008). Modeling energy and development: An evaluation of models and concepts. *World Development*, *36*(12), 2801–2821. doi:10.1016/j.worlddev.2008.01.011

Wells, A. T., & Chadbourne, B. D. (Eds.). (2007). *Introduction to aviation insurance and risk management* (3rd ed.). Malabar: Krieger Publishing Company.

Wilsdon, J., & Willis, R. (2004). *See-through science: Why public engagement needs to move upstream*. London: Demos.

Wynne, B. (1992). Misunderstood misunderstanding: Social identities and public uptake of science. *Public Understanding of Science*, *1*, 281–304.

Wynne, B. (2010). Strange weather, again: Climate science as political art. *Theory, Culture & Society*. Retrieved from http://tcs.sagepub.com/content/27/2-3/289.full.pdf+html

CHAPTER 3

AN INTERNATIONAL DIMENSION: AVIATION

Lucy Budd and Tim Ryley

ABSTRACT

Purpose – *To examine the relationship between aviation and climate change, and the international dimensions of air transport.*

Methodology/approach – *A review of aviation's impacts on the global climate, mitigation strategies to reduce this impact, and the possible consequences of climate change for commercial aviation.*

Findings – *Although a range of mitigation measures have been developed and implemented to reduce aircraft emissions in the short term, with some environmental benefit, there is a real need for the aviation sector to identify the possible impacts of climate change on air travel operations, both to aircraft in flight and to operations at airports. A further challenge will be to devise adaptation plans that will address the vulnerabilities and thus ensure safe aviation-related operations.*

Social implications – *The climate change impacts of aviation will adversely affect society. In addition, some individuals may have to reduce or stop flying as a result of increased taxes and legislation implemented in response to climate change.*

Transport and Climate Change
Transport and Sustainability, Volume 2, 39–64

ISSN: 2044-9941/doi:10.1108/S2044-9941(2012)0000002006

Originality/value of paper – *There is a novel focus on the adaptation challenges for the aviation industry in response to climate change.*

Keywords: Aviation; climate change; mitigation; adaptation

INTRODUCTION

In a little over 100 years between the first tentative 12-second flight of the Wright Brothers' biplane and today, civil aviation has rapidly developed from a dangerous and unreliable form of mobility into a multi-billion dollar commercial enterprise that facilitates the routine international mobility of two billion passengers and tens of millions of tonnes of airfreight every year. Although the current economic downturn has, albeit temporarily, depressed consumer demand for air travel, one of the defining features of mid-twentieth and early twenty-first century commercial aviation has been its seemingly inexorable expansion. Between 1945 and 2000 global air passenger traffic grew at an average rate of 12% per year (Hanlon, 2007), and current forecasts suggest that the industry will grow by 4–5% per annum between now and 2020 (IATA, 2011).

For affluent members of the global economy, flying constitutes a normal and routine mode of long-distance transportation that enables professional and personal relationships to be conducted at a distance and renders even the most far-flung destinations within reach. The aviation industry, furthermore, provides direct and indirect employment to hundreds of thousands of people worldwide, enables individuals, goods, capital and knowledge to flow around the world, and metaphorically brings people and places closer together in time and space. Yet while governments, businesses and individual citizens have variously championed aviation's apparent socio-economic and cultural benefits, critics have drawn attention to the socio-spatial inequalities in the provision and use of air services and climate scientists and environmentalists have highlighted the unpalatable yet inescapable truth that aircraft operations, like so many processes of capitalist production and human activity, impose a significant burden on both the local and the global environment.

The manufacture, transportation, and decommissioning of airframe and engine components, for example, generates greenhouse gases and toxic waste, while routine aircraft operations are known to contaminate groundwater, degrade the acoustic environment and local air quality around airports and perturb the global climate. Growing public awareness of aviation's deleterious environmental effects, combined with consumer unease about airport expansion, has caused aviation to become an increasingly emotive subject with the 'rights' of airline passengers to fly, relatively unhindered, around the world juxtaposed against the rights of others to be protected from the noise, human health implications and global environmental impacts such aerial mobility creates.

In a departure from much of the existing scientific and popular literature on the subject of aviation and the environment (see Bows, 2010; Cairns & Newson, 2006; Daley, 2010; Randles & Bows, 2009), this chapter focuses not only on aviation's impacts on the global climate but also on the mitigation strategies that are being pursued to lessen aviation's climate impact, the possible consequences of climate change for commercial aviation, and the adaptation plans that will be required.

AVIATION AND CLIMATE CHANGE

Although commercial aviation is by no means the largest anthropogenic source of greenhouse gas emissions (current estimates suggest the sector is responsible for generating 2–3% of all carbon dioxide (CO_2) emissions and 12% of all transport-borne CO_2 pollution (IATA, 2011)), continued emissions reductions in other sectors combined with a net increase in flights means that commercial air travel represents one of the fastest growing sources of pollution (*ibid.*, 2011). In the UK alone, increasing demand for air travel means that CO_2 emissions from aviation are forecast to grow from 37.5 million tonnes in 2005 to 59 million tonnes by 2030 despite the introduction of new, more fuel efficient, aircraft (Anderson, Bows, & Footitt, 2007; Department for Transport, 2007). At the global level, as much as 5% of all CO_2 emissions may be attributable to aviation by 2050 (IATA, 2011).

Although concern about the possible effects of aircraft pollution on the environment was first articulated in the late 1960s and early 1970s (Broderick, 1978), it is only within the last few years that scientists have begun to understand not only the complex chemical interactions that occur between different pollution species but also begun to quantify how their presence alters the atmosphere's radiative balance (the proportion of solar radiation that is absorbed by the atmosphere as opposed to being reflected back out into space) and thus contributes to climate change (see Bows, 2010;

Bows, Anderson, & Upham, 2008; Bows, Anderson, & Mander, 2009; Lee et al., 2009; Mendes & Santos, 2008; Randles & Bows, 2009).

When jet fuel is burnt in an aircraft's engine, a range of pollution species, including (but not limited to) water vapour, carbon dioxide, nitrous oxides and particulates, are produced. While the production of certain chemical species, including CO_2, is a direct function of fuel burn, the quantity and proportion of other pollutants that are formed varies according to a range of parameters, including core engine temperature, thrust settings, the age and the type of the engine, the precise chemical composition of the fuel and how well an individual engine has been maintained. Furthermore, the impact these different pollutant species exert on the global climate is understood to vary depending on the altitude at which they are emitted and the prevailing atmospheric conditions at the time of their release (see Cairns & Newson, 2006).

Volumetrically, water vapour is the most significant component of aircraft exhaust yet its contribution to anthropogenic climate change has only recently been recognised. In the cold and rarefied air at 30,000 ft, the water vapour that is contained within the exhaust plume immediately condenses. If the upper atmosphere is relatively dry, the vapour quickly evaporates, but if the air is saturated, evaporation cannot occur and white contrails (condensation trails) form behind the aircraft. Depending on the density of local air traffic and prevailing wind and atmospheric conditions, these contrails can persist for many hours and can seed vast cirrus cloud systems that can cover hundreds of square kilometres of sky.

It is believed that the presence of these anthropogenic cirrus clouds interferes with normal atmospheric thermal exchange processes. Depending on the latitude and the time of day that they form, contrails can variously cool the earth during daylight hours (by reflecting sunlight back out into space) but warm the planet at night by acting as a blanket and trapping heat. Research by Williams, Noland, and Toumi (2002) has suggested that one possible way to reduce the climate change effect of contrails would be to restrict cruise altitudes to prevent contrail formation. However, the balance between the potential climate benefits and dis-benefits of this proposal needs careful consideration. Lower cruise altitude, while hindering contrail formation, would result in higher fuel burn and a net increase in emissions as most jet aircraft are at their most fuel efficient at cruising altitudes of around 35,000 ft.

After water vapour, the second most voluminous component of aircraft exhaust is carbon dioxide (CO_2). Possibly owing to the fact that its production can be easily quantified (3.16 kg of CO_2 is always produced for every 1 kg of fuel that is burnt, irrespective of flight stage, engine setting or altitude) and that its impact on the global climate is well known, the causes and consequences of rising atmospheric concentrations of CO_2 have received considerable scientific and political attention. Although, molecule-for-molecule, CO_2 is a relatively weak greenhouse gas, the quantities with which it is released and the fact that it can remain in the atmosphere for over 100 years render it a considerable cause for concern (Archer, 1993). As a consequence, most of the existing mechanisms for mitigating greenhouse gas emissions, such as the European Union's Emissions Trading Scheme (EU ETS), have focused on carbon dioxide (see Mendes & Santos, 2008, for a discussion of the economic instruments that are being used to address CO_2 emissions in Europe).

In addition to water vapour and carbon dioxide, other non-CO_2 species, including nitrous oxides (which while not greenhouse gases themselves influence the production and destruction of the greenhouse gases ozone and methane), are also believed to exert a significant short-term influence on the radiative balance and further exacerbate aviation's climate impact (Cairns & Newson, 2006; Mendes & Santos, 2008).

Despite a lack of any global scientific or political consensus on what constitutes an 'acceptable' level of greenhouse gases or climate change, many scientists believe that any global temperature rise that exceeds 2°C above pre-industrialisation levels will have 'dangerous' consequences for the earth and its ecosystems (see Bows et al., 2009). Growing awareness of aviation's impact on the global climate has raised challenging questions about our reliance and apparent 'addiction' to flying. Environmental groups, including Greenpeace, enoughsenough.org, Airport Watch and Plane Stupid, local anti-airport expansion campaigns such as HACAN-Clearskies (who oppose expansion at London's Heathrow airport), individual travellers, and even airlines, airports and air transport regulators, have all expressed concern about aviation's environmental impact. Reflecting their different political (and often commercial) standpoints and ideologies, these groups have promoted a range of policy options, from imposing high taxes to dramatically reduce consumer demand, scrapping all proposed airport developments and promoting alternative slower modes of travel, to operating existing aircraft in a more fuel-efficient efficient manner and investing in new and less carbon intensive aircraft technologies, to reduce the sector's present and future climate impact.

CLIMATE CHANGE MITIGATION MEASURES FOR AVIATION

In recognition that aviation contributes to climate change and the urgent need to improve the sector's environmental performance and mitigate some of its most serious environmental effects, a range of new technologies, operating procedures and policy instruments have been introduced. The International Air Transport Association (IATA), for example, has identified a range of possible options, including retrofitting aircraft with winglets to reduce aerodynamic drag, replacing the current fleet of aircraft with new designs and using alternative fuels, which, it is claimed, could reduce aviation emissions by 20–35% per aircraft (IATA, 2009). Elsewhere, individual airports and air navigation service providers have reconfigured the airspace architecture around airports to accommodate more efficient flight paths, while airlines are trialling alternative fuels and less carbon intensive operating practices.

New Operating Procedures

During the last couple of years, a range of revised operating procedures have been introduced by airlines, airports and air navigation service providers to reduce fuel costs and lower emissions. Many airlines, including easyJet at Stansted,[1] have implemented single-engine taxi operations at major airports and instruct their pilots to shut down one engine after vacating the runway on arrival and taxi to the gate using power from the remaining engine. This practice lowers engine wear and reduces fuel consumption, fuel costs and emissions. The technique is, however, only currently certified for use with a limited number of aircraft and it does impose sequencing implications for air traffic control (aircraft taxiing on a single engine need to be given a clear and uninterrupted route to their gate) and is therefore not suited for use at all airports.[2] Other operational improvements include enhanced flight sequencing which helps ground movement controllers reduce the length of time (and associated emissions) aircraft spend queuing on a taxiway before take-off.

As well as improving environmental performance on the ground, new airspace protocols and increasingly sophisticated navigation and communications technologies are enabling more efficient use of airspace. Improved schedule coordination and slot control increasingly facilitates continuous climb departures (CCDs). CCDs expedite an aircraft's climb to its initial cruising altitude and dispense with the need for an aircraft to climb inefficiently in stages and be held at intermediate altitudes. Similarly, continuous descent arrivals (CDAs) enable aircraft to continuously descend from their cruising altitude to the runway without being held at intermediate flight levels. Both CCDs and CDAs improve flight efficiency and reduce fuel burn and emissions, but they may require changes to be made to the airspace architecture around airports and demand that flight crew and air traffic controllers are trained in their use.

In the en-route flight phase, the increased use of user-defined trajectories, the identification and use of optimum cruise altitudes, precision area navigation, flexible use of airspace and functional airspace blocks are helping to maximise en-route capacity, minimise unnecessarily circuitous and inefficient routings and reduce emissions. In Europe, the Single European Skies initiative ultimately aims to eliminate many of the current inefficiencies that result from the continent's fragmented airspace structure (a legacy of the airspace being divided along national boundaries). However, unresolved geopolitical issues have restricted the initiative's progress.

New Technologies

While new operating techniques do confer some environmental benefit, they do not tackle the root cause of the majority of aviation-borne pollution, that is, the aircraft themselves. Despite numerous advances in avionics and automated flight management systems since the end of Second World War, the fuels, propulsion systems and basic airframe configurations (i.e. slender fuselages, jet engines, and swept back wings) have not changed substantially during the last 60 years and a number of new radical aircraft and engine designs have been proposed in the hope that they will confer improved aerodynamic and environmental performance (Marks, 2003). Radical designs, including blended wing airframes, strut-braced wings, and open rotor engines, have been proposed as one way to reduce fuel burn and lower emissions (Daviss, 2007). However, these ideas remain at the experimental stage and it will take many years of testing and development before they are certified for commercial use.

A further area that has seen considerable development in recent years is in the area of aviation alternative fuels, particularly biofuels (see Warwick,

2007). Worldwide, the commercial aviation industry consumes around 5 million barrels of jet fuel per day (around 240 million tonnes per year). Fuel represents a major proportion of an airline's costs and airlines are particularly vulnerable to fluctuations in cost and supply. The dynamic price of oil, combined with concerns about long-term energy security and increasingly stringent international targets for reducing CO_2 emissions, have collectively driven the development of alternative, more environmentally sustainable, aviation fuels.

Although the process of trans-esterification, through which certain types of vegetable oil can be converted into biodiesel, has been performed since the 1940s, it is only very recently that aviation-grade biofuel has been produced. The earliest (so-called 'first generation') biofuels were used to power internal combustion engines in road vehicles, but while they were suitable for use in road transport, the high oxygen content and low freezing point of the fuels rendered them unsuitable for use in aircraft as the fuels were heavy and would congeal in the cold air at 35,000 ft. New production techniques appear to have overcome many of the limitations associated with the earlier fuels and a growing number of major international airlines, including KLM, Virgin Atlantic, Air New Zealand, Iberia, Thomsonfly and Continental Airlines, have performed test flights during 2009, 2010 and 2011 using different combinations and blends of biofuel derived from, amongst other things, algae, jatropha, coconuts and camelina, an inedible green shrub (O'Connell, 2009).

While these flights confirmed that biofuels could be used to safely power large commercial jet aircraft, the challenges of ensuring an adequate supply of feedstock and accurately quantifying the life cycle emissions of the fuels remain, as yet, unresolved. Nevertheless, indications from within the industry suggest that major airlines think that biofuels are viable alternatives to conventional fuel. In February 2010, British Airways (BA) announced plans to construct a biofuel plant in East London that would convert domestic refuse into aviation-grade biofuel that would power BA flights from London City Airport (Arnott, 2010). Later that year, the German national carrier, Lufthansa, announced its intention to use biofuels to power revenue passenger services between Hamburg and Frankfurt (Lufthansa, 2010).

In addition to pursuing biofuels, a number of independent research groups are experimenting with other alternative energy sources for aircraft, including solar power, hydrogen and electricity (Bradbury, 2007; Bremner, 2008; Harris, 2003). However, at the time of writing, these power sources remain at the early stages of development.

Behavioural Change

Despite the industry's focus on alternative technologies and more environmentally efficient operating procedures, one the easiest and quickest ways in which the environmental impact of aviation could be reduced is if individual travellers flew less frequently, people consumed fresh, local produce that is in season and did not demand a ready supply of consumer goods which have been airfreighted half way around the world. Tellingly, however, recent research has indicated that while some individuals are prepared to be more 'green' in certain areas of their lives (by, for example, cycling or walking rather than taking the car for short journeys), the majority of people are unwilling to change their air-dependent lifestyles (see Barr, Shaw, Coles, & Prillwitz, 2010; Becken, 2007; Department for Environment, Food and Rural Affairs, 2008a; Leroux, 2009). Ryley and Davison (2008) discovered that individual propensity to fly is so strong that only a small proportion, 8%, of their survey respondents were trying to reduce the number of flights they took for environmental reasons.

In an attempt to atone for some of the environmental damage a flight causes, a number of airlines and independent providers offer passengers the option to 'offset' the carbon emissions their flight creates by investing in schemes that remove an equivalent amount of CO_2 from the atmosphere, or prevent its release elsewhere by, for example, planting trees or supporting the use of alternative energy (Gössling et al., 2007). While some claim that carbon offsetting is a useful tool for raising consumer awareness, critics have suggested offsetting constitutes little more than a way for profligate polluters to soothe their conscience (Ma'anit, 2006).

Given that many people are apparently unwilling to reduce their personal consumption of air travel and airfreighted goods, attempts to get people out of the air and on to the ground would appear doomed to failure. However, as the development of high-speed rail lines in continental Europe has shown, trains have the potential to complete effectively with aircraft on short-haul routes (Keeley, 2009).

When alternative modes of travel do not represent a viable option, whether because of time or monetary constraints (ironically it can be cheaper to take a domestic flight in the UK than undertake the same journey by rail), airlines and airports are actively trying to reduce the environmental impact of their operations. Airlines, including the UK-based Flybe and easyJet, have used consumer concerns about the environment in their marketing and have stressed that their modern, more fuel efficient aircraft and relatively high passenger load factors mean that the environmental impact of their services is lower than that of many of their competitors. Such strategies may confer significant commercial benefit for the airlines concerned – a survey of passengers at Liverpool John Lennon Airport in the UK revealed that nearly half of the respondents differentiated between airlines based on their perceived environmental 'friendliness' (Mayer, Ryley, & Gillingwater, 2011) – but may ultimately lead to a net increase in flights and emissions if passengers believe that certain services are less environmentally damaging and fly more as a consequence.

A further behavioural issue, and of key concern for airports, relates to surface access and the need to reduce the proportion of trips staff and passengers make by private transport (T. Budd, Ison, & Ryley, 2011). A range of technology and policy options have been proposed, including vehicle sharing, home tele-presence (to reduce the numbers of so-called 'kiss and fly' trips), technologies for facilitating off-airport check-in and baggage drop off and subsidies for public transport. However, while there is a moral (and, increasingly regulatory) imperative for airports to increase public transport use to and from the site, there is also an understandable reluctance to reduce trips by private cars as car parking represents a valuable source of non-aeronautical revenue for airports.

However, all the interventions previously described, while undoubtedly well intentioned, invariably only offer modest or one-off efficiency gains. Consequently much more challenging, politically unpalatable, and demanding management measures, such as higher ticket prices and aviation taxes and a moratorium on global airport expansion, may be required if aviation is to significantly reduce its climate impact.

Taxation/Legislation

Green taxes are one way of restraining demand (and emissions) and encouraging an uptake of 'cleaner' technology. At a national level, different countries have imposed a range of policy and/or financial incentives to either actively encourage or dissuade people from flying. Discussions of the national aviation taxes that are applied across a range of countries revealed that competition between nations regarding domestic airlines, airports and tourism influence national taxes and the provision of financial incentives.

In the UK, a passenger charge, known as Air Passenger Duty (APD), is levied on all passengers flying out of the country. The APD was first introduced in 1994 and has been subsequently revised. It constitutes a flat fee which varies according to the distance flown and class of travel, with long-haul flights in first and business class attracting the highest charges. The APD has been criticised for not considering the age, size, fuel efficiency or passenger load of the aircraft and for the fact that small aircraft (carrying fewer than 20 people) and cargo aircraft are exempt from the charge.

In order to assess the possible impact of APD on consumer behaviour, Mayor and Tol (2007) modelled domestic and international tourist flows in response to different APD charges. They discovered that an increase in APD in 2007 had a small but perverse effect on consumer behaviour because tourist destination choice is driven by relative price. Thus, a departure tax can make distant destinations more appealing, not less, as the charge for longer routes constitutes a much lower proportion of the total holiday cost than it does for short-haul destinations. One possible outcome of this effect is that UK aviation emissions may increase as people choose to fly on long-haul routes which are flown using larger and heavier aircraft, or UK consumers may choose to fly from other airports in Europe which do not impose such charges. Although the levels of APD have since been tweaked so that there is currently a four-destination band structure based on geographical distance, each having two rates of duty depending upon the class of travel (for further details see HM Custom and Excise, 2010), the system arguably requires further refinement. Critics of the scheme have also called for greater transparency about where the APD revenue is spent, with Ryley, Davison, Bristow, and Pridmore (2010) suggesting that it should be spent on aviation-related mitigation measures.

At a European level, much of the focus on aviation-related mitigation measures relates to the EU ETS, a 'cap and trade' system which aims to make the polluter pay by creating a market for carbon in which the 'right' to pollute is a tradable asset (Mendes & Santos, 2008). Phase one of the EU ETS, which represented the largest international multi-sector greenhouse gas emission trading system in the world, was introduced in January 2005 and covered 11,500 energy-intensive installations, including oil refineries, iron and steel plants and power stations, across the EU. In 2006, the European Commission proposed that airlines should be included within the scheme as way to cap carbon dioxide emissions and provide a financial incentive to promote the uptake of new technologies and more environmentally sustainable operating practices.

The resulting directive, which was published in 2008, stipulated that all commercial flights operated by an aircraft with a maximum take-off weight greater than 5,700 kg which operate in to or out of any EU member airport would be included in the ETS from 2012. Under the regime, airlines would be given an initial set of free allowances. They would only be allowed to pollute above their quota if they purchased additional allowances from other airlines or industrial sectors which had not used their full allocation (Albers, Buhne, & Peters, 2009). Using this mechanism, the EU hopes to be able to stabilise carbon emissions. The inclusion of aviation into the ETS is, however, highly controversial, with some non-EU airlines claiming that it contravenes existing international aviation regulations, including the 1944 Chicago Convention on International Civil Aviation.

AVIATION ADAPTATION TO A CHANGING CLIMATE

Despite growing awareness of aviation's impact on the global environment and official recognition that some degree of climate change is now inevitable (IPCC, 2007), relatively little research has explored aviation's vulnerability to climate change. While aviation is, of course, not the only transport mode that will be affected by climate change (see Department for Transport, 2010; Dobney, Baker, Quinn, & Chapman, 2009; Jaroszweski, Chapman, & Petts, 2010; Koetse & Rietveld, 2009; and other Chapters in the book for examinations of climate change's likely impacts on surface transport modes), the globally interconnected nature of contemporary airline networks, combined with rising numbers of flights and growing capacity constraints at major airports, render the sector particularly vulnerable to weather-related disruption and make schedule recovery following a disruptive event very difficult.

In December 2010 and January 2011, prolonged periods of heavy snowfall and sub-zero temperatures affected much of north-western Europe. Snow and ice disrupted power supplies, blocked roads, shut schools and forced the temporary closure of dozens of major airports. Thousands of flights were cancelled, hundreds more re-routed or rescheduled and tens of thousands of passengers had their Christmas and New Year travel plans severely disrupted. The resulting chaos was quickly blamed not on the weather (which had been forecast) but on the failure of airlines and airport operators to be 'proactive' and prepare for such conditions (Milmo, 2011; Rayner, Millward, & Swinford, 2010).

In the UK, the Labour Party Chairwoman of the Commons Transport Select Committee, Louise Ellman, argued that it was unacceptable for politicians and airport operators to claim that that conditions were exceptional as the winter of 2010–2011 represented the 'third bad winter in a row' that the UK had experienced (cited in McKie & Doward, 2010, p. 1). Leaving aside controversial, and perhaps irresolvable, questions concerning the extent to which the wintry conditions were either caused or exacerbated by global changes in weather and precipitation patterns brought about by climate change, the disruption brought aviation's vulnerability to adverse weather conditions into sharp relief. It also highlighted an urgent need for the industry to strengthen resilience and preparedness planning, and to develop effective strategies that will enable it to adapt to, and function in, a changing climate.

In light of climate change predictions, the need to strengthen the aviation industry's resilience to climate change is becoming increasingly acute.

Growing Industry Recognition

In 2008, Eurocontrol, the European organisation responsible for the safety of air traffic within the continent, unequivocally stated that 'climate change *will* affect aviation' (Eurocontrol, 2008, p. 2, our emphasis). In order to understand the nature of these effects and raise awareness that climate change will impact on the sector, the organisation commissioned a series of climate adaptation studies. The resulting report (Eurocontrol, 2011) suggested that climate change may adversely affect the European aviation network by:

- Changing patterns of tourist demand (owing to increased thermal stress and water shortages in certain areas during the summer months and reduced snow cover in others in winter);
- Causing rising sea levels to inundate low-lying airports;
- Resulting in changes in weather patterns that will lead to increased delays, necessitate the re-routing of aircraft, damage airport and air navigation infrastructure and make routine ground handling operations more difficult.

At a national level, governments have also begun to examine the impact that climate change may have on key transportation and utility infrastructures. As a result of the 2008 Climate Change Act, the UK Government's Department for the Environment, Food, and Rural Affairs (DEFRA) has requested that the organisations responsible for operating assets of national importance, including energy suppliers, water companies and transport providers, compile detailed assessments of their vulnerability to the short, medium, and long-term consequences of climate change and describe the steps that they are taking to mitigate or adapt to these threats (DEFRA, 2008b). In the context of aviation, London's Stansted Airport identified 35 separate risks that they consider climate change poses to the safety and integrity of their operations (Jefferson, 2011). Many of these risks are directly related to the consequences of rising surface temperatures and changes to existing weather and precipitation regimes.

Rising Surface Temperatures

Predicted rises in mean global surface temperatures will have a number of significant implications for the aviation industry. Some of these impacts will be felt immediately, whereas others may only begin to manifest themselves after many years. At a local level, the ambient air temperature has a direct and tangible impact on airport and airline operations. Very high temperatures degrade the take-off performance of aircraft, can cause airport infrastructure to buckle or melt and make normal aircraft turnaround operations more hazardous, while sub-zero temperatures can result in airside infrastructure freezing and necessitate the expensive and time-consuming removal of snow and ice.

Ambient air temperature affects air density, and air density affects the lifting capacity of a wing. Broadly speaking, lower air densities generate less lift at any given airspeed than higher air densities, as there are fewer molecules of air moving over the surface of the wing (Sealy, 1957). Consequently, when air density is low, take-off and landing speeds have to be higher to generate sufficient lift to enable aircraft to become, and remain, airborne. Higher surface temperatures, therefore, degrade aircraft performance by lowering the air density and reducing a wing's ability to generate lift and the engine's ability to generate thrust. As a consequence, take-off speeds have to be higher (increasing fuel burn, engine wear, and maintenance costs) and aircraft may require longer take-off runs before they can safely become airborne. In the case of 'hot and high' altitude airports with relatively short runways, higher surface temperatures may ultimately prevent certain larger aircraft type from operating into the facility and may demand that runways are extended so that payload restrictions do not have to be imposed.

In addition to reducing aircraft performance, higher surface temperatures can also damage airport infrastructure by causing it to buckle or melt. While runways are designed to withstand high temperatures and friction, the structural integrity of the tarmac used on aircraft aprons and taxiways begins to degrade when exposed to temperatures of 32°C and above (Jefferson, 2011). If sections of taxiways 'melt' in the heat and become unable to support the weight of aircraft, they must be taken out of service until the surface can be remade.

A related concern about rising surface temperatures and the temperature of airfield tarmac relates to the safety of ground personnel who are involved in aircraft turnarounds. The flashpoint of jet fuel is 38°C. At London's Stansted Airport, temperatures peaked at 37.3°C in the summer of 2003 and the temperature on the tarmac can reach 80°C (Jefferson, 2011). In the event of an overfill or fuel spillage, there is a danger that the spilt fuel could spontaneously ignite when it comes into contact with a hot tarmac. As a result, any increases in surface temperatures will require changes to existing aircraft fuelling arrangements at airports where the 38°C threshold is likely to be reached or exceeded. New costly and potentially time-consuming control measures, such as not allowing passengers to board at the same time as an aircraft is being refuelled and having fire vehicles attend all refuelling activities, may have to be considered to counter this threat. Such interventions, if suddenly imposed, would have serious financial implications for airlines (as aircraft couldn't be turned around as quickly) and airports (as individual aircraft would fly a reduced number of services as a result).

In the medium to longer term, rising temperatures may also lead to an increased fire risk on and around the airfield and alter energy demands in airport buildings. If, as is predicted, winters in the northern hemisphere become milder and wetter and summers hotter and drier (see Hulme et al., 2002), additional energy may be required during sustained periods of high temperatures to cool airport terminal buildings and supply pre-conditioned air to aircraft on stand. Airports will need to ensure a reliable and secure year-round supply of energy to power their facilities, and buildings may have to be redesigned or reconfigured to maximise the use of natural light and ventilation and minimise energy consumption. New terminals will need to be designed to maximise the use of daylight, while operators of older facilities may consider retrofitting them with more energy-efficient lighting, ventilation and heating systems to reduce energy costs and associated emissions.

A further concern about the effects of rising surface temperatures and their effects on aviation relates to a possible relationship between the melting of glaciers and polar ice and increased volcanic activity. The unprecedented closure of parts of northern European airspace that followed the eruption of Iceland's Eyjafallajokull volcano in April and May 2010 demonstrated the extent to which the presence of volcanic ash in the atmosphere can disrupt normal air traffic flows (L. Budd, Griggs, Howarth, & Ison, 2011). The initial six-day shutdown resulted in the cancellation of 108,000 flights, disrupted the travel plans of 10.5 million passengers and cost airlines in excess of $1.7 billion in lost revenue (Eurocontrol, 2010a). Although described as a 'natural' geological event, it has been tentatively suggested that climate change, rising global temperatures and glacial retreat may ultimately promote an increase in volcanic activity in volcanoes that are no longer 'plugged' by a thick layer of ice (see Pagli & Sigmundsson, 2008).

Snow and Ice

The trend towards generally hotter summers and milder winters in the northern hemisphere means that airports in these regions will become less familiar at dealing with colder conditions and thus may be unprepared for sudden periods of colder weather. Snow and ice, as well as being notoriously difficult to forecast, can disrupt normal airport operations by blocking access roads, contaminating runways and necessitating the costly and time-consuming mechanical and/or chemical removal of snow and ice from airside areas and aircraft.

Ice, in particular, represents a significant hazard to aircraft. If it is allowed to form on the wings, ice increases drag and makes the wing less aerodynamic. If, on the other hand, it accumulates on the airframe it can block pitot tubes and result in erroneous altitude and airspeed information being sent to the flight deck. Both scenarios have the potential to cause serious accidents. In January 1982, 74 people died when an Air Florida Boeing 737 crashed into Potomac River shortly after take-off from Washington National Airport. Ice on the wings and airframe increased the weight and drag on the airframe to such an extent that the aircraft was unable to climb (National Transportation Safety Board, 1982). This, and other ice-related accidents, resulted in existing operating procedures being strengthened to ensure that if the air temperature and humidity falls below a defined threshold, aircraft have to be de-iced within a set period of time before take-off. The responsibility of ensuring that de-icing procedures are carried out in a timely and effective manner lies with individual airline operators and their appointed handling agents, and the provision of adequate de-icing capacity is a prerequisite of operating into certain airports.

The process of de-icing usually involves the leading edges of the wings and the vertical and horizontal stabilisers (more commonly known as the 'tail' and the 'tailplane' respectively) being liberally sprayed with a de-icing solution before the aircraft pushes back from its stand and taxis to the runway. While the process ensures the safety of the airframe and its occupants, the solution is expensive (it costs over €7,000 to de-ice a Boeing 747) and contains toxic chemicals (EFM, 2011). In order to prevent de-icing fluids from contaminating local water courses, the run-off must be directed to balancing ponds where it is stored and treated before being discharged. During prolonged periods of cold weather, the capacity of these ponds is placed under considerable pressure and, if climate change is likely to increase the severity and/or length of wintry weather, airports that are likely to be affected may need to invest in larger water storage facilities or alternative de-icing facilities, such as overhead infra-red thermal gantries, which can melt ice without resorting to the use of chemicals.

In addition to endangering aircraft during take-off and in-flight, the presence of snow and ice on runways and taxiways degrades aircraft braking action and can result in a loss of directional control leading to aircraft over-running or veering off paved runway and taxiway surfaces. As a result, it is imperative that all aircraft manoeuvring areas are kept free of snow and ice and mechanical blowers and sweepers and/or specialist anti-icing and de-icing compounds are employed for this purpose. However, such specialist equipment is expensive and airports will only invest in such technology if rigorous cost benefit analyses identify a need for it.

As well as presenting a problem to aircraft, snow and ice also presents a danger to ground personnel by making surfaces slippery, increasing the risk of injuries and raising the risk of vehicle collisions and obscuring safety-critical ramp markings. If water penetrates gaps in the apron surface and then freezes, cracks can appear. This degrades the surface and requires costly and time-consuming repairs to rectify. Clearly, any alteration in the current snowfall regime and/or changes in the frequency or location of freezing weather conditions will have very tangible material and financial impacts on airline and airport operations and revenues. As a result of climate change, some airports may find they have a greater need for de-icing and snow clearance, whereas others may rarely, if ever, be affected by snow and ice.

Consumer Demand and Behavioural Change

Changing surface temperatures may also affect the aviation industry in more subtle ways by changing patterns of consumer demand for air services.

Within the next 10–20 years, it has been suggested that certain summer holiday destinations may be rendered inhospitable due to increased thermal stress and water shortages, while changes in snowfall patterns may necessitate the relocation of winter sport activities to higher latitude and altitude resorts (Eurocontrol, 2008). Such changes would fundamentally alter patterns of demand for air travel and transform the spatial and temporal distribution of air traffic flows. They will also have a profound impact on the socio-economic fortunes of the airports and communities that will lose traffic or receive new inbound flows of tourists.

It has been suggested that growing public awareness of the environmental effects of aircraft operations, together with associated policy responses, may result in changes to consumer demand and travel behaviour (Davison & Ryley, 2010). Some of this may result from the imposition of new environmental taxes or a desire to reduce personal carbon footprints by reducing the number of flights that are taken and by choosing alternative modes of transport. As well as potentially altering flows of international tourism, rising surface temperatures and changing climatic conditions may also alter the spatial demand for airfreight by altering patterns of agricultural productivity.

Sea Level Rise

In addition to affecting airport and airline operations and changing patterns of consumer demand, rising global temperatures are forecast to result in rising sea levels and more frequent incidents of flooding and storm surges. Low-lying coastal airports may be inundated and critical air transport infrastructure damaged. A significant number of major airports around the world are located in low-lying coastal areas or are situated adjacent to tidal estuaries. Cairns airport in Australia, for example, is located only 10 ft (3.05 m) above mean sea level (MSL), while the maximum elevations of Nice's airport in southern France, John F Kennedy airport in New York and Japan's Kansai facility, are only 12 ft (3.66 m), 13 ft (4 m) and 17 ft (5 m) above MSL respectively.

More vulnerable still are airports which are either located below mean sea level (usually on reclaimed land) or which have runways built on artificial peninsulas that stretch out into the sea. Amsterdam's Schiphol airport, for example, is situated on reclaimed land 11 ft (3.4 m) below MSL, while Louis Armstrong International Airport in New Orleans is situated some 4 ft (1.2 m) below MSL. Although a complex system of banks and levees has been constructed to protect the facilities from inundation, the risk of flooding is ever present as any breach in the existing defences would be catastrophic.

Other airports which are at risk from rising sea levels include those serving low-lying islands in the Indian Ocean and the Caribbean. The maximum elevation of Malé airport in the Maldives, for example, is only 6 ft (1.83 m) above MSL, while Princess Juliana International airport, on the eastern Caribbean island of Saint Maarten, is only 13 ft (4 m) above MSL. Consequently, the risks that sea level rise and/or any increase in the frequency and severity of storm surges pose to these facilities are considerable. While it may prove technically feasible and socio-culturally (if not economically) desirable to protect such facilities by constructing sea walls and other defences, any decision to protect or relocate an airport will be highly controversial.

Wind and Convective Weather Activity

While increased surface temperatures and rising sea levels are two of the most frequently cited consequences of climate change, they are not the only ones.

Anticipated changes to the distribution, frequency, duration and/or intensity of severe weather activity, including storms, ice, snow, severe turbulence and wind shear, as a result of climate change may cause considerable delays disruption to aircraft operations (see Pejovic, Williams, Noland, & Tourni, 2009), in addition to potentially endangering lives and damaging aircraft and airport infrastructure. Of all the weather phenomena that aircraft encounter, wind has the greatest potential to impact on the safety of the air transport system. In order to generate maximum lift in the shortest possible time, aircraft need to take-off and land into the wind and runways are aligned in accordance with the prevailing wind direction. If, however, changing weather patterns cause the prevailing wind direction to alter, airports may be left with runways that are not optimally aligned and flight crew will be obliged to perform more dangerous and cognitively demanding crosswind landings.

Wind also plays an important role during the en-route flight phase as wind speed and wind direction can hinder or expedite a flight's progress. Where possible, pilots will try and take advantage of favourable high-altitude winds and avoid strong headwinds which will slow their aircraft down and increase fuel consumption. For example, on transatlantic routes, the jet stream, a ribbon of fast moving air which blows from west to east across the northern Atlantic Ocean, enables transatlantic flights from North America to Europe to cross the Ocean more quickly than westbound services. While the strength and location of certain high-altitude winds, including the jet stream, can be predicated and forecast with considerable accuracy and air routes planned accordingly, other phenomena resulting from variable atmospheric pressure are much less predictable. Regions of clear air turbulence, which are usually found at the boundaries of neighbouring air masses, and thunderstorms, can produce violent downdrafts and moderate to severe turbulence that can damage aircraft and result in fear and injury to people onboard.

Thunderstorms, and associated convective weather activity, whether encountered at the airport or in the en-route flight phase can result in the delay, rerouting, diversion, or cancellation of flights and lead to a serious deterioration in normal schedule performance (Eurocontrol, 2010a; Janic, 2005; Pejovic et al., 2009; Sasse & Hauf, 2003). The atmospheric environment in and around a thunderstorm represents one of the most hazardous airborne environments an aircraft may encounter (Collins, 2002; Godwin, 2006). Microbursts, icing, wind shear, severe turbulence, hail, electrical discharges, heavy precipitation, reduced visibility and sudden and unpredictable changes in atmospheric pressure are just some of the hazards that have been reported (Machol & Barnett, 1988). Hailstones can damage the airframe, crack flight deck windows and destroy paintwork, causing a safety hazard and costing tens of thousands of pounds' worth of damage, while electrical discharges have the potential to interfere with aircraft navigation and communication equipment (Godwin, 2006).

Although the sophistication of modern meteorological forecasts and weather radar means that the location and intensity of storms can be predicted, located and often avoided with increasing accuracy, numerous fatal aircraft crashes have been attributed, at least in part, to aircraft encountering adverse weather conditions in the take-off, en-route, or final approach phases of a flight (Collins, 2002; Owen, 2006). At the time of writing, it is believed that turbulence associated with severe thunderstorm activity may have contributed to the loss of Air France flight AF447 over the Atlantic Ocean in June 2009, which killed 228 people (BEA, 2009). If such weather events become more common and more severe, there may be a need to enhance the minimum design standards for aircraft to ensure that the airframes can withstand the higher structural stresses that may be imposed on them.

While in-flight structural failures remain a mercifully rare occurrence, news reports attest to the injury, fear and distress severe turbulence can cause (BBC News, 2010, 2011; Friedman, 2009) and pilots will endeavour to avoid known areas of convective activity by requesting alternative headings and/or flight levels and will often accept significant deviations from their filed flight plans (Gillingwater, Budd, Caves, & Reynolds, 2009). This practice almost inevitably results in longer flight times, higher emissions and greater operating costs for the airlines concerned. Convective weather already represents a significant cause of en-route air traffic delay (as much as 89% of all delays recorded in one sector of central European airspace in 2009 were attributable to adverse weather) has the potential to reduce the capacity of a sector of airspace by 15–20%, close airports, reduce runway capacities for landings and take-offs and hinder or prevent normal ground servicing and flight turnaround operations (Eurocontrol, 2010b, 2010c). Given that rising incidents of severe convective weather are a possible consequence of a changing climate, there is an urgent need to quantify the disruptive impacts of convective weather on air traffic flows and examine the extent to which a changing climate may affect the occurrence of such phenomena.

CONCLUSIONS

Commercial aviation is a known contributor of greenhouse gases and one of the few industrial sectors where emissions are predicted to increase as demand for air travel outstrips our ability to deliver emissions reductions through technological and operational innovation. Although uncertainties remain concerning both aviation's exact contribution to climate change and also how climate change will affect the aviation industry, recent research from both academia and the aviation industry itself indicates that the potential for disruption is real and considerable.

In light of this, a range of mitigation measures have been developed and implemented to reduce aircraft emissions in the short-term. These include more efficient use airspace and the introduction of new aircraft technologies and alternative fuels which, it is hoped, will confer short- to medium-term efficiency gains when they are introduced. Other proposals, including the use of economic instruments to reduce consumer demand and thereby lower emissions have, unsurprisingly, been criticised by passenger groups and airlines for being unworkable, unduly bureaucratic and/or ineffective.

Yet while these interventions may, either singularly or collectively, confer some environmental benefit, there is growing recognition from within both government and the air transport industry of the urgent need for the aviation sector to identify all the possible impacts of climate change on air transport operations, assess their severity and devise appropriate preparedness plans to address these vulnerabilities and ensure its safe and continued operation. The key challenge will be to successfully align short-term priorities (including efficiency, regulatory demands, value for money and shareholder returns) with the need for longer-term resilience measures. Many global airports and airlines are private enterprises and many national regulatory regimes do not explicitly require climate adaptation policies to be incorporated within their planning cycles (although the financial imperative for them to remain operational under all but the most extreme weather conditions will almost certainly drive investment in this field).

While the sector will not be able to eliminate all of its deleterious climate impacts in the short term, a better understanding both of how aviation affects the global climate and how a changing climate will affect aviation will hopefully drive the creation of a more equitable and sustainable commercial aviation industry that is responsive to the needs and challenges of the twenty-first century.

NOTES

1. Airside operations manager, London Stansted Airport, personal communication, 2011.
2. Air traffic controller, London Stansted Airport, personal communication, 2011.

REFERENCES

Albers, S., Buhne, J., & Peters, H. (2009). Will the EU ETS instigate airline network reconfigurations? *Journal of Air Transport Management*, *15*, 1–6.

Anderson, K., Bows, A., & Footitt, A. (2007). *Aviation in a low carbon EU*. A Research Report by the Tyndall Centre, University of Manchester. Friends of the Earth. Retrieved from http://www.foe.co.uk/resource/reports/aviation_tyndall_07_main.pdf

Archer, L. J. (1993). *Aircraft emissions and the environment: CO_x, SO_x, HO_x, and NO_x*. Oxford: Oxford Institute for Energy Studies, Parchment Press.

Arnott, S. (2010). British Airways to fly jets on green fuel made from London's rubbish by 2014. *The Independent*, 16 February, p. 40.

Barr, S. W., Shaw, G., Coles, T., & Prillwitz, J. (2010). 'A holiday is a holiday': Practicing sustainability, home and away. *Journal of Transport Geography*, *18*, 474–481.

BBC News. (2010). *United Airlines plane hits turbulence injuring 30*. Retrieved from http://www.bbc.co.uk/news/world-us-canada-10716960. Accessed on 13 April 2011.

BBC News. (2011). *Turbulence led to spinal injury for plane stewardess*. Retrieved from http://www.bbc.co.uk/news/uk-england-london-12695584. Accessed on 13 April 2011.

BEA. (2009). *Interim report on the accident on 1st June 2009 to the Airbus A330-203 registered F-GZCP operated by Air France flight AF447 Rio de Janerio-Paris*. Paris, Bureau d'Enquêtes et d'Analyses pour la sécurite de l'aviation civile. Retrieved from www.bea.aero/docspa/2009/f_cp090601e1.en/pdf/f-cp090601e1.en.pdf. Accessed on 13 April 2011.

Becken, S. (2007). Tourists' perception of international air travel's impact on the global climate and potential climate change policies. *Journal of Sustainable Tourism*, *15*(4), 351–368.

Bows, A. (2010). Aviation and climate change: confronting the challenge. *The Aeronautical Journal*, *114*(1158), 459–468.

Bows, A., Anderson, K., & Mander, S. (2009). Aviation in turbulent times. *Technology Analysis and Strategic Management*, *21*(1), 17–37.

Bows, A., Anderson, K., & Upham, P. (2008). *Aviation and climate change: Lessons for European policy*. London: Routledge.

Bradbury, D. (2007). Blue sky thinkers. *The Independent Extra*, 14 February, pp. 10–11.

Bremner, C. (2008). Air travel switches to electricity. *The Times*, 3 January, p. 39.

Broderick, A. J. (1978). Stratospheric effects from aviation. *Journal of Aircraft*, *15*(10), 643–653.

Budd, L., Griggs, S., Howarth, D., & Ison, S. (2011). A fiasco of volcanic proportions? Eyjafjallajökull and the closure of European airspace. *Mobilities*, *6*(1), 31–40.

Budd, T. M. J., Ison, S. G., & Ryley, T. J. (2011). Airport ground access: Issues and policies. Airport surface access in the UK: A management perspective. *Research in Transportation Business & Management*, *1*, 100–117.

Cairns, S., & Newson, C. (2006). *Predict and decide. Aviation, climate change and UK policy*. Oxford, UK: University of Oxford, Environmental Change Institute.

Collins, R. L. (2002). *Thunderstorms and airplanes* (2nd ed.). Newcastle, WA: Aviation Supplies and Academics.

Daley, B. (2010). *Air transport and the environment*. Farnham: Ashgate.

Davison, L. J., & Ryley, T. J. (2010). Tourism destination preferences of low-cost airline users in the East Midlands. *Journal of Transport Geography*, *18*(3), 458–465.

Daviss, B. (2007). Green sky thinking: Eight ways to a cleaner flying future. *New Scientist*, 22 February, pp. 32–38.

Department for Environment, Food and Rural Affairs. (2008a). *A framework for environmental behaviours: A report*. London: HMSO.

Department for Environment, Food and Rural Affairs. (2008b). *Climate change bill*. London: HMSO.

Department for Transport. (2007). *UK air passenger demand and CO_2 forecasts*. London: HMSO.

Department for Transport. (2010). *Climate change adaptation plan for transport 2010–2012 enhancing resilience to climate change*. London: DfT.

Dobney, K., Baker, C. J., Quinn, A. D., & Chapman, L. (2009). Quantifying the effects of high summer temperatures die to climate change on buckling and rail related delays in south-east United Kingdom. *Meteorological Applications*, *16*, 245–251.

EFM. (2011). *Gesellschaft für Enteusen und Flugzeugschleppen am Flughafen München*. Price List Effective as of 1st October 2010. Retrieved from www.munich-airport.de/media/download/bereiche/efm/preisliste.1.pdf

Eurocontrol. (2008). *Challenges of growth*. Brussels: Eurocontrol.

Eurocontrol. (2010a). *Ash-cloud of April and May 2010: Impact on air traffic*. STATFOR/Doc394. Brussels: Eurocontrol.

Eurocontrol. (2010b). *PRR 2009. Performance review report. An assessment of air traffic management in Europe during the calendar year 2009*. Performance Review Committee. May 2010. Brussels: Eurocontrol.

Eurocontrol. (2010c). *First look at delays in December 2010*. Central Office for Delay Analysis. Brussels: Eurocontrol. Retrieved from http://www.eurocontrol.int/coda/gallery/content/public/docs/coda_reports/2010/FLAD1210.pdf

Eurocontrol. (2011). *Challenges of growth environmental update study-climate adaptation case studies: Eurocontrol commentary*. June 2011. Brussels: Eurocontrol.

Friedman, E. (2009). *Texas woman breaks neck, back in airplane turbulence*. Retrieved from http://abcnews.go.com/health. Accessed on 21 April 2009.

Gillingwater, D., Budd, L., Caves, R., & Reynolds, T. (2009). *Environmental effects of aircraft operations and airspace charging regimes*. Final Report. OMEGA, Manchester.

Godwin, P. D. (2006). *The air pilot's manual Volume 2. Aviation law and meteorology* (7th ed., amended). Cranfield: Air Pilot Publishing.

Gössling, S., Broderick, J., Upham, P., Ceron, J.-P., Dubois, G., Peeters, P., & Strasdas, W. (2007). Voluntary carbon offsetting schemes for aviation: Efficiency, credibility and sustainable tourism. *Journal of Sustainable Tourism*, *15*(3), 223–248.

Hanlon, P. (2007). *Global airlines. Competition in a transnational industry* (3rd ed.). Oxford: Butterworth-Heinemann.

Harris, P. (2003). Solar aircraft heralds revolution in the sky. *The Observer*, 29 June, p. 9.

HM Custom and Excise. (2010). *Notice 550 air passenger duty*. Retrieved from http://customs.hmrc.gov.uk/channelsPortalWebApp/channelsPortalWebApp.portal?_nfpb=true&_pageLabel=pageExcise_InfoGuides&propertyType=document&id=HMCE_CL_000505#downloadopt

Hulme, M., Jenkins, G. J., Lu, X., Turnpenny, J. R., Mitchell, T. D., Jones, R. G., … Hill, S. (2002). *Climate change scenarios for the United Kingdom: The UKCIP02 scientific report*. Norwich: Tyndall Centre for Climate Change Research, School of Environmental Sciences, University of East Anglia.

Intergovernmental Panel on Climate Change. (2007). In S. Soloman, D. Qin, Z. Chen, M. Marquis, K. B. Avery, M. Tignor & H. L. Miller (Eds.), *Climate change 2009: The physical science basis. Contribution of work group one to the intergovernmental panel on climate change*. Cambridge: University Press.

International Air Transport Association. (2009). *A global approach to reducing aviation emissions*. Montreal: IATA.

International Air Transport Association. (2011). *Air transport market analysis June 2011*. Retrieved from www.iata.org/economics. Accessed on 27 July 2011.

Janic, M. (2005). Modeling the large scale disruptions of an airline network. *Journal of Transportation Engineering*, *131*(4), 249–261.

Jaroszweski, D., Chapman, L., & Petts, J. (2010). Assessing the potential impact of climate change on transportation: The need for an interdisciplinary approach. *Journal of Transport Geography*, *18*, 331–335.

Jefferson, A. (2011, May). *London stansted airport climate change adaptation plan*. London: BAA, Stansted Airport.

Keeley, G. (2009). Spain's ailing airlines suffer significant bullet wound. *The Times*, 2 December, p. 50.

Koetse, M. J., & Rietveld, P. (2009). The impact of climate change and weather on transport: An overview of empirical findings. *Transportation Research Part D*, *14*, 205–221.

Lee, D., Fahey, D., Forster, P., Newton, P., Wit, R., Lim, L., ... Sausen, R. (2009). Aviation and global climate change in the 21st century. *Atmospheric Environment*, *43*, 3520–3537.

Leroux, M. (2009). Yes, the planet's future is a worry, but we'll carry on flying, consumers tell poll. *The Times*, 30 April, p. 59.

Lufthansa. (2010). *World premiere: Lufthansa first airline to use biofuel on commercial flights*. Press release dated 29/11/2010. Retreived from http://presse.lufthansa.com/en/news-releases/singleview/archive/2010/november/29/article/1828.html

Ma'anit, A. (2006). CO_2 conned carbon offsets stripped bare. Keynote. *New Internationalist*, *391*, 2–6.

Machol, R. E., & Barnett, A. (1988). Thunderstorms and aviation: A dialogue. *Interfaces*, *18*(2), 20–27.

Marks, P. (2003). The shape of wings to come. *New Scientist*, 13 December, pp. 28–33.

Mayer, R. M. J., Ryley, T. J., & Gillingwater, D. (2011). Passenger perceptions of the 'green image' associated with airlines. *Journal of Transport Geography*, *22*, 179–186.

Mayor, K., & Tol, R. S. J. (2007). The impact of the UK aviation tax on carbon dioxide emissions and visitor numbers. *Transport Policy*, *14*(6), 507–513.

McKie, R., & Doward, D. (2010). Britain in gridlock as big freeze heralds Christmas travel misery. *The Observer*, 19 December, p. 1.

Mendes, L. M. Z., & Santos, G. (2008). Using economic instruments to address emissions from air transport in the European Union. *Environment and Planning A*, *40*, 189–209.

Milmo, D. (2011). Heathrow chaos down to 'wrong amount of snow'. *The Guardian*, 9 March, p. 15.

National Transportation Safety Board. (1982, 10 August). Air Florida, Inc., Boeing 737-222, N62AF, collision with 14th Street Bridge, near Washington National Airport, Washington, DC, January 13. NTSB Report No. AAR-82-08. NTSB, Washington, DC.

O'Connell, D. (2009). Your next holiday may be powered by weeds. *The Sunday Times*, 29 November, p. 13.

Owen, D. (2006). *Air accident investigation* (3rd ed.). Sparkford, Somerset: Patrick Stephens Ltd.

Pagli, C., & Sigmundsson, F. (2008). Will present day glacier retreat increase volcanic activity? Stress induced by recent glacier retreat and its effect on magmatism at the Vatnajokull ice cap, Iceland. *Geophysical Research Letters*, *35*(9), L09304.

Pejovic, T., Williams, V. A., Noland, R. B., & Tourni, R. (2009). Factors affecting the frequency and severity of airport weather delays and the implications of climate change for future delays. *Transportation Research Record: Journal of the Transportation Research Board*, *2139*, 97–106.

Randles, S., & Bows, A. (2009). Aviation, emissions and the climate change debate. *Technology, Analysis and Strategic Management*, *21*(1), 1–16.

Rayner, G., Millward, D., & Swinford, S. (2010). Airport chaos for 1m as snow returns. *The Daily Telegraph*, 18 December, p. 1.

Ryley, T. J., & Davison, L. J. (2008). UK air travel preferences: Evidence from an East Midlands household survey. *Journal of Air Transport Management*, *14*(1), 43–46.

Ryley, T. J., Davison, L. J., Bristow, A., & Pridmore, A. (2010). Public engagement on aviation taxes in the United Kingdom. *International Journal of Sustainable Transportation*, *4*(2), 112–118.

Sasse, M., & Hauf, T. (2003). A study of thunderstorm-induced delays at Frankfurt Airport, Germany. *Meteorological Applications*, *10*, 21–30.

Sealy, K. (1957). *The geography of air transport*. London: University Hutchinson Library.

Warwick, G. (2007). Fuel's gold. *Flight International*, 11–17 September, pp. 36–38.

Williams, V., Noland, R., & Toumi, R. (2002). Reducing the climate change impacts of aviation by restricting cruise altitudes. *Transportation Research Part D: Transport and Environment*, *7*(6), 451–464.

CHAPTER 4

AN INTERNATIONAL DIMENSION: SHIPPING

Kevin Cullinane

ABSTRACT

Purpose – *The shipping industry is generally recognised as having better fuel efficiency than other transport modes. In many regions of the world, therefore, policy has promoted shipping as the preferred freight transport mode of choice. In recent years, however, environmental problems associated with shipping have emerged. Several influential analyses have revealed the impact of shipping on air quality, particularly in the form of emissions of sulphur, nitrogen oxides and particulate matter, all of which have adverse consequences for human health.*

Methodology/approach – *An extensive environmental profile of shipping is provided, focusing specifically on the atmospheric pollution that is directly attributable to shipping operations.*

Findings – *It is important, however, to place the environmental profile of the shipping industry into the context of exactly how much transport work it does. This makes it clear that where shipping is a viable modal alternative then, in relative terms and most contexts, it still retains significant environmental advantages over other modes. The industry and its regulators have been slow, however, to improve its environmental profile and maintain its inherent advantage. Technical and operational*

Transport and Climate Change
Transport and Sustainability, Volume 2, 65–104

ISSN: 2044-9941/doi:10.1108/S2044-9941(2012)0000002007

measures which the industry may implement unilaterally are analysed, but these are deemed insufficient to stem the adverse tide of environmental concerns. Regulation is a necessity. Recently implemented regulatory measures are analysed, together with possible scenarios for the future regulation of greenhouse gas emissions. The IMO approach of global regulation is supported in preference to regionally based regulatory policies. There is also a danger that regulatory intervention may distort mode choice contexts.

Originality/value – *The provision of an extensive environmental profile of shipping and an examination of this profile in relation to the importance of this transport mode to the global economy.*

Keywords: Shipping; environment; IMO; carbon; SO_x; NO_x

INTRODUCTION

The shipping industry plays a pivotal role in facilitating international trade and the economic development of nations. Although it is notoriously difficult to precisely measure the modal split of international freight transport, Hoffmann (2008) reports that, in 2006 and excluding intra-European Union trade, seaborne trade accounted for 89.6% of global trade in volume terms (tons) and 70.1% in value terms. For the same year, air accounted for 0.27% of trade volume and 14.1% of trade value, while land-based modes (including pipelines) accounted for 10.2% of trade volume (tons) and 15.8% of trade value.[1] Including intra-European Union trade, most of which is obviously land-based, in 2009 total world trade amounted to 7.8 billion tons, down from 8.2 billion tons recorded in 2008 (UNCTAD, 2011). Hoffmann (2008) also reports that, after taking into account intra-European trade, the Lloyds Maritime Intelligence Unit have estimated that, on average, about 77% of total world trade volume is seaborne. Given the immense distances involved in seaborne transport, however, it is clear that in terms of transportation activity (as measured in ton-miles or tonne-kms), the domination of shipping in the carriage of international trade is even greater. Indeed, the oft-quoted claim that 'shipping moves more than 90% of world trade' (e.g. ECSA/ICS, 2008, p. 2; Mitropoulos, 2010, p. 5) may even be vindicated.

In particular, the shipping industry has been central to the phenomenon of globalisation, where locations for the production and consumption of a product are distributed widely around the world. As such, the shipping mode has emerged as a fundamental component in the global supply chains which link production locations to points of consumption. It might be tempting to attribute this simply to being an inevitable outcome of world geography. However, other characteristics of the shipping industry have also exerted an influence; most notably that it is the least expensive mode of freight transport per tonne-kilometre moved and also possesses the most spatially extensive network coverage.

The relationship between the shipping industry and the environment is a complex one. Climate change will bring about rising sea levels that are likely to have a differential impact on the world's ports, with certain of the lowest lying major ports, including Rotterdam, New York, New Orleans, Miami, Osaka, Kobe and Nagoya already identified as highly vulnerable in terms of exposed asset values (Nicholls et al., 2007). At the same time, other ports will benefit from deeper channels and increased depth at berth. The potentially dramatic impact on these ports complements the comparatively mundane, but widely dispersed, expectation of more flooding and inundations causing disruptions to coastal supply chains, higher temperatures requiring greater expenditure on refrigeration, more frequent freeze and thaw cycles causing faster deterioration of paved areas and the increased mobility of sediment creating a greater requirement for dredging (UNCTAD, 2009). A greater propensity for extreme weather events will almost certainly have an adverse impact on the navigability of some established trade routes, thereby increasing costs, transit times and potentially inducing modal switching under certain circumstances. At the same time, rising temperatures in the Arctic could, however, open up new opportunities for the shipping industry through the increased navigability of both the North-West Passage and the Northern Sea Route. The impact of this phenomenon could be felt not only in terms of faster transit times between Asia and the Americas and Europe respectively, but is also likely to facilitate the exploitation of new sources of natural resources which could also have an important impact on the shipping industry and, indeed, on global trade patterns (Vernya & Grigentinc, 2009).

Shifting focus to the opposite direction of influence, the environmental effects of any logistics industry are numerous. For the shipping industry specifically, there are or have been concerns over deleterious environmental impacts such as: the pollution caused by oil spills and the discharge of gases, solids or liquids from ships; noise; vibration and; the visual intrusion and land-take associated with land-based infrastructure such as ports and ship-breaking facilities.

These categories of environmental concerns associated with the shipping industry are obviously very important in their own right. However, there can be no doubt that the issue of greenhouse gas emissions, as well as certain other noxious forms of atmospheric pollution, has emerged in the past five years or so as the most urgent and topical for debate, policy development and regulatory action. It is for this reason that the remainder of this chapter will focus exclusively on the atmospheric pollution from ships and its impact on the environment. To this end, section 2 presents an environmental profile of the shipping industry, detailing the recent trends in the industry's fuel consumption and the main characteristics and trends relating to the atmospheric pollution produced as a direct result of that fuel consumption. Section 3 places this environmental profile into context by relating it to the amount of transport activity which is undertaken by the shipping industry. The environmental damage per unit of freight moved which is attributable to shipping is then compared to that of other alternative modes of freight transport. In section 4, possible approaches to mitigating the adverse environmental effects of the shipping industry are enumerated in terms of technical, operational and regulatory measures. Conclusions are then drawn in section 5.

AN ENVIRONMENTAL PROFILE OF THE SHIPPING INDUSTRY

The Intergovernmental Panel on Climate Change lists 27 substances as greenhouse gases (IPCC, 2000). The six most significant of these are: Carbon dioxide (CO_2 – by far the most abundant greenhouse gas), Methane (CH_4), Nitrous Oxide (N_2O), Hydro fluorocarbons (HFCs), Per fluorocarbons (PFCs) and Sulphur Hexafluoride (SF_6). It is important to recognise, however, that shipping emissions are not covered by the Kyoto Protocol, where legally binding commitments are established for the reduction of greenhouse gases produced by industrialised (i.e. 'Annex I') countries. This is because in the preparations for the Kyoto Protocol no agreement could be reached about how to allocate shipping or aviation emissions to national carbon budgets (CE Delft, MMU, & MNP, 2006). Instead, Article 2.2 of the Kyoto Protocol calls on Annex I countries to 'pursue limitation or reduction of emissions of greenhouse gases not controlled by the Montreal Protocol from aviation and marine bunker fuels, working through the International Civil Aviation Organization and the International Maritime Organization, respectively'. With shipping now carrying over 80% of world trade by volume, in recent years the industry's greenhouse gas emissions have increasingly become the subject of public attention (UNCTAD, 2009). The issue of greenhouse gas emissions from shipping was, for example, a subject for consideration and debate at the United Nations Climate Change Conference in Copenhagen (COP 15) in December 2009 (UNFCCC, 2010).

Fundamentally, the amount of CO_2 and other emissions produced by the shipping industry is directly related to the amount and type of marine fuel burned.[2] In the past, estimates of fuel consumption for the industry were derived from a top-down analysis of sales figures for international marine fuels. In a highly influential and significant work, Corbett and Koehler (2003) employed a bottom-up approach to estimate the fuel consumption and vessel activity for the world shipping fleet. Using this approach, they derived an estimate for global marine fuel consumption which was more than twice the quantity reported as 'international fuel'. The authors argued that this implied the presence of systematic model bias in any top-down approach based solely on the sales of international marine fuels and that, furthermore, this served to undermine the validity of estimates based on such an approach.

Endresen, Sorgard, Bakke, and Isaksen (2004), however, criticised Corbett and Koehler (2003) for using some unrealistic assumptions and input values. Using the same fundamental bottom-up approach, but different assumptions and exogenous input data, they produced a reduced estimate of marine fuel consumption and associated ship emissions than that of Corbett and Koehler (2003). Indeed, Endresen et al. (2004) claimed to substantiate a lower estimate for marine fuel consumption that more closely agreed with published fuel statistics and previous studies that assumed that reported marine fuel statistics were essentially accurate and complete. In a rebuttal of these assertions, Corbett and Koehler (2004) responded by incorporating the assumptions of Endresen et al. (2004) into their activity-based model to confirm, albeit in a slightly diluted form, their earlier conclusion that the international shipping fleet uses more fuel than reported in international marine fuel statistics.

Irrespective of the level of historical precision in estimating fuel consumption and air pollution inventories for the shipping industry, it is certainly the case that bottom-up activity-based methodologies now hold sway. However, the degree of variability in the estimates derived from using such an approach is still quite significant. As shown in Fig. 1, however, all the estimates derived using this bottom-up approach have been significantly

2007	Low bound	Best	High bound
Total fuel consumption	279	333	400

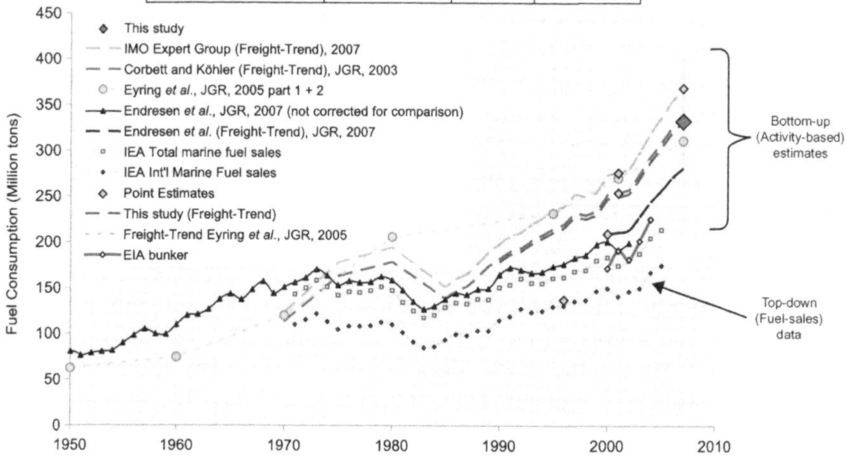

Fig. 1. World Fleet Fuel Consumption. *Note*: The large dark diamond shows the IMO consensus estimate and the whiskers the high and low bound estimates. *Source*: Buhaug et al. (2009).

larger than what is yielded by applying a top-down approach based on fuel sales figures. Indeed, it is this corpus of empirical work which has helped stimulate a greater sense of urgency with respect to shipping emissions in policy and regulatory circles.

Based on the range of bottom-up empirical estimates of the industry's fuel consumption, international shipping is estimated to account for somewhere between 1.6% and 4.1% of world CO_2 emissions from fuel combustion, with the emissions of the global shipping industry in 2007 estimated at 943.5 million tonnes of CO_2 by Psaraftis and Kontovas (2009). By way of validating this figure, MARINTEK (2008) independently arrived at an estimate of 954 million tonnes of CO_2 for the same year, which excludes fishing vessels and is split into 111 million tonnes for domestic shipping and 843 million tonnes for international shipping. Davidson and Faber (2009) point out that the latter figure is about the same order of magnitude as the total national CO_2 emissions of Germany, and twice the total national CO_2 emissions of Mexico.

As can be seen in Fig. 2, the industry's regulatory body (the International Maritime Organization or IMO) is working on the basis that international shipping accounts for 2.7% of all global CO_2 emissions, with domestic

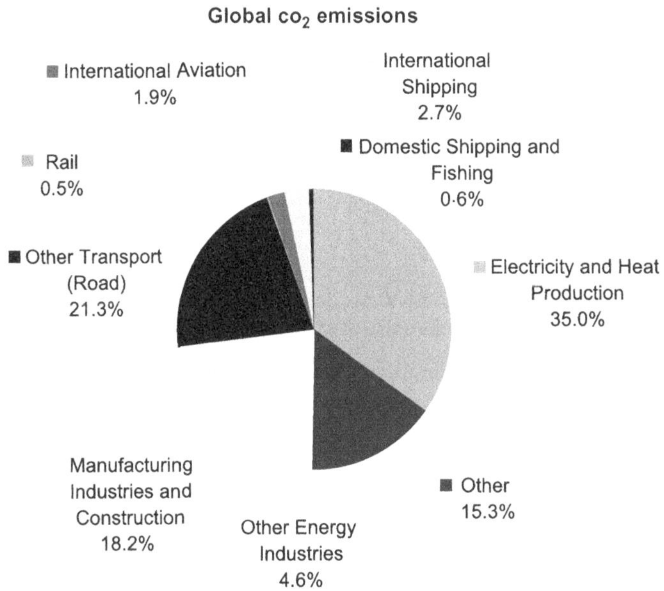

Fig. 2. CO_2 Emissions from Shipping Compared with Global Total Emissions. *Source*: Buhaug et al. (2009).

shipping and fishing accounting for a further 0.6% (Buhaug et al., 2009). These figures compare starkly to that of 1.9% for international aviation, an industry which over the past decade has not only attracted significant attention from both news agencies and international policy makers, but which has also received a fair degree of vilification from some parts of the general public, for its perceived high level of greenhouse gas emissions (Bows & Anderson, 2007; Mayor & Tol, 2007; Olsthoorn, 2001; Upham, 2003). The absolute volume of CO_2 emissions from the shipping industry in 2005 can be seen in comparison to all of aviation, road and rail in Fig. 3.

More than 95% of the world's shipping fleet is powered by diesel engines (Deniz, Kilic, & Civkaroglu, 2010) and, primarily as a result of the low quality of fuels used,[2] even the most modern of marine engines produces higher emissions per power output than regulated on-road diesel engines (Corbett & Farrell, 2002). Indeed, the bunker oil used in ocean-going ships has been estimated to produce over 100 times per unit volume the amount of sulphur of on-road diesel (Bailey & Solomon, 2004; California Air Resource Board, 2003; US EPA, 1999). Thus, while some of other main greenhouse

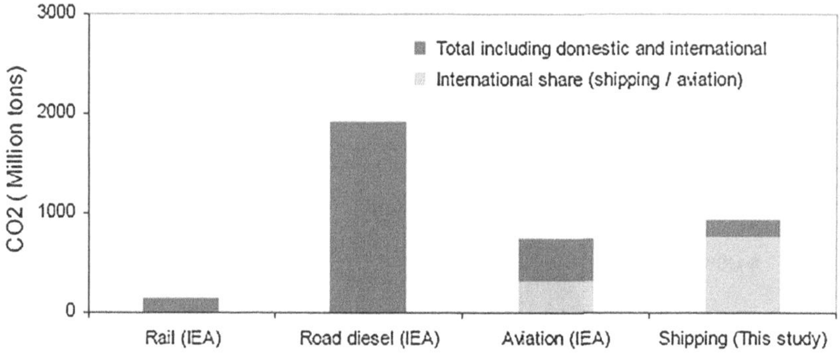

Fig. 3. CO_2 Emissions from Shipping Compared to Other Modes of Transport in 2005. *Source*: Buhaug et al. (2009).

gases apart from CO_2 are also of relevance to the shipping industry (e.g. carbon monoxide), there are other atmospheric pollutants which are of more particular concern because of the relatively poor quality of the fuel which is typically used in ships. These include:

- SO_x (oxides of sulphur) – The emission of this shipping pollutant is related directly to the relatively high sulphur content of marine fuel in comparison to other fuels.
- NO_x (oxides of nitrogen) – Volumes of this emission relate directly to the temperature at which a fuel combusts and because of the low energy efficiency of marine fuels, they need to be heated to higher temperatures.
- VOC (volatile organic compounds) – The cargoes of tankers emit VOC during cargo loading/discharge, crude oil washing operations and during sea voyages.
- PM (particulate matter, typically measured in terms of PM_{10} and $PM_{2.5}$) – This is produced during combustion and mainly takes the form of soot and ash. Again, this pollutant is especially prevalent from ships because of the poor quality of fuel used in comparison to other transport modes.

In addition to an understanding of the forms of atmospheric pollution produced by the shipping industry, it is also important to recognise the different geographical impacts of these emissions. As is shown in Table 1, not only can the impact of greenhouse gases be either direct or indirect, but also certain greenhouse gases have both a global and a regional impact. In addition, Table 1 reveals that a large range of pollutants have a very specific

Table 1. Geographical Impact of Pollutants.

Effect	PM	HM	NH_3	SO_2	NO_x	NMVOC	CO	CH_4	CO_2	N_2O
Local										
Health and air quality										
Regional										
Acidification										
Photochemical										
Global										
Greenhouse – indirect										
Greenhouse – direct										

Note: PM, particulates; HM, heavy metals; NH_3, ammonia; SO_2, sulphur dioxide; NO_x, oxides of nitrogen; NMVOC, volatile organic compounds (hydrocarbons); CO, carbon monoxide; CH_4, methane; CO_2, carbon dioxide; N_2O, nitrous oxide.
Source: Hickman (1999).

impact on the local environment, affecting not only local air quality but also potentially the health of the proximate human population.

It has been found that ships contribute around 55–77% of total emissions in port regions (Hulskotte & Gon, 2010; Lowles, 1998). Even while stationary at berth (in 'hotelling' mode), ships typically still need to run their engines in order to provide power for basic ship operations that are required while in port. Standing at three to five times higher than from other activities (e.g. manoeuvring and cruising), ship emissions at berth are the fundamental determinant of the concentration of exhaust emissions in ports (Deniz et al., 2010; Tzannatos, 2010). As described above and suggested by the data presented in Fig. 4, three of the pollutants with a very specific local impact (SO_x, NO_x and PM) are of particular concern when considering the environmental credentials of the shipping industry. This is particularly the case given the influential findings of Corbett et al. (2007) linking the emission of PM from ships to cardiopulmonary mortality in and around the world's major port locations (see Fig. 5).

THE SHIPPING INDUSTRY CONTEXT

Irrespective of the negative nature of shipping's environmental profile presented in the previous section, as the introduction has already suggested, it is generally accepted that more than 90% of global trade in volume terms is carried by sea (IMO, 2011a). In recent decades, in tandem with a general trend of increasing total trade, world seaborne trade figures have also grown

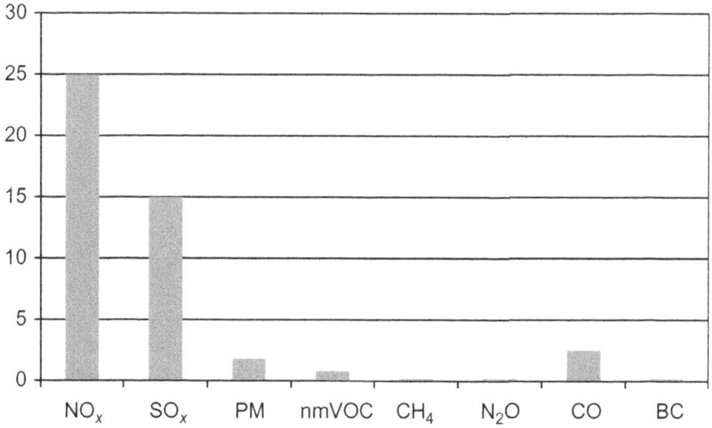

Fig. 4. Estimates of Other Emissions from Shipping (Million Tonnes).
Source: Derived by the author from data contained in Table 3.7 of Buhaug et al. (2009). *Note*: The data is based on estimates of the total shipping industry fuel consumption for 2007. As such, there is uncertainty of $\pm 20\%$ associated with these emissions estimates.

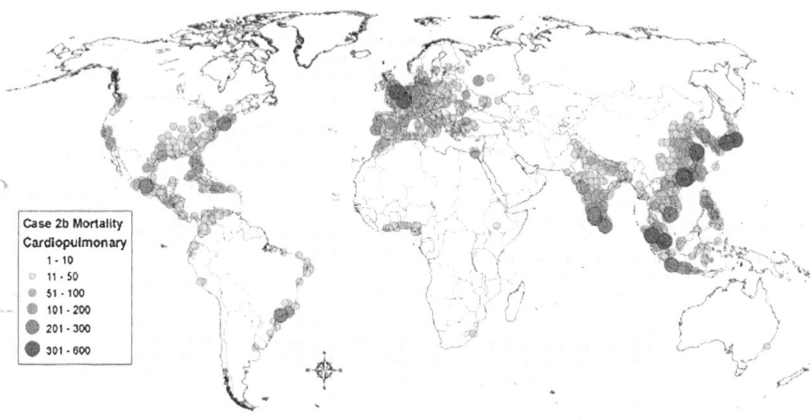

Fig. 5. Cardiopulmonary Mortality Attributable to Ship $PM_{2.5}$ Emissions.
Source: Reprinted from Corbett et al. (2007). © 2007 American Chemical Society.

Table 2. Development of World Seaborne Trade (Selected Years, Millions of Tons Loaded).

Year	Oil	Main Bulks	Other Dry Cargo	Total (All Cargoes)
1970	1,442	448	676	2,566
1980	1,871	796	1,037	3,704
1990	1,755	968	1,285	4,008
2000	2,163	1,288	2,533	5,984
2006	2,698	1,849	3,135	7,682
2007	2,747	1,972	3,265	7,983
2008	2,732	2,079	3,399	8,210
2009	2,649	2,113	3,081	7,843

Note: The 'main bulks' are iron ore, grain, coal, bauxite/alumina and phosphates.
Source: UNCTAD (2011).

considerably (see Table 2), reaching a record of 8.2 billion tons of goods loaded in 2008.

As with many other industries facing global recession, shipping experienced a contraction in 2009 when total goods loaded fell to 7.8 billion tons (UNCTAD, 2011). However, a continuing general trend towards increasing international trade and greater demand for sea transport is irrevocable. Indeed, Stopford (2010) has suggested that if the historical trend in trade growth continues into the future, then by 2060 the 8 billion tonnes of cargo that is currently carried in ships will have expanded to 23 billion tonnes and that, in the absence of any corrective action, this will increase the carbon footprint of shipping by 300%.

Nevertheless, it is the sheer volume of freight moved in ships that has earned shipping its recognition as the least environmentally damaging mode of freight transport. For all the environmental problems posed by the industry, its level of activity is massive and far outweighs that of any other mode of freight transport. The corollary of this in terms of the volume of emissions per unit of activity can clearly be imputed from Figs. 6 and 7.

As well as supporting the assertion that shipping is the least environmentally damaging mode of freight transport, what Figs. 6 and 7 also suggest is that there are significant economies of scale to be reaped in terms of CO_2 emissions, a fact which is explicitly portrayed in Fig. 8 and which spells good news in terms of the current direction of the highly significant containership sector, in particular, where maximum and average ship size has been increasing continuously since 1995 (Cullinane & Khanna, 1999, 2000).

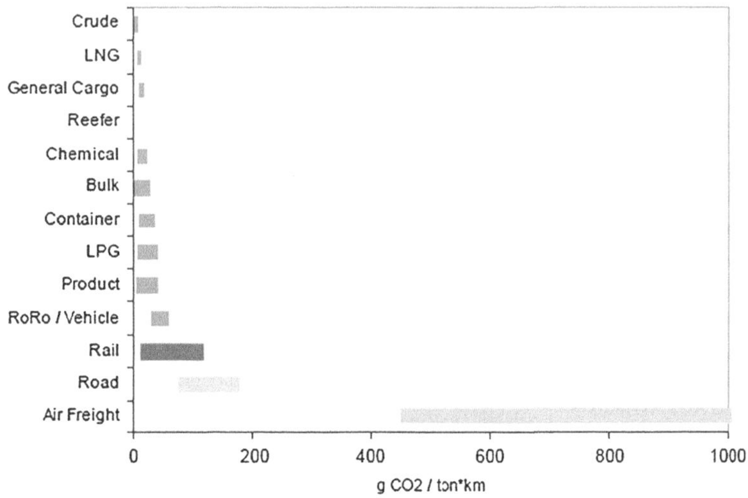

Fig. 6. Typical Range of Ship CO_2 Efficiencies Compared to Rail, Road and Airfreight. *Note*: The ranges of CO_2 efficiency of various forms of transport were estimated using actual operating data, transport statistics and other information. Efficiency is expressed as mass of CO_2 per tonne-kilometre, where the mass of CO_2 expresses the total emissions from the activity and 'tonne-kilometre' expresses the total transport work done. The ranges shown represent solely the typical average range for each mode of transport; absolute maximum or minimum efficiencies that may be observed in practice are not represented. *Source*: Buhaug et al. (2009).

Advances in engine efficiency, hull forms and propulsion systems through the years have been responsible for a significant downward trend in the fuel consumption of individual ships. In 1960, the average specific fuel consumption of a ship was about 220 g/kWh, while in 2010, the equivalent figure is about 170 g/kWh (Clausen, 2009). However, in aggregate across the whole shipping fleet, the growth in the volume of shipping which has taken place over this period has swamped these efficiency benefits. From a more positive perspective, this might also be alternatively interpreted as these efficiency gains having facilitated burgeoning international trade and that shipping's environmental profile would be much worse than it is currently, if such technological progress had not been made. What this does serve to highlight is that history provides us with evidence that incremental technological improvements will not, in isolation, help resolve the environmental problems associated with shipping; instead, a whole gamut

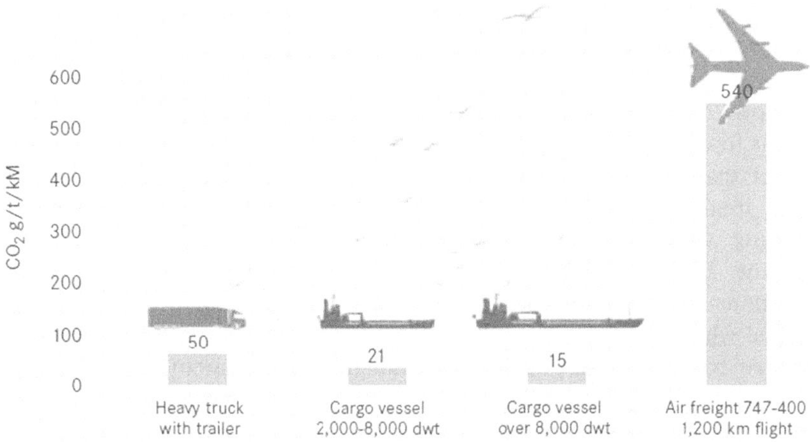

Fig. 7. CO_2 Emissions by Mode of Freight Transport. *Source*: Swedish Network for Transport and the Environment (cited in World Shipping Council, 2009).

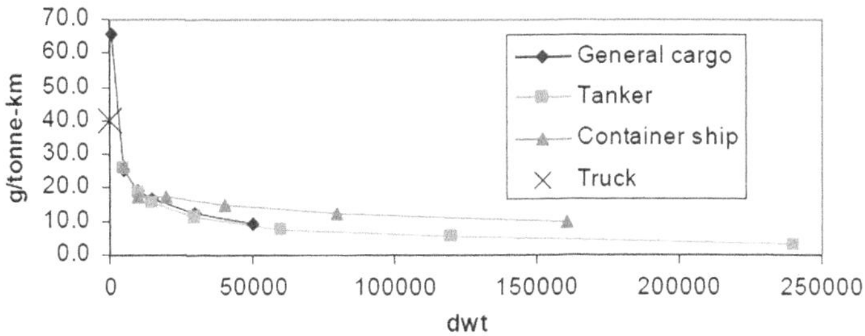

Fig. 8. Economies of Scale in CO_2 Emissions by Mode of Transport. *Source*: Fridell (2011).

of measures should be deployed to ensure that progress is made and to buy time before major technological innovations can be found and deployed.

POTENTIAL MITIGATION MEASURES

The shipping industry is fundamentally driven and controlled by private sector organisations pursuing profit-oriented objectives. It might be thought, therefore, that environmental considerations would be purely incidental to achieving such objectives. Perhaps very fortunately, because of the inextricable link between fuel consumption and particularly airborne pollution, many of the measures introduced to reduce pollution also amount to cost-saving measures. The environment benefits, therefore, from the fact that greater fuel efficiency not only reduces emissions across the board, it also reduces costs. It is also clear that for certain companies operating within the sector, a second motivation for unilateral action lies with the trade-off between cost increases incurred in implementing environmental measures and the possible, but ill-defined, 'green image benefits' which may accrue; for example, in meeting the requirements of a customer base pursuing policies of sustainable procurement. In the field of logistics, these situations are generically referred to as 'green gold', since the achievement of environmental objectives is coincident with enhanced profitability. Of course, this is not true of all the many possible measures through which purely environmental objectives might be achieved and, in many cases, such measures are not deployed simply because it is considered uneconomic to do so.

Technical Measures

These all revolve around reducing fuel consumption or using alternative sources of energy that produce less harmful emissions per unit of output or work. They include the following measures.

Greater Engine Efficiency

Over the past few decades, the development of more efficient marine engines with lower Specific Fuel Oil Consumption (SFOC) has progressed steadily. Typical incremental improvements have revolved around attempting to optimise the combustion process by raising the ratio of the maximum (firing) pressure and the mean effective pressure and by deploying electronic engine control systems with more parameters for adjustment compared to a standard camshaft (or mechanical) control system (Clausen, 2009; Crist, 2009). As well as many truly novel ideas that have not yet been fully explored, there remain several, more conventional, avenues for further development and implementation, where their cost and/or complexity are deterring the investment of research funding (ECSA/ICS, 2008). Nitrogen oxides (NO_x), specifically, are formed during combustion and, according to Clausen (2009), can be reduced by incorporating the following methods into the engine system design: (a) Regulation of fuel injection (rate shaping), (b) Water in Fuel emulsion (WIF), (c) Humidification of the scavenge air (Scavenge Air Moistening, SAM), (d) Exhaust Gas Recirculation (EGR) and (e) Selective Catalytic Reduction (SCR).

Waste Heat Recovery
The exhaust gas and cooling water from ships contains substantial energy that could be utilised, thereby improving the overall thermal efficiency of the engine system, in many cases by 5–10%. Maersk Line is the first shipping company to install waste heat recovery systems as standard on all its new ships. Their system works by utilising the very high heat potential contained in the exhaust gas which leaves the engine; it is converted to steam in an exhaust gas boiler, which is then supplied to a turbine connected to a generator which then recovers electrical energy. In this way, the engine's waste heat has been changed into valuable electric power, thus reducing CO_2 and other emissions from the ship. Maersk estimates that, using this system, its Triple-E fleet saves about 9% in fuel consumption and emissions when operating within a speed range of 18–23 knots. At $10 million per ship for installing the system, they suggest this represents a payback period of somewhere between 5 and 10 years depending on fuel price and service speed (Maersk Line, 2011). Similarly, one of the leading marine engine manufacturers, Wärtsila, has developed a total heat recovery system for which it claims a 12% saving (Aabo, 2009) in fuel consumption and emissions. In a comprehensive review, Wang, Zhang, Peng, and Shu (2011) report that average efficiency gains for such systems typically range from 10% to 30%.

Improved Hull Design and Performance
To make it easier to propel, the fundamental objective here is to either reduce the weight of a ship by selecting appropriate lighter design materials which do not compromise hull strength or to reduce the resistance of the ship in the water, so that less power (and, therefore, less fuel) is required to achieve a specific ship speed. The shape and form of the hull is clearly an important aspect of minimising resistive forces, but there are other less obvious considerations which have received attention. For example, both animals (e.g. barnacles, molluscs, polychaete worms, encrusting hydroids, bryozoa and sea squirts) and plants (e.g. green, red and brown algae and diatoms) can find a home on the hull of a ship. Such fouling communities can significantly increase the frictional resistance of a ship, leading to greater fuel consumption and exhaust emissions. As a consequence, a whole

international industry has grown up around the development of anti-fouling paints and coatings that reduce a ship's proclivity to hull fouling. Compared to an untreated hull, an effective antifouling paint can produce fuel savings and emission reductions of at least 15% due to reduced drag. Although there are also supplementary environmental benefits in terms of reducing the likelihood of transporting invasive marine species, these paints and coatings have created their own environmental problems through their previous reliance on tributyltin (TBT) which itself is known for its highly toxic and endocrine disrupting properties. Although the IMO banned TBT from use in anti-fouling paints and coatings in 2008 in its *International Convention on the Control of Harmful Anti-Fouling Systems on Ships*, full compliance remains a long way from being attained. With a similar objective in mind, innovative research has found that the pumping of compressed air bubbles over a ship's hull surface while in motion can significantly reduce vessel drag, particularly where the hull is polymer-coated to reduce surface tension (Truong, 2001; Winkler, 2009).

More Efficient Propellers and Rudders
In general, the larger the propeller diameter, the higher the propeller efficiency and the lower the optimum propeller speed. Thus, achieving a propeller speed which is as low as possible (within the design restrictions of a ship) is the desired objective, although some fine tuning can be instigated without significantly affecting propulsion efficiency (Clausen, 2009). When the design speed of a ship is reduced, the corresponding propulsion power and propeller speed will also be reduced, which again may have an influence on propeller and main engine parameters; aspects which are very germane to the operational strategy of 'slow steaming' which is discussed below. With optimum hull design, together with an appropriate choice of engine and propeller, emissions reduction potential for a new ship has been evaluated at up to 30% (Eyring, Köhler, Lauer, & Lemper, 2005).

Low Sulphur Fuel
Obviously, this is primarily aimed at reducing sulphur emissions directly, but also has an indirect ameliorative impact on PM and CO_2 emissions. In order to produce low sulphur fuel in the large quantities that may be required by the shipping industry, the residual oil which is currently used as marine fuel will need to be further processed. Alternatively, distillate fuels will need to be burned in ships and engines modified accordingly. In both cases, this results not only in greater production emissions but also in a higher priced fuel which means that there is no economic motivation for its

adoption, apart from the possibility of 'green marketing' benefits. The shipping industry is extremely concerned about the physical capacity of the global oil refining industry to supply their possible future demand for low sulphur fuel (CE Delft, Germanischer Lloyd, MARINTEK, & Det Norske Veritas, 2006).

Seawater Scrubbers and Filters

A scrubber is a packed tower on a ship to where the exhaust gas is directed and where it makes contact with seawater which is pumped into the scrubber. When SO_2 comes into contact with seawater there is a fast and very efficient reaction between the SO_2 and calcium carbonate ($CaCO_3$) in the seawater, to form CO_2 and calcium sulphate (gypsum), a major constituent of ordinary seawater (Andreasen & Mayer, 2007). The treated seawater is then discharged into the sea without causing any harm to the marine environment, though it can be filtered if required. The reaction is complete in a very short time, so the equipment can be compact and is, therefore, relatively inexpensive, while still achieving high reduction efficiencies of about 95% (Andreasen & Mayer, 2007).

The use of seawater scrubbers will also reduce PM (caused by the nucleation and subsequent agglomeration of carbon during the combustion of Heavy Fuel Oil (HFO) or Marine Diesel Oil (MDO)), and has been reported to be a more effective means of reducing PM than lowering the sulphur content of marine fuel. Also, PM filters and NO_x adsorbers can be fitted to marine diesel engines to achieve emission reductions of 90% or more. However, for the largest marine diesel engines, the requirement for extremely large filters means that the technology is not yet fully developed, but is quickly emerging. Because of the severe impacts of fine PM on human health, in terms of economic welfare there is no doubt that the benefits of reducing these emissions will far outweigh the costs of implementing these systems. Indeed, in seeking to comply with emissions regulations (see later) and because of the difficulties attendant upon the latter, the shipping industry may be moving towards the installation of seawater scrubbers as a preferred approach to the burning of low sulphur fuel (Agrawala, Malloy, Welch, Miller, & Cocker, 2008; Cooper, 2003; Fridell, Steena, & Peterson, 2008; Lin, 2002).

Shore-Based Power ('Cold Ironing')

This relates to the use of shore-based power to provide electrical energy to a ship while at berth. Rather than use its auxiliary engines (as is normally the case) to supply the electricity which a ship requires while docked at berth, utilising this shore-based source of energy means that all engines can be shut down. The practice has been comprehensively implemented in all ports within the states of Alaska and California in the United States and is increasingly being adopted by ports in Europe, sometimes as part of a 'Green Port' initiative (Arduino, Murillo, & Ferrari, 2011). Although the reduction of greenhouse gases is often cited as the raison d'être behind implementing the practice, these benefits are very variable and are heavily dependent upon the type of fuel displaced, as well as the source of the power provided from the shore; benefits are only meaningful if the on-shore electricity is derived from renewable sources of energy (Green, Winebrake, & Corbett, 2008). Also, given the current fuel mix for electricity generation aboard ships at berth, the use of auxiliary engines at berth contributes only a small proportion of total marine transport greenhouse gases. Thus, cold ironing reduces total shipping-related greenhouse gases by less than 0.5% (Frey, 2008). Of far greater importance, in fact, are the benefits in terms of SO_x, NO_x and PM reductions and the improvement in local air quality achieved by the shutting down of ships' engines while in port. For example, in a study of the Rotterdam–Dordrecht area where there are extensive port facilities, Hammingh et al. (2007) concluded that the large-scale application of shore-side electricity could significantly reduce the remaining number of poor air quality hotspots. Based on a multi-annual meteorological dispersion model and taking into account a context of impending new regulations on the sulphur content of fuel for ships in the North Sea (see section 'Regulation'), the authors estimated a reduction in NO_2 of between 1.5 and 3 $\mu g/m^3$ in 2020 and a reduction in PM_{10} of 0.4 $\mu g/m^3$ in the same year. Hammingh et al. (2007) also pointed to the potential additional advantage of shore-side electricity being a reduction in noise. The major barriers to the widespread take-up of cold ironing, however, are the high costs and long-term efforts associated with the installation or retrofitting of appropriate power systems on ships and the expansion of electricity lines, substations, etc. ashore, both in port and beyond. Moreover, in the absence of any international standards for cold ironing, the risk of investing in the required infrastructure is immeasurably increased. Thus, to promote the application of this approach, international standards and phase-in schedules are urgently needed.

Alternative Sources of Energy

There are several possibilities for replacing conventional fossil fuels with less environmentally damaging energy sources.

Biofuels. In the form of biodiesel, this works well in ship engines. It can provide 100% of a ship's fuel or it can be blended with conventional fuel so that, for example, 5% biodiesel content in the fuel results in a CO_2 reduction of about 4% (ECSA/ICS, 2008). An additional positive factor is that biodiesel does not contain sulphur. Problems currently associated with its use include the limited available production capacity and the ethical problems surrounding the agricultural land-take required for producing first generation biofuels. While the development of second generation biofuels is still in its infancy (using agricultural waste, industrial waste such as sawdust and paper pulp, woody biomass from forestry, municipal solid waste including household food, garden and paper waste), it does not suffer from the same ethical problems, though the capacity problem may be even more acute. Because of the current limited supply of biofuels, there exists a clear disadvantage in its use due to its very high price. Even if production capacity can be increased, there are concerns in maritime circles that priority in supply will be given to fulfil the likely high demand from land transport, notably cars.

Nuclear Power. Although having a proven track record as a marine fuel in military vessels, there are obvious attendant problems related to public attitudes. This creates its own political sensitivities, especially in relation to the acceptance and reception of such ships in ports and the complex legal issues which may arise upon their deployment; for example, in relation to the role of the International Atomic Energy Agency in maritime affairs (ECSA/ICS, 2008). Even if political and social barriers were to be overcome, crew competency and training needs are likely to be a significant barrier to the commercial application of nuclear power, with the controversial issue of disposal of nuclear waste being a further complicating factor.

Gas (LNG). There are very large reserves of natural gas and, in consequence, it is regarded as a fuel of the future. However, there are significant issues surrounding its storage on board ship as it takes up a lot of space (Einang, 2007). Current technologies suggest, therefore, that its immediate use as a marine fuel will be limited to auxiliary engines, main engines when used on shorter voyages and LNG ships (where relatively minute proportions of the cargo can be used to power the ship). While LNG may not be immediately relevant to ocean-going vessels, it has a lot of potential for use in short sea shipping, including the ferry sector. Due to the nature of its different chemical composition, the use of LNG as a substitute for HFO is likely to lead to a reduction in CO_2 emissions of approximately 23% (Clausen, 2009).

Fuel Cells. These are electrochemical devices which convert the chemical energy within a fuel directly into electricity and heat and do so more efficiently than conventional combustion-based technologies. In contrast to batteries, this energy source will not deplete. The preferred fuel for use with fuel cells is hydrogen, largely because it is non-toxic, abundant, has greater energy production potential than the alternatives and it is non-polluting (Sorensen, 2005). The vast majority of hydrogen production is currently from fossil fuels. The US National Renewable Energy Laboratory have found that this form of production is very energy inefficient, with every 1 MJ of fossil energy consumed producing only 0.66 MJ of energy in the form of hydrogen (Spath & Mann, 2001). It is also the case that for every ton of hydrogen produced, there are 8.9 tons of CO_2 emissions. Some high temperature fuel cells can internally reform fossil fuel into hydrogen without the need for an external heat energy source, thereby significantly improving efficiency and CO_2 emission performance. However, to have any overall practical environmental benefit with widespread application, hydrogen for use with fuel cells would need to be generated from alternative energy sources and there are, indeed, some applications to ships where this has been done. Deployment is currently minimal and limited solely to highly specialised vessels largely in Scandinavia (particularly Iceland), but research is ongoing into various fuel cell technologies, particularly the use of LNG for use with fuel cells and, ultimately, the possibility of ships creating their own hydrogen from the seawater on which they travel. Overall, the inefficiency of upstream production precludes any contemporary proliferation of this technology, as does its cost, but it may yet prove to be a viable avenue for future development.

Wind and Solar. A few shipping companies have experimented with wind and solar energy. In 2008, for instance, the largest Japanese shipping company, NYK, launched its car-carrier, the Auriga Leader, equipped with 328 solar panels costing $1.68 million. The actual ship cost was also high since an innovative design had to be implemented to ensure maximum exposure of the deck-mounted panels to sunlight. Following trials over the first year of voyages, the power generated by the 40 kW system was found to represent the equivalent of only 0.05% of the ship's propulsion power and 1% of its onboard electricity consumption (Eason, 2009). In Denmark, the Windships project seeks to utilise new technologies and material developed in the aerospace sector to fit high performance steel masts and fibreglass sails with a profile like an aircraft wing to bulk carriers of up to 50,000 tons. Unlike traditional sails, these are controlled hydraulically from the bridge. One important disadvantage of sails, however, is that they exert aerodynamic drag which negates some of the fuel savings which accrue from having them there in the first place. A German company, SkySails, is seeking to mitigate this disadvantage by developing innovative inflatable aerofoils which are winched up and down and which act like a large kite that pulls a ship across the ocean by utilising winds at a height of 100–500 metres; a height where winds are often stronger and more dependable. The main disadvantage of this system is that there is a fairly significant fixed cost associated with winching the sails up and down and also the world's prevailing winds are unidirectional. The ECSA/ICS (2008) suggest that both solar and wind power could become important supplementary sources of energy on selected routes, but do not consider them realistic options even in this role for the foreseeable future.

Operational Measures

Energy Optimal Fleet Operation

This relates to minimising CO_2 emissions in relation to the transport work that is undertaken. In theory, this can be achieved in a variety of ways. For example, minimising the backhauls (i.e. return voyages) in bulk shipping that are undertaken in ballast and utilising the hub-and-spoke and load centre concepts in container shipping to minimise aggregate shipping distance travelled and to maximise capacity utilisation on board ships. This objective is further supported by improving logistics systems and enhancing port efficiency (Cullinane and Khanna, 2000; Wang and Cullinane, 2006). All these strategies and approaches can lead to significant reductions in fuel consumption, and thus in CO_2 emissions. However, because of the relationship to costs and the fact that the profit motive drives the vast majority of strategic decisions taken in the shipping sector, there has always been a focus on this aspect in order to obtain maximum returns from assets. The scope for further improvement is, therefore, probably quite limited.

Vessel Speed Reduction

There is no doubt that reducing ship speed does reduce fuel consumption and the CO_2 emissions per unit of transport work. In the bulk shipping markets, where time sensitivity of cargo delivery is not much of an issue, so-called 'slow steaming' (defined as operating a vessel below 60% of maximum engine load) has for decades been applied as a standard cost-reduction strategy in times of poor freight rates. Maersk (2010) reports that reducing speeds by 20% leads to fuel consumption savings of 40% and CO_2 emission reductions of about 7%. However, the feasibility of this as a viable option within an environmental strategy remains highly questionable. For the container trades, in particular, for competitive reasons (not least with land-based modes where they provide a feasible modal alternative), such a strategy would require either the consent or the acceptance of major customers that would have to wait longer to receive their goods; continuity of supply and time of delivery are often essential determinants of competitiveness for the customers of container shipping. Their acceptance of slow steaming would have the immediate effect of their having to increase the volume of buffer stocks held. It would also increase the value of inventory in transit on ships and mean that they have to surmount the associated problem of having to finance this increase in working capital. For ferries too, travel time for passengers and goods is a key influencing factor in the competition with other transport modes.

Ultimately and very critically, with all other things being equal, the systematic adoption of slow steaming will reduce shipping supply to the market. This will mean either higher prices or, more likely, that in order to maintain the same transport capacity, more ships will have to be built and put into operation. This will require not only an additional use of energy in the production process and thus more CO_2 and other emissions, but also the larger fleet which emerges will offset a large proportion of the environmental and cost benefits of speed reductions. Also, additional crews will have to be recruited to an industry where finding suitable human resources has already become an almost insoluble problem. Another consideration relates to the engine design speed referred to earlier, in that operating a ship at a lower speed than that for which it was designed could lead to higher maintenance costs and longer downtime. On the other hand, Maersk (2010) is so convinced of the benefits of reduced speeds that it is ordering 10 new 18,000 TEU ships, actually built to a new design to operate at lower speeds. The company suggests that with the double benefits of economies of scale and lower speeds, CO_2 emissions will be reduced by 50% per moved container. While such a strategy will clearly improve Maersk's, already market leading, green credentials and this may well bring marketing advantages in the short-term, the true test of speed reductions as a viable and sustainable environmentally motivated measure will come when the shipping industry emerges from its current market slump and there is pressure to increase the productivity of ships as an immediate and relatively inexpensive capacity-expanding response to increasing market demand.

Improved Routeing and Less Waiting
This involves attempting to optimise the routeing of ships, so as to avoid impending weather systems that are likely to adversely affect the time it takes to complete a voyage. In the same vein, minimising the amount of time spent waiting in port to either load or discharge cargoes will also reduce emissions and is most readily achieved through increased port efficiency (Cullinane, Wang, Song, & Ji, 2006).

Enhanced Fleet Management
Better planning of ship deployment across a fleet can help minimise the time that a ship, and the fleet in aggregate, spends on ballast voyages. This can also contribute to minimising down time for both routine maintenance and technical problems, so as to ensure the greater productivity of a fleet.

A joint position statement of the European Community Shipowners' Association and the International Chamber of Shipping concludes that the remaining scope for further improvements in either technology or operations is perceived by the shipping industry to be diminishing (ECSA/ICS, 2008). In addition, in order to properly appeal to the industry as potentially implementable measures, any potential measure needs to be commercially attractive and this is not always the case. In such a context, if the environment is to take priority over commercial considerations, then regulation is the last remaining recourse for mitigating the environmental externalities of the shipping industry.

Regulation

Although the IMO has a long history of regulating ship pollution through its International Convention on the Prevention of Pollution from Ships (known as the MARPOL Convention), air emissions have been one of the last major ship pollutants to be regulated. Work began on addressing air emissions from ships in the late 1980s. This led to the amendment of the MARPOL Convention by the '1997 Protocol' to include an Annex VI which was entitled 'Regulations for the Prevention of Air Pollution from Ships'. MARPOL Annex VI prohibits the deliberate emission of ozone depleting substances in line with the Montreal Protocol and sets limits on NO_x and SO_x emissions from ship exhausts.

IMO emission standards are commonly referred to as the Tier I, Tier II and Tier III standards. The Tier I standards were originally defined in the 1997 Protocol which entered into force on 19 May 2005 and apply retrospectively to new engines greater than 130 kW installed on vessels constructed on or after 1 January 2000 or which have undergone a major conversion after that date. In anticipation of the ratification of the Annex VI provisions, most marine engine manufacturers have been building engines compliant with these Tier I standards since 2000.

Following cumulative pressure for more urgent action, the work conducted on this issue by the IMO's Marine Environment Protection Committee (MEPC) culminated in amendments to Annex VI of the MARPOL Convention which were adopted in 2008 and have only relatively recently come into force on 1 July 2010. The amendments define the Tier II and Tier III standards which introduced new fuel quality requirements and Tier II and Tier III NO_x emission standards for new engines, as well as a revised Tier I NO_x requirement for existing pre-2000 engines. The issue of energy efficiency, CO_2 and other greenhouse gas emissions is not covered in Annex VI.

SO_x Regulations

The latest revised version of MARPOL Annex VI sets the following global caps on the sulphur content of a ship's fuel:

- A reduction to 3.50% (35,000 ppm), effective 1 January 2012.
- A reduction to 0.50% (5,000 ppm), effective 1 January 2020 but subject to a review to be completed no later than 2018. If this review is negative, the effective date defaults to 1 January 2025.

Specific 'Sulphur Emission Control Areas' (SECAs) were defined as the English Channel, Baltic Sea and North Sea (see Fig. 9) where more stringent caps on the sulphur content of fuel were established as follows:

- A reduction to 1.00% (10,000 ppm), effective 1 March 2010
- A reduction to 0.10% (1,000 ppm), effective 1 January 2015

In addition to these new global and SECA caps, the new version of Annex VI reinforces the EU Marine Fuel Directive by specifying a sulphur cap of 0.10% at berth within the whole of the EU. It also specified the future creation of North American and US Caribbean ECAs (for both SO_x and NO_x) in 2012 and 2014 respectively (see Figs. 10 and 11) and allowed for the use of abatement technology, rather than low sulphur fuel, to achieve the same outcome in terms of the sulphur content of exhaust gas. The expected successful impact of the IMO's regulations on SO_x emissions is very clearly

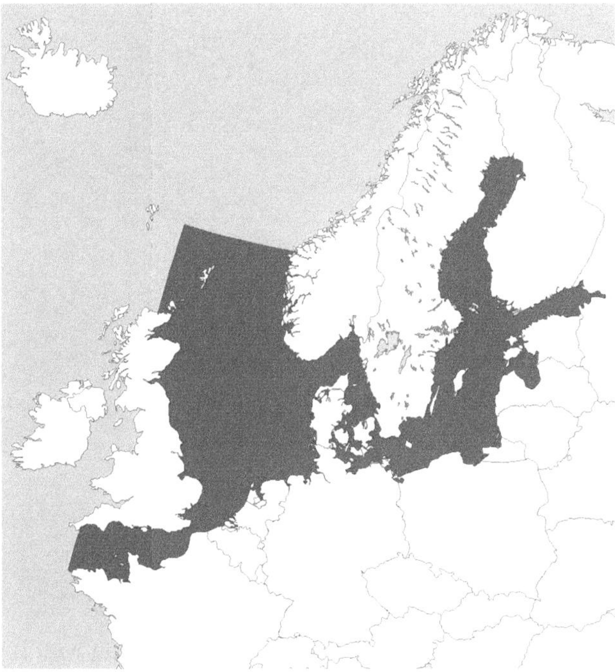

Fig. 9. Europe's Existing Sulphur Emissions Control Area (SECA).

shown in the forecast figures for 2020 in Fig. 12. This also clearly shows that, due in large part to the considerable reductions made in shore-based industry and other terrestrial sources within the EU and in the absence of any regulation on the part of the IMO or the EU itself, by 2020 the shipping industry is expected to be a larger source of sulphur pollution than all other sources combined.

NO_x Regulations

The IMO NO_x emission standards are based on an engine's maximum operating speed (in rpm) because, in general terms, the lower the rpm, the greater the propensity to generate NO_x. As shown in Table 3, the limits are expressed in g/kWh, with Tier I and Tier II limits being global, while Tier III standards are to apply only in NO_x Emission Control Areas.

Tier II standards are expected to be met by designing the engine to optimise the combustion process through fuel injection timing, pressure and rate (rate shaping), fuel nozzle flow area, exhaust valve timing, and cylinder

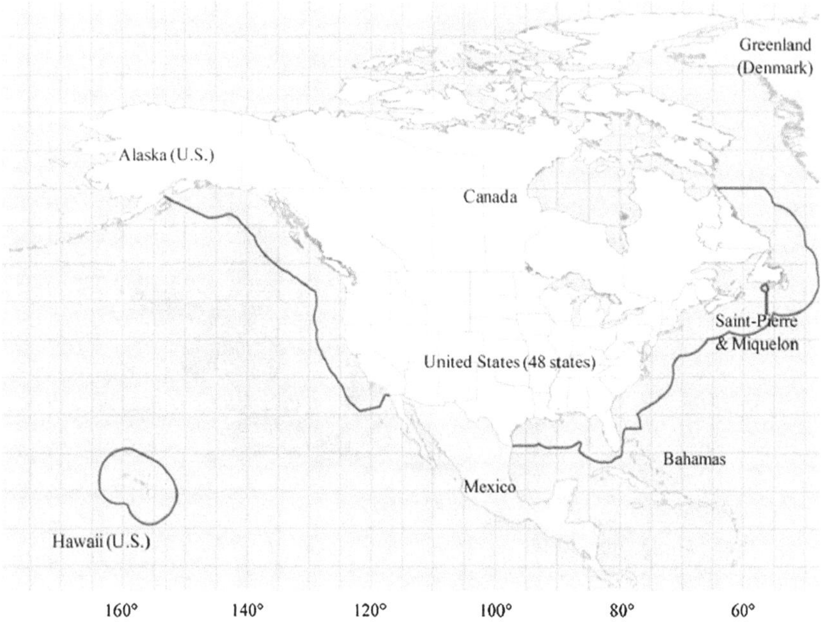

Fig. 10. The Future North American Emissions Control Area (ECA).
Source: FMC (2012).

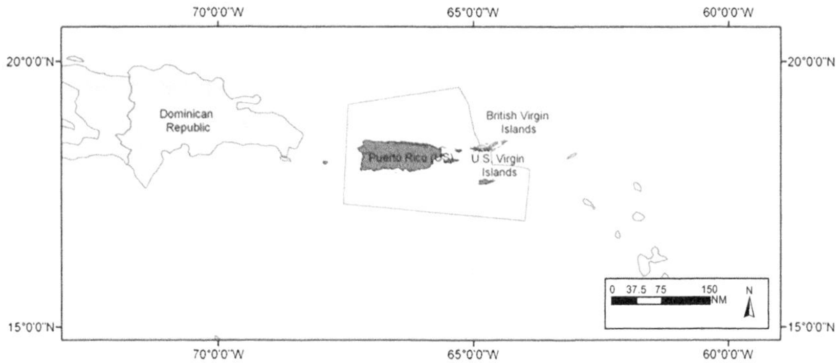

Fig. 11. The Future US Caribbean Emissions Control Area (ECA).
Source: FMC (2012).

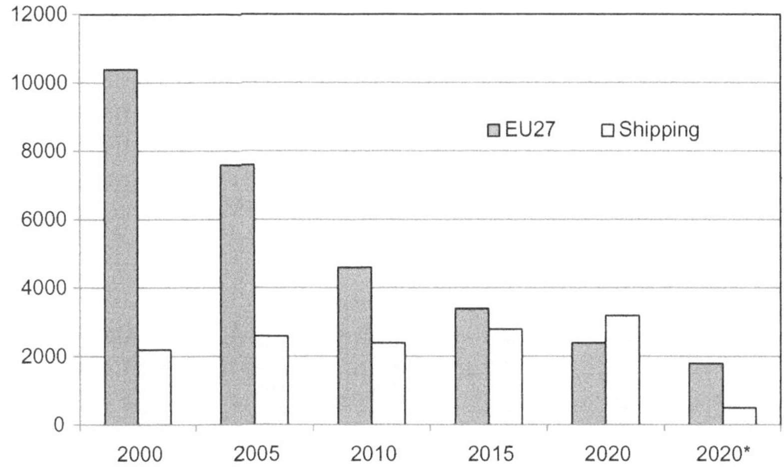

Fig. 12. Forecast of SO_2 Emissions in the EU (2000–2020) (ktonnes). *Notes*: *This is the TSAP – EU Target for the 'EU27' and the IMO Expected Outcome after Regulation for 'Shipping'. The forecast methodology is a scenario analysis employing an extended version of the RAINS model called GAINS that allows the analysis of interactions between air pollution control and greenhouse gas mitigation. The methodology of the GAINS model and the differences to the RAINS methodology are summarised in Amann et al. (2006), with the different optimisation approaches documented in Wagner, Schöpp, and Heyes (2006) and Wagner, Amann, and Schoepp (2007). Emission figures include updated projections for shipping for 2020, based on the IMO-MEPC decision from 9 October 2008 on a revised MARPOL Annex VI. *Source*: Derived by the author from data contained in IIASA (2008) and AirClim (2011).

Table 3. MARPOL Annex VI NO_x Emission Limits.

Tier	Year	NO_x Limit (g/kWh)		
		rpm < 130	130 ≤ rpm ≤ 2,000	rpm ≥ 2,000
I	2000	17.0	$45 \times rpm^{-0.2}$	9.8
II	2011	14.4	$44 \times rpm^{-0.23}$	7.7
III	2016	3.4	$9 \times rpm^{-0.2}$	1.96

Source: IMO (2011b).

compression volume. Tier III standards are expected to require dedicated NO_x emission control technologies such as various forms of water induction into the combustion process (with fuel, scavenging air or in-cylinder), exhaust gas recirculation, or selective catalytic reduction (Burgel, 2007; Durán Grados, Uriondo, Clemente, Jiménez Espadafor, & Gutiérrez, 2009). A technical review in 2012 will determine if the Tier III limits are really feasible but, in any case, Fig. 13 reveals that the IMO's own expectations with respect to the efficacy of their regulations are not that positive.

Greenhouse Gas Regulations and Guidelines
Although not all yet formally embodied within MARPOL Annex VI, the IMO has adopted a 2-pronged approach to reducing the greenhouse gases of the shipping industry. In the first instance, it is attempting to formulate a

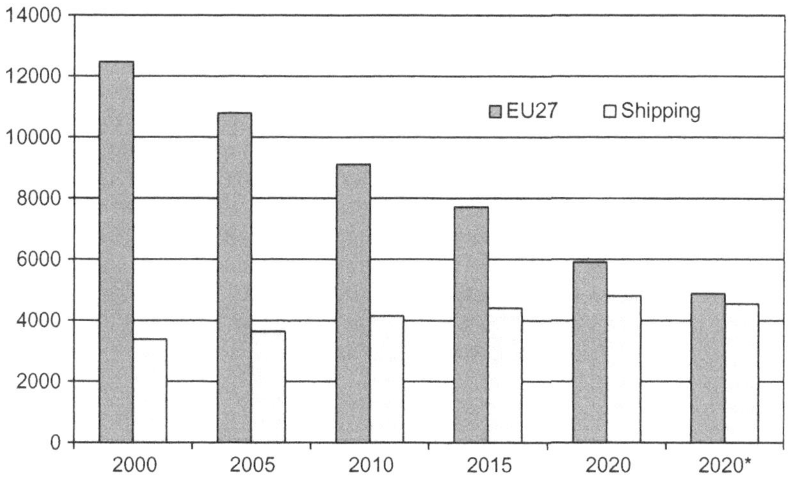

Fig. 13. Forecast of NO_x Emissions in the EU (2000–2020). *Notes*: *This is the TSAP – EU Target for the 'EU27' and the IMO Expected Outcome after Regulation for 'Shipping'. The forecast methodology is a scenario analysis employing an extended version of the RAINS model called GAINS that allows the analysis of interactions between air pollution control and greenhouse gas mitigation. The methodology of the GAINS model and the differences to the RAINS methodology are summarised in Amann et al. (2006), with the different optimisation approaches documented in Wagner et al. (2006) and Wagner et al. (2007). Emission figures include updated projections for shipping for 2020, based on the IMO-MEPC decision from 9 October 2008 on a revised MARPOL Annex VI. *Source*: Derived by the author from data contained in IIASA (2008) and AirClim (2011).

range of 'Technical and Management Strategies' which will either increase the fuel efficiency of the sector or reduce greenhouse gas emissions in some other manner. In a study conducted on behalf of the IMO, however, Buhaug et al. (2009) concluded that even if the whole gamut of available 'Technical and Management Strategies' were to be implemented, they would not alone achieve the level of greenhouse reductions for the shipping industry that is required by the UNFCCC. As a parallel initiative, therefore, the IMO is also pursuing the development of a range of what are generically referred to as 'Market Based Measures', but these remain very much in their infancy.

The IMO's Energy Efficiency Design Index (EEDI) is a formula that is intended to enable ship designers and builders to design and construct ships of the future for maximum fuel efficiency and, thus, minimum greenhouse gas emissions. Conceptually, the index boils down to a ratio of environmental costs over the benefit to society. Alternatively, in practical terms, this can be interpreted as the CO_2 emissions being the environmental cost and the cargo capacity and potential transport work of the ship as the benefit to society. Broadly speaking, the EEDI is to be made operational by applying the following general formula although, in order to accommodate different specific ship types and sizes, there also exist much more complex variations on this theme:

$$\frac{\text{Installed power} \times \text{Specific fuel consumption (SFC)} \times \text{Fuel carbon content factor}}{\text{Capacity} \times \text{Design speed}}$$

The IMO's Marine Environment Protection Committee (MEPC) adopted the EEDI at its meeting in July 2011 as a mandatory element within a newly amended MARPOL Annex VI, with the intention that the EEDI will apply to all new ships as from 2019. It is envisaged that a baseline maximum limit for greenhouse gas emissions according to ship type, size, age, etc. will gradually be reduced over time to ensure a declining trend in such emissions. A major criticism in the development of this regulatory approach is that it has never been the intention for it, or anything similar, to be applied retrospectively to existing ships. In other words, the objective for developing the EEDI has never been to hasten the removal of the worst polluting ships from the existing world fleet; a decision which is anathema to many industry participants and observers as, with such a long-lived asset as a ship, it will bring about only incremental improvements in the environmental footprint of the shipping fleet once it comes into operation.

The IMO's Energy Efficiency Operational Index (EEOI) has been designed to measure the operational efficiency of all new and existing ships by allowing efficiency comparisons between similar ships on similar routes and enables the operator to introduce further efficiency measures. It is basically the ratio of actual fuel consumption to the amount of transport work done and is purely voluntary in nature:

$$\frac{\text{Fuel consumption in operation}}{\text{Cargo loaded} \times \text{Distance travelled}}$$

In practice, the EEOI amounts to a purely voluntary voyage-by-voyage comparator and, despite original intentions, is not really suitable for comparing ships; even sister ships have quite different characteristics. The IMO recommends that the EEOI should be used in conjunction with a Ship Energy Efficient Management Plan (SEEMP). In July 2011, the mandatory possession of such a plan was adopted for inclusion in MARPOL Annex VI and will come into force as from 2019. The SEEMP was developed in close collaboration with the shipping industry and is intended to make ship crews and owners think about how energy is used on board ship. It encompasses a guidance document from the IMO setting out best practice on shipboard environmental measures that provides ship operators with practical advice as to the technical and operational means at their disposal to make their ships more efficient. These include aspects related to improving voyage planning (weather routeing/routing for just-in-time), speed and power optimisation, optimised ship handling (ballast/trim/use of rudder and autopilot), improved fleet management, improved cargo handling and better energy management. A major criticism of the SEEMP as a future regulatory measure is that it will be couched in terms of being a requirement simply to have one, rather than to actually use it.

Taken together with the industry's self-imposed measures that are motivated by the desire to reduce costs through better fuel consumption, the IMO's regulated 'Technical and Management Strategies' and associated guidelines may be enough to engender a much improved environmental profile for the shipping industry if the demand for international shipping were to simply stop growing. However, world trade continues to increase, even during a global economic crisis. Also, emerging economies, such as China, India, Brazil and others generate a growing need for shipping. In addition, many developing countries depend on sea transport for food and basic commodities and the globalisation phenomenon shows no signs of retracting. Thus, the reductions in greenhouse gas emissions from shipping that are achieved by applying technical, operational and management measures alone, irrespective of whether they are regulatory or not, are highly likely to be offset by an increase in shipping activity over time. It is for this reason that the IMO and a number of other industry stakeholders argue fervently in favour of the need for what the IMO refers to as 'market-based measures'. However, there exists considerable opposition to the introduction of market-based measures as many industry players feel that this is would be an unnecessary step.

Although there are a number of subtly different options which have been proposed (including a 'do-nothing' option), there are, essentially, just three types of market-based mechanisms under consideration by the MEPC:

1. An international fund for greenhouse gas emissions from ships, proposed and supported primarily by Denmark and Japan. In effect, this is a 'carbon offset' scheme, where flag states would make a contribution to the fund on the basis of the total emissions from their flag fleet. Either a target line for emissions is established where contributions from highly polluting flag fleets are used to compensate states whose fleet emissions fall below the target line or, alternatively, the fund is utilised to invest in measures to improve ship efficiency.
2. A global emissions trading scheme for shipping, proposed and supported primarily by France, Germany and Norway. The UK government has also explicitly stated that it favours a global emissions trading scheme for shipping (Gilbert, Bows, & Starkey, 2010). This is also the case for the shipping chambers of Australia, Belgium and Sweden. This would be an open cap and trade scheme involving full auctioning, as successfully utilised in other industries (e.g. within the EU Emissions Trading System, the industries currently covered include energy activities, production and processing of ferrous metals, mineral industries and the pulp and paper industries).
3. A trading scheme using energy efficiency credits that is based on the EEDI, proposed and supported primarily by the United States, which represents a hybrid form of the first two main alternatives.

Although discussions continue as to if, and how, market-based measures should be implemented, a fundamental principle which has been agreed is that any surplus funds generated by a market-based instrument under the auspices of the IMO should be used for ameliorating climate change effects in developing countries. However, it would seem that the IMO remains a long way from achieving a consensus amongst its members on the issue of market-based measures. Even if a particular approach were to be adopted in the foreseeable future, its implementation would then inevitably require a new IMO instrument, which could well take some considerable time to develop and to bring to full deployment. In the meantime, the EU is currently exploring potential mitigation options for shipping, given its intention to include shipping emissions in its greenhouse gas reduction commitment if the IMO has not agreed the implementation of a market-based mechanism by 31 December 2011. In the absence of any effective IMO regulation of greenhouse gas emissions from ships, unilateral regulatory action has been threatened by a number of nations and regions (such as Australia, the EU and Japan). This, of course, would undermine the IMO's efforts to achieve a global and generally applicable solution that encompasses both regulatory and market-based measures. A recent report by CE Delft (2009) highlights four potential policy instruments that could be used by the EU and which would appear, in any case, to be quite similar to those already under consideration at the IMO:

- A cap-and-trade system for maritime transport emissions.
- An emissions tax with hypothecated revenues.
- A mandatory efficiency limit for ships in EU ports.
- A baseline and credit system based on an efficiency index.

CONCLUSIONS

There has obviously been a clear recognition that, in absolute terms, ships are a major source of air pollution and associated climate change, global warming, acidification and eutrophication, all of which can have significant adverse effects on human health (Buhaug et al., 2009; Capaldo, Corbett, Kasibhatla, Fischbeck, & Pandis, 1999; Corbett, Fischbeck, & Pandis, 1999; Deniz and Durmusoglu, 2008; Eyring et al., 2005; Kasibhatla et al., 2000; Lawrence and Crutzen, 1999; Richter et al., 2004; Streets, Carmichael, & Arndt, 1997; Tzannatos, 2010). At the same time, it is also important to recognise that, in a relative sense that takes account of the amount of transport work in which it engages, shipping can generally be regarded as the least environmentally damaging mode of freight transport, particularly in terms of the emission of greenhouse gases. Once IMO regulations for the reduction of shipping's SO_x and NO_x footprint (and by default the PM associated with these emissions) begin to have a meaningful impact, then it will generally be the case that an informed and neutral observer would want shipping to be the preferred mode of transport for the carriage of freight, where this is at all feasible. Taking account of work done, no other form of transport can approach shipping's level of performance with respect to energy consumption though, admittedly, some degree of trade-off with SO_x, NO_x and PM emissions does currently exist under certain circumstances. Despite this, it is most certainly the case, however, that the shipping industry has done itself no favours with respect to the public perception of its environmental credentials. With some notable exceptions, shipping industry decision-makers have generally been late and probably reluctant to engage in corporate social responsibility, while regulators (in particular, the IMO) have lacked anticipation with respect to environmental concerns and have been very late to take action to improve the industry's environmental performance and image.

As maritime transport continues to grow in tandem with enhanced economic activity and international trade (particularly as the world economy emerges from its current trough), there is a critical need for the fuel consumption from shipping to be rigorously analysed and effective strategies implemented for greater fuel efficiency and, as a consequence, for the reduction of air pollution from ships. Largely because of the cost saving incentive, examples of unilaterally implemented good environmental practice do exist within the sector. This will no doubt continue into the future as the anticipated rising cost of fuel proves to be an influential driver. In addition, the customers of shipping companies (i.e. the owners of the cargoes carried) will increasingly demand more sustainable transport for the movement of their products, especially as they seek to influence, gain control over and integrate their supply chains, ultimately in response to the demands and environmental concerns of the retail consumer. As with all industries, 'first movers' in responding to this imperative (e.g. Maersk) will gain considerable market momentum and competitive advantage.

The evidence suggests, however, that market forces alone will not resolve the environmental problems attributable to the shipping industry and that intervention is most definitely required. Seemingly rather late in the day, the IMO has embarked on developing a new regulatory regime to control the air pollution of ships. While SO_x regulation is likely to prove highly effective within the foreseeable future, the effect of NO_x regulation remains questionable. With respect to energy efficiency, CO_2 and other greenhouse gas emissions, the introduction of the EEDI and the SEEMP will have little effect in the short term. Achieving a workable consensus on a way forward for the development of market-based measures seems a long way off. A major stumbling block to progress has been a conflict between the ethos of 'common but differentiated responsibility' (Rajamani, 2000), which implies a differential treatment between developed and developing nations and which underpins climate change discussions at the UNFCCC (United

Nations, 1997), and the IMO stance that shipping industry regulation requires a global level of implementation and should, in consequence, apply equally to all flag-states.

In designing a regulatory framework for the mitigation of air pollution from ships, it is vitally important that the supply of international shipping for the transportation of goods around the world should not be curtailed in any way. To do so would cause severe disruption to global trade and economic development. Davidson and Faber (2009), for instance, suggest that the costs of international maritime transport may increase as a result of international climate mitigation policies and that even a marginal increase in costs could have a disproportionate affect on developing countries, particularly small island states and landlocked countries. In order to avoid such an outcome from the imposition of market-based measures, the IMO has committed to using the resources generated from whatever system might be implemented to finance adaptation and mitigation measures in developing countries.

In addition, care should be taken to avoid undermining the inherent environmental advantage of shipping versus other modes of transport. In this respect, it is also important to adopt a holistic view on the regulation of shipping in order not to stimulate a modal shift from waterborne transport to a less environmentally credible mode of transport such as road, rail or air transport. For example, the IMO's SECA proposals for the English Channel, North Sea and Baltic Sea were received with a barrage of criticism with respect to their potential adverse impact (a) on the competitiveness of short-sea shipping in North-West Europe due to the higher costs that the new regulations would impose and (b) in relation to inter-continental shipping, on the competitiveness of ports in the region compared to those in the Mediterranean. Indeed, the Chairman of the EU transport Committee is reported as stating that 'It's unusual for the IMO to take such a decision .... normally it takes them 50 years, but when they take it quickly they get it wrong' (Stares, 2010a, 2010b).

In order to achieve a holistic solution to the atmospheric pollution from ships, there is clearly a need for cooperation and coordination between the IMO and the world's various regional powers. However, as the IMO prevaricates with respect to a decision on the regulation of greenhouse gas emissions, regional and unilateral national actions have been threatened by, for instance, the EU, the United States, Japan and Australia. These may yet come to fruition and even proliferate; resulting in a very uneven playing field that does not deliver environmentally because of the inherent mobility of shipping assets. The EU, in particular, has set a deadline of the end of 2011 for agreement to be reached within the IMO on a strategy for imposing market-based measures. It remains to be seen whether this threat prompts swift action on the part of the IMO or whether the EU follows through and, for example, opts to bring shipping into the fold of the EU Emissions Trading System.

NOTES

1. Figures are based on the mode of transport on which imports arrive at a country's border, seaport or airport.
2. The three main types of marine fuel (bunkers) are most usually referred to as Heavy Fuel Oil (HFO), Intermediate Fuel Oil (IFO) and Marine Diesel Oil (MDO). HFO is the highly viscous residual fuel oil that remains after the more valuable cuts (or fractions) of crude oil have been boiled off during the refining process. MDO is the highest quality common marine fuel, being a gasoil distillate which, as distinct from ordinary diesel, contains some HFO. IFO is the medium quality of the three standard marine fuels and contains blended volumes of HFO, MDO and other lighter, higher quality fuel oils.

ACKNOWLEDGEMENTS

The author is grateful to the editors and two anonymous referees for their helpful comments on an earlier draft of this chapter.

REFERENCES

Aabo, K. (2009). *2-stroke exhaust gas emission control.* 31st Annual Motorship Conference: Saving fuel for lower costs and lower emissions, 26–27 May, Copenhagen.

Agrawala, H., Malloy, Q. G. J., Welch, W. A., Miller, J. W., & Cocker, D. R., III. (2008). In-use gaseous and particulate matter emissions from a modern ocean going container vessel. *Atmospheric Environment, 42*(21), 5504–5510.

AirClim (2011). *Air pollution fom ships.* Retrieved from http://www.airclim.org/sites/default/files/documents/Air%20pollution%20from%20ships_Nov_2011.pdf

Amann, M., Asman, W., Bertok, I., Cofala, J., Heyes, C., Klimont, ... Wagner, F. (2006). *Emission control scenarios that meet the environmental objectives of the Thematic Strategy on Air Pollution* (NEC Scenario Analysis Report No. 2). Laxenburg, Austria: International Institute for Applied Systems Analysis (IIASA). Retrieved from http://www.iiasa.ac.at/rains/reports/NEC2full.pdf

Andreasen, A., & Mayer, S. (2007). Use of seawater scrubbing for SO_2 removal from marine engine exhaust gas. *Energy Fuels, 21*(6), 3274–3279.

Arduino, G., Murillo, D. C., & Ferrari, C. (2011). *Key factors and barriers to the adoption of cold ironing in Europe.* Retrieved from http://ww2.unime.it/sefisast/SEFISAST/Conference_Paper_files/Arduino_Carrillo_Murillo_Ferrari.pdf

Bailey, D., & Solomon, G. (2004). Pollution prevention at ports: Clearing the air. *Environmental Impact Assessment Review*, *24*, 749–774.

Bows, A., & Anderson, K. L. (2007). Policy clash: Can projected aviation growth be reconciled with the UK Government's 60% carbon-reduction target? *Transport Policy*, *14*(2), 103–110.

Buhaug, Ø., Corbett, J. J., Endresen, Ø., Eyring, V., Faber, J., Hanayama, S., … Yoshida, K. (2009). *Prevention of air pollution from ships, Second IMO GHG study 2009*. London: International Maritime Organization (IMO). Retrieved from http://www.imo.org/includes/blastDataOnly.asp/data_id%3D26047/%20INF-10.pdf

Burgel, A. P. (2007). Air pollution from ships: Recent developments. *WMU Journal of Maritime Affairs*, *6*(2), 217–224.

California Air Resource Board. (2003, January). *Draft state and federal element of South Coast state implementation plan*, II-F. California Environmental Protection Agency, Sacramento, CA.

Capaldo, K., Corbett, J. J., Kasibhatla, P., Fischbeck, P. S., & Pandis, S. N. (1999). Effects of ship emissions on sulphur cycling and radiative climate forcing over the ocean. *Nature*, *400*, 743–746.

CE Delft. (2009). Technical support for European action to reducing greenhouse gas emissions from international maritime transport. Tender DG ENV.C3/ATA/2008/0016, CE Delft.

CE Delft, Germanischer Lloyd, MARINTEK, & Det Norske Veritas. (2006). *Greenhouse gas emissions for shipping and implementation guidance for the marine fuel sulphur directive*. Delft: CE Delft.

CE Delft, MMU, & MNP. (2006). *Aviation and maritime transport in a post 2012 climate policy regime*. CE Delft, Delft, The Netherlands.

Clausen, N. B. (2009). Marine diesel engines: How efficient can a two-stroke engine be? *STG ship efficiency conference*, Hamburg, 28–29 September. Retrieved from http://www.ship-efficiency.org/onTEAM/pdf/Clausen.pdf

Cooper, D. A. (2003). Exhaust emissions from ships at berth. *Atmospheric Environment*, *37*(27), 3817–3830.

Corbett, J. J., & Farrell, A. (2002). Mitigating air pollution impacts of passenger ferries. *Transportation Research Part D*, *7*, 197–211.

Corbett, J. J., Fischbeck, P. S., & Pandis, S. N. (1999). Global nitrogen and sulfur inventories for oceangoing ships. *Journal of Geophysical Research*, *104*, 3457–3470.

Corbett, J. J., & Koehler, H. W. (2003). Updated emissions from ocean shipping. *Journal of Geophysical Research*, *108*(D20).

Corbett, J. J., & Koehler, H. W. (2004). Considering alternative input parameters in an activity-based ship fuel consumption and emission model: Reply to comment by Oyvind Endresen et al. on "Updated emissions from ocean shipping". *Journal of Geophysical Research*, *109*(D23303).

Corbett, J. J., Winebrake, J. J., Green, E. H., Kasibhatla, P., Eyring, V., & Lauer, A. (2007). Mortality from ship emissions: A global assessment. *Environmental Science & Technology*, *41*, 8512–8518.

Crist, P. (2009, May). *Greenhouse gas emissions reduction potential from international shipping* (Discussion Paper 2009-11). International Transport Forum, OECD, Paris.

Cullinane, K. P. B., & Khanna, M. (1999). Economies of scale in large container ships. *Journal of Transport Economics and Policy*, *33*(2), 185–208.

Cullinane, K. P. B., & Khanna, M. (2000). Economies of scale in large containerships: Optimal size and geographical implications. *Journal of Transport Geography*, *8*(3), 181–195.

Cullinane, K. P. B., Wang, T.-F., Song, D.-W., & Ji, P. (2006). The technical efficiency of container ports: Comparing data envelopment analysis and stochastic frontier analysis. *Transportation Research A: Policy and Practice*, *40*(4), 354–374.

Davidson, M. D., & Faber, J. (2009). Impacts of climate policies on air and maritime transport in developing countries. *Trade and Environment Review: Promoting poles of clean growth to foster the transition to a more sustainable economy*, United Nations Conference on Trade and Development, Geneva, 129–142.

Deniz, C., & Durmusoglu, Y. (2008). Estimating shipping emissions in the region of the Sea of Marmara, Turkey. *Science of the Total Environment*, *390*, 255–261.

Deniz, C., Kilic, A., & Civkaroglu, G. (2010). Estimation of shipping emissions in Candarli Gulf, Turkey. *Environmental Monitoring and Assessment*, *271*, 219–228.

Durán Grados, C. V., Uriondo, Z., Clemente, M., Jiménez Espadafor, F. J., & Gutiérrez, J. M. (2009). Correcting injection pressure maladjustments to reduce NO_X emissions by marine diesel engines. *Transportation research Part D: Transport and environment*, *14*(1), 61–66.

Eason, C. (2009). Solar panels efficient but have a long way to go. *Lloyd's List*, September 15. Retrieved from http://www.lloydslist.com/ll/sector/Environment/article24728.ece

ECSA/ICS (2008). *Climate change and shipping* (Position Paper). Brussels: European Community Shipowners' Association and International Chamber of Shipping.

Einang, P. M. (2007). Gas-fuelled ships, CIMAC paper 261, *Proceedings of the 25th CIMAC World congress on combustion engines*, Vienna, Austria, 21–24 May.

Endresen, O., Sorgard, E., Bakke, J., & Isaksen, I. S. A. (2004). Substantiation of a lower estimate for the bunker inventory: Comment on "Updated emissions from ocean shipping" by James J. Corbett and Horst W. Koehler. *Journal of Geophysical Research*, *109*, D23302.

Eyring, V., Köhler, H., Lauer, A., & Lemper, B. (2005). Emissions from international shipping: 2. Impact of future technologies on scenarios until 2050. *Journal of Geophysical Research*, *110*, D17306.

FMC (2012). *Emissions control areas*. Retrieved from http://www.fmc.gov/maritime_environmental_issues/maritime_environmental_news.aspx. Accessed on April 5, 2012.

Frey, H. C. (2008). Identification and evaluation of potential best practices for greenhouse gas emissions reductions in freight transportation. Retrieved from www.arb.ca.gov/research/seminars/frey/frey.pdf

Fridell, E. (2011, September 9). *Emissions to air from shipping and future abatement techniques*. Reducing the Impact of Emissions from Aviation and Shipping Conference, Edinburgh.

Fridell, E., Steena, E., & Peterson, K. (2008). Primary particles in ship emissions. *Atmospheric Environment*, *42*(6), 1160–1168.

Gilbert, P., Bows, A., & Starkey, R. (2010). *Shipping and climate change: Scope for unilateral action*. Manchester: University of Manchester.

Green, E. H., Winebrake, J. J., & Corbett, J. J. (2008, July 30). *Opportunities for reducing greenhouse gas emissions from ships*. Report Prepared for the Clean Air Task Force, Boston. Retrieved from http://www.ceoe.udel.edu/CMS/jcorbett/CATF_GHG_Reduction_Memo_20080801_external.pdf

Hammingh, P., Aben, J. M. M., Blom, W. F., Jimmink, B. A., de Vries, W. J., & Visser, M. (2007). *Effectiveness of international emission control measures for North Sea shipping on Dutch air quality*. MNP Report 500092004/2007. The Hague: Netherlands Environmental Assessment Agency.

Hickman, A. J. (1999). *Methodology for calculating transport emissions and energy consumption.* Final report of the EC MEET project. Transport Research, 4th framework programme DGVII. Crowthorne: Research for sustainable mobility, TRL-© TRL (Transport Research Laboratory).

Hoffmann, J. (2008). The modal split of international goods transport, *UNCTAD Transport Newsletter, 38*, UNCTAD/SDTE/TLB/MISC/2008/1, United Nations Conference on Trade and Development, Geneva.

Hulskotte, J. H. J., & Gon, H. A. C. (2010). Fuel consumption and associated emissions from seagoing ships at berth derived from an on-board survey. *Atmospheric Environment, 44*, 1229–1236.

IIASA. (2008, July). *National emission ceilings for 2020 based on the 2008 climate and energy package* (NEC Scenario Analysis Report No. 6). Laxenburg, Austria: International Institute for Applied Systems Analysis. Retrieved from http://www.iiasa.ac.at/rains/reports/NEC6-final110708.pdf

IMO (2011a). *International convention for the prevention of pollution from ships (MARPOL).* London: International Maritime Organization. Retrieved from http://www.imo.org/About/Conventions/ListOfConventions/Pages/International-Convention-for-the-Prevention-of-Pollution-from-Ships-(MARPOL).aspx. Accessed on August 22, 2011.

IMO (2011b). *Maritime knowledge centre.* London: International Maritime Organization. Retrieved from http://www.imo.org/KnowledgeCentre/ShipsAndShippingFactsAndFigures/TheRoleandImportanceofInternationalShipping/Documents/International%20Shipping%20Facts%20and%20Figures%20final.pdf

IPCC (2000). *Special report on emissions scenarios.* Arendal, Norway: United Nations Environment Programme, Intergovernmental Panel on Climate Change. Retrieved from http://www.grida.no/publications/other/ipcc_sr/?src=/climate/ipcc/emission/

Kasibhatla, P., Levy, H., II., Moxim, W. J., Pandis, S. N., Corbett, J. J., Peterson, M. C., … Ryerson, T. B. (2000). Do emissions from ships have a significant impact on concentration of nitrogen oxides in the marine boundary layer? *Geophysical Research Letters, 27*(15), 2229–2233.

Lawrence, M. G., & Crutzen, P. J. (1999). Influence of NOx emissions from ships on tropospheric photochemistry and climate. *Nature, 402*, 167–170.

Lin, C.-Y. (2002). Reduction of particulate matter and gaseous emission from marine diesel engines using a catalyzed particulate filter. *Ocean Engineering, 29*(11), 1327–1341.

Lowles, I. (1998). *Emissions of sulphur dioxide from ships, in coasts and benefits of controlling sulphur dioxide emissions from ships in the North Sea and Seas to the West of Britain.* A report to the Department of the Environmental, Transport and the Regions, London.

Maersk. (2010). *Slow steaming here to stay.* Retrieved from http://www.maersk.com/AboutMaersk/News/Pages/20100901-145240.aspx. Assessed on September 1, 2010.

Maersk Line. (2011). *Technology harnesses waste heat for energy.* Retrieved from http://www.maerskline.com/link/?page=news&path=/news/story_page/11/Technology_harnesses

MARINTEK. (2008). *Propulsive tests database.* Internal documentation, Trondheim, Norway.

Mayor, K., & Tol, R. S. J. (2007). The impact of the UK aviation tax on carbon dioxide emissions and visitor numbers. *Transport Policy, 14*(6), 507–513.

Mitropoulos, E. E. (2010) A message to the world's seafarers. *IMO News* (Issue 1, p. 5). London: International Maritime Organisation.

Nicholls, R. J., Hanson, S., Herweijer, C., Patmore, N., Hallegatte, S., Corfee-Morlot, J., Chateau, J., & Muir-Wood, R. (2007). *Ranking of the world's cities most exposed to coastal flooding today and in the future*. ENV/WKP/(2007)1, OECD, Paris.

Olsthoorn, X. (2001). Carbon dioxide emissions from international aviation: 1950–2050. *Journal of Air Transport Management*, *7*(2), 87–93.

Psaraftis, H. N., & Kontovas, C. A. (2009). CO_2 emission statistics for the world commercial fleet. *WMU Journal of Maritime Affairs*, *8*, 1–25.

Rajamani, L. (2000). The principle of common but differentiated responsibility and the balance of commitments under the climate regime. *Review of European community and international environmental law*, *9*(2), 120–131.

Richter, A., Eyring, V., Burrows, J. P., Bovensmann, H., Lauer, A., Sierk, B., & Crutzen, P. J. (2004). Satellite measurements of NO_2 from international shipping emissions. *Geophysical Research Letters*, *31*, L23110.

Sorensen, B. (2005). *Hydrogen and fuel cells: Emerging technologies and applications*. Amsterdam: Academic Press.

Spath, P. L. & Mann, M. K. (2001). *Life cycle assessment of hydrogen production via natural gas steam reforming*. (National Renewable Energy Laboratory Report NREL/TP-570-27637) Golden, CO: National Renewable Energy Laboratory.

Stares, J. (2010a). Full steam ahead for Baltic and North Sea low-sulphur rules. *Lloyd's List*, December 7. Retrieved from http://www.lloydslist.com/ll/sector/regulation/article351675.ece

Stares, J. (2010b). Europe's emissions control areas a big mistake. *Lloyd's List*, September 27. Retrieved from http://www.lloydslist.com/ll/sector/ship-operations/article345779.ece

Stopford, M. (2010, September 7). *How shipping has changed the world and the social impact of shipping*. Global Maritime Environmental Congress, SMM Hamburg.

Streets, G. D., Carmichael, G. R., & Arndt, R. L. (1997). Sulfur dioxide emissions and sulfur deposition from international shipping in Asian waters. *Atmospheric Environment*, *31*, 1573–1582.

Truong, V.-T. (2001). *Drag reduction technologies* (Report No. DSTO-GD-0290). Australia: DSTO Aeronautical and Maritime Research Laboratory.

Tzannatos, E. (2010). Ship emissions and their externalities for the port of Piraeus-Greece. *Atmospheric Environment*, *44*, 400–407.

UNCTAD. (2009, February 16–18). *Multi-year expert meeting on transport and trade facilitation: Maritime transport and the climate change challenge – Summary of proceedings* (UNCTAD/DTL/TLB/2009/1), United Nations Conference on Trade and Development, Geneva. Retrieved from http://www.unctad.org/en/docs/dtltlb20091_en.pdf

UNCTAD (2011). *Review of maritime transport 2010*. United Nations Publication No. E.11.II.D.4. United Nations Conference on Trade and Development, Geneva.

UNFCCC. (2010, February 12). *Report of the Subsidiary body for scientific and technological advice on its thirty-first session held in Copenhagen from 8th to 12th December 2009*. Document FCCC/SBSTA/2009/8, United Nations Framework Convention on Climate Change, Copenhagen.

United Nations. (1997, December 10). *Kyoto protocol to the United Nations framework convention on climate change*. Conference of the parties on its third session, FCCC/CP/1997/L.7/Add.1, Kyoto.

Upham, P. (2003). *Towards sustainable aviation*. Oxford: Earthscan.

US EPA. (1999, August). *In-use marine diesel fuel*. United States Environmental Protection Agency. Retrieved from http://www.arb.ca.gov/planning/sip/scsip03/scsip03.htm

Vernya, J., & Grigentinc, C. (2009). Container shipping on the Northern Sea Route. *International Journal of Production Economics*, *122*(1), 107–117.

Wagner, F., Amann, M., & Schoepp, W. (2007). *The GAINS optimization module as of 1 February 2007* (Interim Report IR-07-004). Laxenburg, Austria: International Institute for Applied Systems Analysis (IIASA). Retrieved from http://www.iiasa.ac.at/Admin/PUB/Documents/IR-07-004.pdf

Wagner, F., Schöpp, W., & Heyes, C. (2006). *The RAINS optimization module for the Clean Air For Europe (CAFE) programme* (IR-06-29). Laxenburg, Austria: International Institute for Applied Systems Analysis (IIASA).

Wang, T., Zhang, Y., Peng, Z., & Shu, G. (2011). A review of researches on thermal exhaust heat recovery with Rankine cycle. *Renewable and Sustainable Energy Reviews*, *15*(6), 2862–2871.

Wang, T.-F., & Cullinane, K. P. B. (2006). The efficiency of European container terminals and implications for supply chain management. *Maritime Economics & Logistics*, *8*(1), 82–99.

Winkler, J. (2009, May 26–27). *Efficient ship designs can regulate CO_2 now*. Saving fuel for lower costs and lower emissions, 31st Annual motorship conference, Copenhagen.

World Shipping Council (2009, September). *Liner shipping and carbon emissions policy*. Retrieved from http://www.worldshipping.org/pdf/liner_shipping_carbon_emissions_policy_presentation.pdf

CHAPTER 5

THE IMPACTS OF CLIMATE CHANGE ON NATIONAL ROAD AND RAIL NETWORKS

Elizabeth Hooper and Lee Chapman

ABSTRACT

Purpose – *To investigate the potential impacts of future climate change in the United Kingdom on its road and rail networks.*

Methodology/approach – *The climate change impacts of increasing summer temperatures, decreasing winter temperatures, increased heavy precipitation, greater numbers of extreme weather events and rises in sea level are reviewed.*

Findings – *Surface transportation is the most exposed element to the localised impacts of climate change. High summer temperatures will result in road rutting, rail buckling and decreased thermal comfort, whereas more intense winter precipitation will cause flooding, landslips and bridge scour across all modes. For all impacts, it is the extreme events (e.g. heat waves and storms) that are potentially the most devastating. As shown, there are some positive climate change impacts. For example, in the case of winter maintenance, all transport networks stand to benefit.*

Originality/value – *In order for transport to react appropriately to the potential changes in climate, it is essential to understand how the road and*

Transport and Climate Change
Transport and Sustainability, Volume 2, 105–136

ISSN: 2044-9941/doi:10.1108/S2044-9941(2012)0000002008

rail networks may be affected and to build strategies for both adaptation and mitigation into plans for future developments for both modes.

Keywords: Climate change; national; road; rail; impact

INTRODUCTION

The UK transport sector has already been identified as one of the six key sectors – energy; business; transport; domestic; public services; and agriculture, forestry and fisheries – which will be most affected by climate change (McKenzie Hedger, Gawith, Brown, Connell, & Downing, 2000). It is crucial to understand the potential impacts of future climate change on each transport mode in order to be able to prepare for these effects. Transport networks are highly valuable assets – the UK road network alone was valued as the government's single most expensive asset in 2005 with major trunk roads and motorways having an approximate value of £62 billion (Department for Transport, 2005). With this in mind, it is clear that the impacts of climate change on transport networks need careful consideration to allow the networks to continue operating effectively in the future.

The need to mitigate the impacts of climate change is highlighted in several reports (e.g. Garnaut, 2008; The Royal Academy of Engineering, 2011) as a result of the wide range of effects that may be endured by the various transport modes. Impacts on the road network are not limited to pavements but also long-term assets on the network such as bridges so it is important that adaptation strategies can be incorporated into upgrade and future development works to prevent more extensive costs in a few decades time (Arkell & Darch, 2006). The need for flexibility in infrastructure designs to cope with the uncertainty of climate change and prevent increasing adaptation costs is identified by the UK's Adaptation Sub-Committee in the 'How well prepared is the UK for climate change?' report (Krebs et al., 2010). For both road and rail, it is unlikely that any single aspect of climate change will have a single impact on each mode. As a result, understanding needs to be developed of the combination of factors which impact on each mode and the corresponding infrastructure to allow strategies to be developed to help mitigate for and also to adapt to the potential effects.

Table 1 provides an overview of the predicted changes in climate, as defined by the UK Climate Projections (UKCP09), and the potential

Table 1. The Effects of Climate Change on Transport.

Increased numbers of hot days	1. Increased thermal loading on road pavements a. Melting tarmac b. Roadway buckling c. Expansion/buckling of bridges d. Increased numbers of tyre blow-outs 2. Increased rail track buckling 3. Expansion/buckling of railway bridges 4. Increased heat exhaustion of maintenance and operations staff 5. Effects of higher density air on aviation a. Reduced engine combustion efficiency b. Increased runway lengths required
Decreased numbers of cold days	1. Reduced winter maintenance costs for road and rail 2. Improved working conditions for personnel in cold environments 3. Permafrost problems: a. Unable to rely on 'frozen roads' b. Infrastructure problems caused due to settlement when permafrost thaws c. Increased subsidence and landslides on slopes and embankments 4. Positive effects on marine transportation: a. Less de-icing required and freezing fog b. Less icebreaking required c. Potential opening of new sea passages in polar regions 5. Reduction in icing problems for electric rail systems
Increased heavy precipitation	1. Road and rail submersion and underpass flooding 2. Increased landslides and undercutting 3. Poor visibility 4. Exceedance of existing 100 year flood
Seasonal changes	1. Longer summers/shorter winters will mean changes in timing of a. Leaf-fall for railways b. Winter maintenance regimes c. Shift in ice/snow belts 2. Reduction in frozen precipitation – significant improvements in road safety
Drought	1. Navigation problems on inland waterways 2. May lead to increased failure of earthworks due to changes in the water table
Sea-level change	1. Locations of ports may be inappropriate 2. Other infrastructure – many airports are built 10 m above sea level 3. Localised problems, e.g. storm surges 4. Possible increases in coastal erosion causing problems for coastal transport routes (e.g. Dawlish railway, A55 along the north Wales Coast)

Table 1. (*Continued*)

Extreme events	1. Increased numbers of tropical storms? 2. Increased lightning effects on aviation
	No clear projections are available for wind

Source: Based on Jaroszweski, Chapman, and Petts (2010) and Peterson et al. (2008).

impacts of these changes on the UK road and rail transport networks. Each predicted change in climate impacts upon the transport network in a number of ways, and there are a number of consequences for road and rail which need to be considered and incorporated into adaptation strategies. This chapter provides a detailed account of the impacts of climate change on national road and rail transport subdivided into the categories outlined in Table 1.

Links between different modes of transport are also important as one mode may be more resilient to one particular impact of climate change than another, for example the rail network may be more resilient to precipitation changes than the road network with fewer implications for safety and delay times unless there are extreme occurrences of heavy precipitation which cause infrastructure damage.

TEMPERATURE

The UKCP09 predict that there will be an increase in the number of days with high temperatures in the summer months and a decrease in the number of cold days during the winter. Table 2 shows the predicted changes in temperatures for winter, summer and daily temperatures across the UK under the different emission scenarios.

Increased Number of Hot Days

The UKCP09 forecast that the number of days with high temperatures will increase in the future under all emission scenarios. Road and rail transport will both suffer some common and individual consequences as the number of days with high temperatures increases Fig. 1 shows the possible changes in mean daily maximum summer temperatures by the 2080s under

Table 2. Highest and Lowest Changes in Mean Daily, Mean Daily Maximum and Mean Daily Minimum Temperature (°C) in Winter and Summer, by the 2080s, Relative to 1961–1990.

Variable		Mean Temperature, Winter			Mean Temperature, Summer			Mean Daily Maximum Temperature, Winter			Mean Daily Maximum Temperature, Summer			Mean Daily Minimum Temperature, Winter			Mean Daily Minimum Temperature, Summer		
Probability Level		10%	50%	90%	10%	50%	90%	10%	50%	90%	10%	50%	90%	10%	50%	90%	10%	50%	90%
High emissions	Highest change in UK	2.2	3.8	5.8	2.9	5.3	8.4	1.6	3.4	6.1	3.0	6.8	11.7	2.0	4.2	7.0	2.8	5.3	8.8
	Lowest change in UK	1.0	2.1	3.5	1.6	3.1	5.0	1.1	2.3	3.9	1.2	3.5	6.3	0.8	2.4	4.3	1.7	3.3	5.6
Medium emissions	Highest change in UK	1.7	3.1	4.8	2.2	4.2	6.8	1.3	2.9	5.1	2.2	5.4	9.5	1.5	3.5	5.9	2.0	4.1	7.1
	Lowest change in UK	0.8	1.8	3.1	1.2	2.5	4.1	0.8	2.0	3.4	1.1	2.8	5.0	0.6	2.1	3.7	1.3	2.7	4.5
Low emissions	Highest change in UK	1.5	2.7	4.1	1.4	3.1	5.3	1.3	2.6	4.3	1.4	4.1	7.5	1.4	2.9	4.8	1.4	3.2	5.6
	Lowest change in UK	0.8	1.7	2.7	0.8	1.9	3.2	0.9	1.8	3.0	0.7	2.1	3.9	0.7	2.0	3.3	0.9	2.1	3.5

Source: UKCP09 – Murphy et al. (2009).

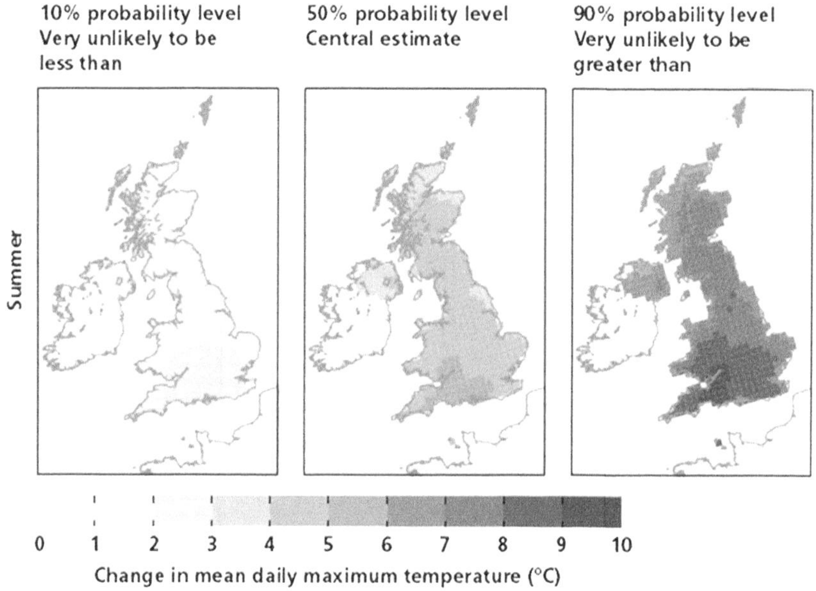

Fig. 1. Changes to Mean Daily Maximum Temperature in Summer by the 2080s at 10, 50 and 90% Probability Levels, Under the Medium Emissions Scenario. *Source*: UKCP09 – Murphy et al. (2009).

the medium emissions scenario at the 10, 50 and 90% probability estimates. The central estimate, at the 50% probability level, shows that there is likely to be a temperature increase of between 3°C and 6°C which would have a significant impact on both road and rail transport.

Impacts on the Road Network

During the summer months, prolonged periods of higher temperatures may increase the amount of rutting on roads across the UK (Clarke et al., 2002a). Current literature suggests that there may be a need to adapt the specifications of road surfaces to enable roads to withstand more prolonged periods of higher temperatures. Following the hot summer of 1995 and extensive damage to the roads as a result, road surfacing specifications were amended to prevent such damage recurring in future (Clarke et al., 2002a). At present there is no indication as to whether these specifications will enable roads to cope with predicted future summer temperatures. Road surface specifications in the United Kingdom may be an issue which needs extensive attention and adjustment to cope with the predicted temperature increases.

Road surfaces in the south west of England have been identified as especially vulnerable with more incidences of road rutting likely to occur as temperatures in the region rise and as the number of visitors and hence the volume of traffic increases (McKenzie Hedger et al., 2000). The summer of 2003 showed the vulnerability of roads in south east England with reports of tarmac 'melting/bleeding' and minor roads becoming more slippery as loose stones rose to the surface because of the heat (Standley et al., 2009). If summer temperatures are to increase and sustain over longer time periods, then this will become a more commonplace problem across the UK.

Climate analogues use climate regimes currently experienced which may be representative of future climates in a given region (Mearns et al., 2001). For example, the hot summer of 2003 experienced in the United Kingdom may be used as a temporal analogue for future climate change and as a result is an instance from which lessons can be learned about the reactions of transport across the UK to increasing temperatures. Learning from the effects of the summer of 2003 will help to prepare both road and rail networks for possible impacts from increasing summer temperatures and allow national transport bodies such as the Highways Agency and Network Rail to be better prepared for future events.

The impact of repair work needed following road rutting in hot weather must also be considered. Unless road surfaces are adapted and specifications increased to withstand higher temperatures over more prolonged periods, there will be more instances of rutting. Naturally it follows that there will need to be more extensive repair work, and this may in turn increase journey times and delays as more service and maintenance work is carried out (Hunt et al., 2006).

The impact of higher summer temperatures on road safety also needs consideration. The *Climate Change and London's Transport Systems* report raises this issue and indicates that problems may arise as a result of reduced skid resistance on road surfaces following prolonged periods of higher temperatures (London Climate Change Partnership, 2005). This report also highlights the potential for problems on the road network in London which may not be apparent under present day hot weather conditions or are currently only mild impacts which will become more serious as a result of the predicted changes in climate. The UK will not yet have experienced the full extent of the range of effects that may be incurred as the full scale of climate change has not yet been felt. Therefore, using the UKCP09 in future design codes is paramount to ensure that the UK is well equipped to cope with future changes and pressures on the road networks and keep the UK traffic flowing.

Impacts on the Rail Network

Like rutting on the road network, the rail network is also likely to suffer severe buckling as a result of higher temperatures over more prolonged periods of time (Arkell & Darch, 2006). The number and severity of buckles on rail tracks is strongly linked to the maintenance regimes and conditions of the track (Hunt et al., 2006). As a result, this makes it more difficult to explicitly determine the relationship between rail buckling and hot weather (Dobney, Baker, Chapman, & Quinn, 2010). In this instance, mitigation strategies which address the issues of track condition and maintenance would be highly beneficial – improving maintenance of the tracks would help to reduce the number of buckles occurring during the summer months in future, reducing costs of further repairs and delays across the railway network. The impacts of high temperatures on the UK rail network have already been observed during the summers of 2002, 2003, 2006 and 2008 where there were increased instances of track buckling (Standley et al., 2009).

Incidents such as rail buckling cause long delays and are expensive to repair, so solutions have to be found to reduce money lost during delays and the cost of repairs (Dobney et al., 2010). With a likely increase in the number of days with high temperatures, there needs to be significant investment in attempting to increase the strength of rail tracks and the thresholds at which buckling will occur (Dobney et al., 2010; London Climate Change Partnership, 2005).

The need to adapt and improve the conditions of railways due to higher temperatures is already being addressed by Network Rail as a result of damage caused following the summer of 2003 (Arkell & Darch, 2006). Although this work is already taking place to improve the network's resilience, there is still more that needs to be done in order for the network to be able to cope with an increased number of days with high temperatures during the summer months. Speed restrictions on rail tracks have to be applied at air temperatures of between 36°C and 41°C (London Climate Change Partnership, 2005). At higher track temperatures, it is likely that buckle harvesting will occur – where the critical rail temperature has previously been exceeded and the rail has consequently buckled, it will not buckle again until the previous highest temperature is exceeded (Chapman et al., 2008). These restrictions are imposed where there is either a risk of

Table 3. Critical Rail Temperature Values for the United Kingdom.

Track Condition	On Standby	Impose 30/60 mile/h Speed Restriction	Impose 20 mile/h Speed Restriction
Good condition	SFT + 32	SFT + 37	SFT + 42
Inadequate ballast	SFT + 10	SFT + 13	SFT + 15

Source: From Dobney, Baker, Quinn, and Chapman (2009).
Note: SFT = stress-free temperature is the temperature at which track is laid and is the temperature where no thermal forces are acting upon the rail. In the United Kingdom, it is normally 27°C.

buckling occurring or where a buckle has previously occurred. Table 3 shows the critical rail temperatures at which speed limits have to be imposed on rail lines in the United Kingdom. Understandably these restrictions cause lengthy delays and prolong journey times. With predictions of between 25 and 45 days per summer where air temperatures will exceed 25°C in London alone and rail temperatures being between 15°C and 18°C higher than the air temperature (Hunt et al., 2006), then the need to improve the resilience of the railway network to temperatures such as these is critical.

In the summer of 2003, there were 165,000 delay minutes on the UK's railways which could be attributed to hot weather between 14th May and 18th September, compared to only 30,000 delay minutes for the same period in 2004 (Hunt et al., 2006). These figures highlight the need for the rail network to be more resilient to higher temperatures which will become more common under climate change. Better maintenance of tracks is needed to cope with higher temperatures, with another possible adaptation being to increase the stress-free temperatures at which rail tracks are laid (Dobney et al., 2010). The stress-free temperature normally used in the United Kingdom is 27°C (Dobney, Baker, Quinn, & Chapman, 2009), but if daily temperatures are set to increase, then the stress-free temperature will also need to be increased so that tracks are able to withstand higher temperatures and reduce the risk of buckling. In implementing higher maintenance standards, damage to tracks will be less likely and hence there will be fewer costs entailed as a result. Higher maintenance standards are likely to be a good investment and adaptation strategy for the rail network.

Increased instances of lineside railway fires are another potential impact of climate change for the rail network. However, care needs to be taken when investigating this issue as it is not fully understood what the causes of such incidents are and so it can be difficult to attribute instances of lineside fires to the weather conditions alone (Hunt et al., 2006). However, during the heat wave of 2006 a lineside fire at Clapham Junction in London occurred when undergrowth caught fire and the cause of this was determined to be the high temperatures at the time (Standley et al., 2009). As this relationship is not fully understood at present, it warrants further investigation to try and determine exactly how instances of lineside fires may be affected as a result of climate change and increasing numbers of days with high temperatures.

High temperatures are already experienced by passengers on the London Underground in hot weather especially at peak times (Arkell & Darch, 2006) and so passenger discomfort as a result of more days with high temperatures is an issue which has started to receive more attention. Many reports which detail the findings of research into climate change and its impacts on London identify passenger comfort on the Underground as a key issue which has to be addressed in order for the Underground to continue operating effectively and providing its current service without compromising passenger comfort and safety (Clarke et al., 2002b). In future, thermal comfort for passengers on the Underground will be determined by changes in temperature (Jaroszweski, Chapman, & Petts, 2010) and with over 3 million journeys made on the Underground every week day, keeping the Underground fully operational is likely to be one of the biggest challenges to the transport network in London (London Climate Change Partnership, 2005). Stakeholders have already expressed concern about rising air temperatures on the Underground (Clarke et al., 2002a). Temperatures on the Underground cannot solely be determined by the air temperature above ground but also depend on the number of trains running, the number of passengers and also the station design. It may not just be passengers' discomfort which is affected by higher future temperatures – ventilation and cooling equipment on the underground system will also be under considerable pressure as the systems struggle to cope with increases in temperatures. A further issue to be taken into consideration is that of the urban heat island – the combined impact of regional climate change and an intensified urban heat island will significantly affect the London Underground system (Clarke et al., 2002a).

Increased Congestion

An additional impact of an increasing numbers of days with higher temperatures which has received little attention is that of increased congestion in the south east of England due to the movement of people looking to enjoy the warmer weather and improving climate. There are mentions of this possible impact in a number of reports (Clarke et al., 2002a), but there is presently no in-depth exploration of potential impacts on the UK's transport networks. Increased movement in this region is of particular concern as the existing transport infrastructure already experiences significant pressure during the peak tourism season (McKenzie Hedger et al., 2000) and further increases in tourist numbers would impose further pressure on these already stretched transport networks. It is likely that passenger trips on both road and rail networks will increase as more people from other parts of the UK head to the south east to take advantage of the climate, escalating demands on the transport network as a whole (Hunt et al., 2006). At present, it is assumed that tourism in the south east of England will increase as a result of more favourable summer weather conditions but further research is needed to determine how increasing temperatures in the United Kingdom are likely to affect people's travel decisions. Some degree of congestion around London has already been observed on a number of occasions where summer temperatures were higher than average. The high temperatures and heat waves of July 2002, August 2003, July 2006 and July 2008 led to traffic jams on many of the roads and motorways around London as people left the city to make the most of the good weather on the coast (Standley et al., 2009). It is likely that if summer weather improves as a result of climate change, then instances such as these will increase as more people head for the coast, not only from London but across the UK.

Decreased Number of Cold Days

With the expected reduction in the number of cold days during the winter under the current UKCP09, it is expected that there will be a reduction in the level of winter maintenance needed to maintain the roads and railways during these cold periods. Fewer incidences of frost and snow will mean that councils do not need to invest as much in winter maintenance and treatment as at present (Clarke et al., 2002a). Improved safety across all modes of transport is considered to be an advantage as a result of fewer days with cold temperatures and fewer occurrences of frost and ice (Peterson, McGuirk, Houston, Horvitz, & Wehner, 2008). Fig. 2 shows the possible changes in the average daily mean temperature in winter by the 2080s under the medium emissions scenario. It is clear that there will be some degree of

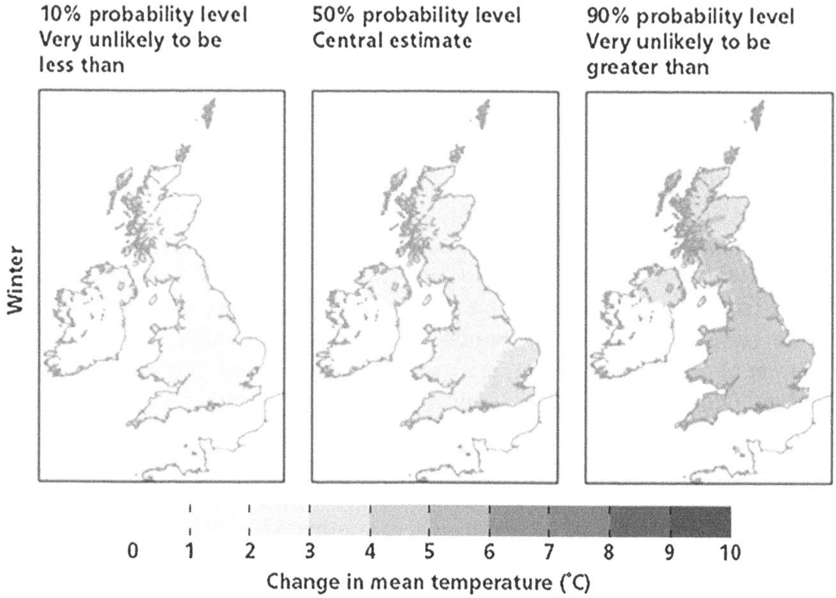

Fig. 2. Change to the Average Daily Mean Temperature (°C) of the Winter by the 2080s at 10, 50 and 90% Probability Levels, Under the Medium Emissions Scenario. *Source*: UKCP09 – Jenkins et al. (2009).

warming during the winter, and therefore this will impact on winter maintenance and treatment.

Impacts on the Road Network

As the number of snowy days is expected to decrease across the UK, it is expected that money spent on both road treatment and maintenance throughout the winter will decrease (Alcamo et al., 2007). The suggestion that road safety will also be improved as a result of the reduction in the number of snowy days (Peterson et al., 2008) is an idea that has been raised in current literature but which warrants further investigation for the future (Keay & Simmonds, 2006). The work of Andersson and Chapman (2011) investigating the impact of climate change on traffic accidents in the West Midlands identifies that with a decrease in the number of frost days, there will theoretically be a decrease in the number of winter accidents occurring.

Transport in Scotland currently experiences significant disruption due to low temperatures in winter and the need for gritting and salting of road surfaces during prolonged periods of cold weather. With a decrease in the number of cold days predicted by the UKCPs, it is expected that there will be less disruption to transport in Scotland due to less need for winter road maintenance (McKenzie Hedger et al., 2000). It is not just Scotland where the need for winter road maintenance will be reduced; London is also expecting to reduce the level of resources which have to be committed to winter road maintenance particularly due to fewer incidences of frost and snow as well as higher winter temperatures (Clarke et al., 2002a). This will result in savings for local authorities as spending on gritting and salting activities is reduced. With both London and Scotland anticipating a reduction in the need for winter road maintenance activities, it follows that the rest of the UK will see a reduction in these activities too.

It is likely that with a decrease in the number of cold days in the future, then safety on the road network will be less heavily impacted by snow occurrences. It is possible that the decrease in road safety which may occur as a result of increased heavy precipitation throughout the remainder of the year may somewhat be offset by an improvement in road safety resulting from fewer snowy days (Keay & Simmonds, 2006). Although this idea has been proposed in a number of studies recently, there is little explicit research into what is likely to happen to congestion and road safety as a result of climate change and as this is an issue of great importance, it is an area which will need further research (Metroeconomica Limited, 2005–2006). Whilst a reduction in the number of cold days during winter months may lead to improved road safety, the familiarity of road users with winter weather conditions may also impact on safety across the road network. With fewer instances of cold weather, road users will become less familiar with the weather conditions and as a result, when cold weather conditions do occur, then it is possible that there may be an increase in accidents as road users fail to adapt their driving accordingly.

The most critical temperatures during winter weather are those occurring around the freezing temperature. Road surfaces are most slippery at temperatures around 0°C (Thornes, 1991). The change in occurrences of temperatures around freezing will need investigation in future to determine how such changes are likely to affect road and rail safety and maintenance. Road temperatures are affected by weather conditions and local micro climate variations along a stretch of road (Andersson, Gustavsson,

Bogren, & Holmer, 2007) and so any changes in future weather patterns as a result of climate change will impact on future road temperatures.

Impacts on the Rail Network

Research into how rail transport is affected by current weather or is likely to be affected by climate change is scarce but the few existing studies show that icing is a major problem for the UK's railways and causes considerable delays (Koetse & Rietveld, 2009). It can be assumed that as a result of fewer cold days and occurrences of frost and ice then there will be less need for winter maintenance of the tracks. Consequentially, with fewer snowy and cold days, there will be a reduction in treatment costs as the number of instances of points freezing decreases (Wilbanks et al., 2007). A study by the Rail Safety and Standards Board investigating the impacts of climate change in the East Midlands identified that warmer winters are likely to benefit the rail system by reducing the demand for winter maintenance and the associated costs (Eddowes et al., 2003). A similar study for railways in Wales suggested that warmer winters would potentially reduce the frequency of snowfalls which currently prevent services running (Eddowes et al., 2003).

Although there will be fewer days with cold temperatures and instances of frost and ice, the impact of increasing winter precipitation needs to be considered in greater detail. If there is increased winter precipitation then there is a chance that when ice does occur on rail tracks then it will be more extensive than current instances of ice forming due to the increased volumes and extent of precipitation (Eddowes et al., 2003). It is likely that changes in snow, sleet and ice will need to be monitored over the coming years to determine how resources currently used for winter maintenance will be able to be reallocated in future (Eddowes et al., 2003).

PRECIPITATION

The costs to the UK transport network associated with flooding are likely to be substantial and will continue to increase if there are further instances of flooding as a result of increased heavy precipitation under future climate change. Flooding in London, for example would affect major transport interchanges (London Climate Change Partnership, 2005) with consequences not just for travellers within London but from further afield as well – both nationally and internationally. The UKCP09 projected changes in precipitation on the wettest day of the season for both winter and summer

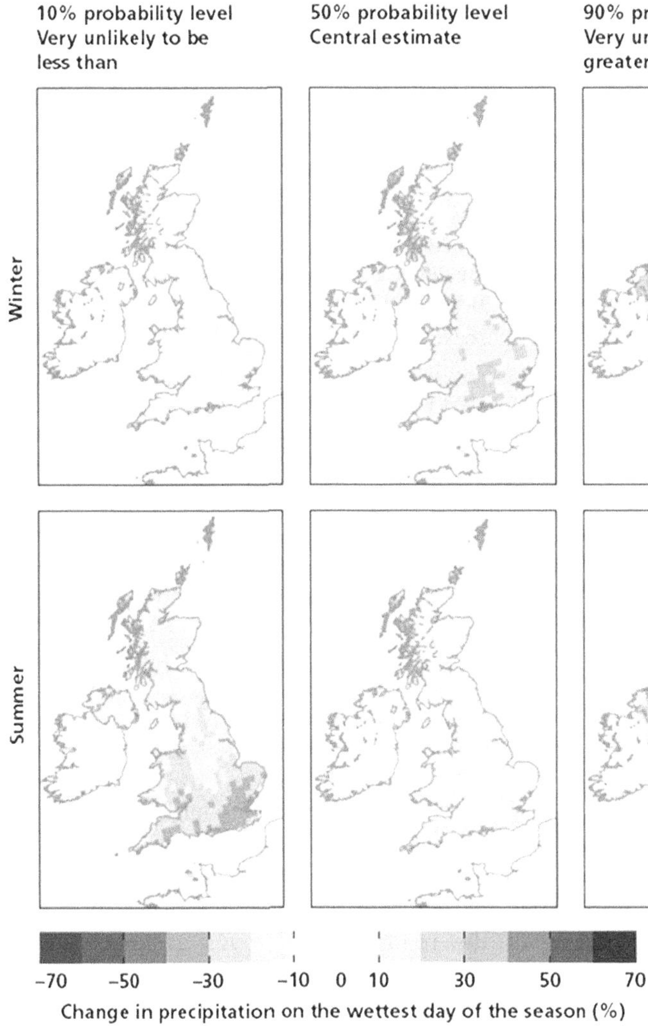

Fig. 3. Changes to Precipitation on the Wettest Day of the Winter (top) and of the Summer (Bottom) at the 10, 50 and 90% Probability Levels, for the 2080s Under the Medium Emissions Scenario.
Source: Murphy et al. (2009).

shown in Fig. 3 indicate that, at the central estimate, there is likely to be an increase in precipitation on the wettest day during both seasons.

Impacts on the Road Network

One of the greatest threats of increased heavy precipitation to the road network is flooding. The estimated cost of flood-related traffic disruption on roads is at least £100,000 per hour delay on each main road affected during peak periods (Arkell & Darch, 2006). If there are more instances of flooding in the future, as a result of increased heavy precipitation, then it is likely that these costs will increase significantly. If the number and extent of flooding events increases, there are likely to be increases to journey times as traffic jams build due to obstructions on roads from flood waters. In July 2007, flooding in and around London caused lengthy delays as traffic jams built up around roads blocked by flooding (Standley et al., 2009). Events such as these are widely reported because of the impact on traffic flows – the floods in July 2007 received widespread media coverage because of the disruption to transport (BBC News, 2007a). Instances like this are likely to become more common unless efforts are made to improve drainage and adapt the road network to be able to cope with increased heavy precipitation.

With increased instances of heavy precipitation, it is likely that road surfaces will need to undergo increased amounts of maintenance work to repair flood damage. Flooding of road surfaces causes road scouring and road washout (Peterson et al., 2008) and if these effects increase in future, then more money will have to be invested either in repairing road surfaces after flooding events have occurred, or in preventing damage – by means of improving road specifications or implementing infrastructure which will help to prevent flooding of the roads. Where roads are particularly susceptible to flooding then it may be that investment in preventing flooding is more worthwhile than continually repairing sections of road after each flood event.

Flooding of the road network has an adverse impact on road safety for all road users (Keay & Simmonds, 2006). Precipitation has been identified as the weather variable which has the greatest impact on road safety. Whilst higher temperatures, stronger wind speeds and occurrences of fog will all also have an adverse impact on road safety, precipitation greatly increases the likelihood of a road accident occurring (Koetse & Rietveld, 2009). As a result, an increase in the number of heavy precipitation events due to climate change prompts the need for careful consideration of the safety of the road network and measures to improve safety for road users in an attempt to reduce the number of accidents occurring when there is precipitation. There is little explicit research into what is likely to happen to congestion and road safety under climate change and as this is an issue of great importance, it is an area which will need further research (Metroeconomica Limited, 2005–2006).

Reduced visibility is known to cause clear reductions in traffic speed. For example, Maze et al. (2006, in Koetse & Rietveld, 2009) discovered a 12% reduction in speed in the Minneapolis/St. Paul area over the four year study period. During heavy precipitation events, visibility on the roads is severely restricted both as a result of the view through the windscreen being obscured and also as there is increased spray and splash from other vehicles on the road (Edwards, 2002). An increase in the number of heavy precipitation events will lead to an increase in the number of instances where visibility is significantly reduced and hence will have an adverse impact on road safety. At present, it is highly uncertain as to what the impact of climate change will be on fog, mist and hence visibility (Koetse & Rietveld, 2009) and so it is difficult to predict exactly how visibility will be affected and the impact this will have on road users and safety.

Impacts on the Rail Network

It is likely that physical damage to railway lines will increase as a result of increased heavy precipitation in the future. Damage to the infrastructure of the rail network is likely to increase as flooding can cause damage to tracks, the ground beneath and also to the lineside equipment. Flooding causes extra debris to be deposited on railway tracks and following particularly large flooding events, extra resources have to be used to clear tracks of any debris and make them safe for use once flood waters have cleared (Standley et al., 2009). In extreme circumstances, heavy precipitation can lead to the closure of sections of track or even stations – for example in April 2004, half of the average monthly rainfall fell in 2 hours causing tube stations to be closed and extensive delays to be experienced on the railways (Standley et al., 2009). Bridge scour during flooding events is another problem for the rail network with long delays occurring until bridges have been assessed and deemed safe after an event, with even greater delays where repairs need to be carried out. Instances of extreme flooding may even lead to ballast stones being washed away, as happened on the Blaenau Ffestiniog tracks in February 2011 following prolonged heavy rain (Daily Post, 2011). Events such as this lead to extended track closures and service disruptions whilst repairs are carried out (BBC News, 2005) and if future occurrences of flooding cause similar problems then there are likely to be more extensive disruptions and increasing repair costs unless measures are implemented to prevent such damage occurring during flooding events.

There are not only physical aspects to the damage which may be caused by flooding on the UK transport network. The indirect effects caused by flooding and indeed other elements of climate are also of great importance and it is here that the true costs of a changing climate and its effects on transport become apparent. Increased instances of flooding as a result of more heavy precipitation events will increase delays to journeys on all modes of transport and some journeys may even be cancelled (Kirshen et al., 2004). As a result, the economic costs will be highly significant – it is only once costs are considered for all transport modes that the true costs of a changing climate can be realised. During July 2007, flooding in London lead to the inundation of railway lines and in order to be able to resume services, extra resources had to be used to clear the lines of debris left behind by flood waters to make them fit for use again (Standley et al., 2009). Such events are not isolated incidents – there appears to be more and more occurrences of such floods during the past 10 or 15 years with notable floods in London in November 2001, August 2002, April 2004, July 2007 and August 2008 which significantly impacted on the transport network and so the need to prepare for or prevent these events is ever growing (Standley et al., 2009).

The possibility of increased occurrences of heavy precipitation events also brings the possibility of increased numbers of landslides and undercutting. Railway lines in particular are at risk from landslips as a result of less frequent, more intense rainfalls – the study of the Rail Safety and Standards Board found that East Midlands and Welsh railway lines are likely to be particularly susceptible to landslips (Eddowes et al., 2003). Increased winter rainfall intensities in particular may increase the instability of slopes and embankments and this has been identified as a possible impact of climate change on railway lines in London (Clarke et al., 2002a). The railway network in London has already experienced some disruption due to landslips as a result of heavy precipitation – in July 2006, the Boston Manor Underground station had to be closed and passengers evacuated from a train as 4 tonnes of earth slid onto the track (Standley et al., 2009).

There are a number of implications of landslips and undercutting including closure of railway lines and extensive delays to journey times. Where there are unknown landslips on railway lines, there is an increased safety risk for trains on susceptible lines as trains meeting a landslip may be derailed (Eddowes et al., 2003) and is something which needs careful monitoring during the future.

The London Underground will face a number of challenges under the predicted climate change scenarios. Not only will higher temperatures affect comfort levels but the potential increased number of heavy precipitation events may lead to increased and more frequent flooding of the Underground system. There have already been a number of flooding incidents recorded on the Underground network – over 1,200 between 1993 and 2002, and around 200 station closures in the same period due to flooding (Arkell & Darch, 2006). With the predicted increase in heavy precipitation events, it is expected that this number will rise unless adaptation and mitigation strategies are implemented. The economic consequences of such incidents are substantial and so any action to reduce these costs now will provide long-term savings as well as improving the resilience of the Underground network to weather almost immediately (London Climate Change Partnership, 2005). There is also the possibility that rising water levels which parts of London are currently experiencing may exacerbate the effects of flooding on the Underground leading to further equipment failure and more delays and cancellations (Clarke et al., 2002a).

Coastal areas are likely to be particularly susceptible to flooding and hence the transport infrastructure in these regions will be highly affected (McKenzie Hedger et al., 2000). Not only will more frequent heavy precipitation events lead to flooding of coastal infrastructure but the possibility of sea level rise during the next century also poses the risk of flooding, especially for coastal railway lines (Eddowes et al., 2003). Existing research into the vulnerability of the rail network found that the Welsh coastline is likely to experience flooding as a result of sea level rise due to the flat terrain on which most of the railway lines are built and as a result, it is possible that corrosion on the lines will be accelerated requiring higher levels of maintenance than at present (Eddowes et al., 2003; Metroeconomica Limited, 2005–2006).

The Rail Safety and Standards Board have identified a need to look specifically at the possible impacts of climate change on coastal rail infrastructure and Dawlish in Devon has been used as an initial case study area. This area is likely to experience increased inundation of the railway line as a result of sea level rise and the need to minimise the risk to train operations has already been identified with flood action and storm response procedures in place to minimise disruption (Rail Safety and Standards Board, 2008). Storm surges and rain associated with these storms have the potential to cause severe problems for coastal railway lines such as in Dawlish and so mitigation and adaptation is crucial. The impacts of increased sea level rise when combined with storm surges must also be considered as these two events combined are likely to cause more significant damage leading to greater repair times and costs.

Embankment and Slope Stability

Changes in precipitation patterns and intensities will have implications for the stability of embankments on both road and rail networks. The arising problems have to be addressed to ensure that the operability of these networks is not compromised. Events such as landslips require extensive recovery work, often over prolonged time periods (Lindgren, Jonsson, & Carlsson-Kanyama, 2009), and so it is essential to understand how these events may occur and what can be done to prevent them or minimise impacts when they do occur. Previous studies have already identified that there is likely to be an increase in the failure of slopes along the motorway if no preventative measures are implicated to cope with changes in slope stability as a result of climate change (Perry, 1999, in Clarke, Hughes, Barbour, & Sivakumar, 2006).

The risk of embankment subsidence and heave will increase due to wetter winters, drier summers and changing vegetation (Department for Transport, 2005). Dry summer periods followed by intense precipitation events will cause significant problems and lead to landslips along both railway lines and the road network. Prolonged periods of dry weather during the summer months cause embankment materials to become desiccated and the structure of the embankment is altered. It is this alteration in structure which eventually causes problems giving the potential for landslides to occur more frequently during intense precipitation events following the drier periods (Eddowes et al., 2003; Manning, Hall, Kilsby, Glendinning, & Anderson, 2008).

The Rail Safety and Standards Board report (Eddowes et al., 2003) on the *Safety Implications of Weather, Climate and Climate Change* identified a number of possible scenarios which may lead to increased embankment instability across the railway network and hence increased numbers of landslides as a result of climate change. Hotter and drier summers have been recognised as a cause of shrinkage in soils leading to subsidence and therefore an increased need for work to reduce the possibility and severity of associated problems. The issue of changing vegetation on embankments is also raised in the report. Changes in climate may lead to changes in vegetation presence along the railway. If vegetation cannot adapt to future climate change and as a result dies, then the absence of vegetation may lead to increased embankment instability, contributing to the possibility of landslips occurring where the embankment is also exposed to other elements of climate change.

There is still the need for further research especially along specific sections of the road and rail networks particularly where there are high usage levels of the network in order to be able to minimise future disruption. Subsidence of embankments is a critical issue for both road and rail across the country. Existing studies conducting research into effects in specific areas have identified that the impacts of subsidence of embankments on London's road network during hot weather is not fully understood and has not fully been experienced to date (London Climate Change Partnership, 2005) and so further work is required to better understand the full implications of such events. Likewise, clay shrinkage in London has been identified as having significant effects on embankments and the infrastructure which is central to railway operations, including bridges and tunnels (Clarke et al., 2002a). This shrinking of clays in London coupled with increased winter precipitation intensities is likely to cause increased disruption on the railways and as a result, structures affected will need reinforcement to prevent this.

Whilst many studies have identified embankment stability as being particularly susceptible to climate change, further research is needed to understand fully the implications of changes in precipitation in future for embankment stability. The Rail Safety and Standards Board report (Eddowes et al., 2003) suggests some primary themes which need in-depth investigation to address the potential hazards that may be caused by climate change and one of these themes is excess rain-related impacts including embankment stability. Understanding how antecedent conditions of slopes and embankments along both road and rail networks will change as a result of climate change is crucial. Findings from future research, along with improved understanding of processes affecting slope stability, will enable improved forecasting of slope stability as a result of climate change (Dijkstra & Nixon, 2010). Developing understanding of potential future problems associated with embankment stability through research such as this will help to inform future planning and management.

SEASONAL CHANGES

One predicted effect of climate change which will have some impact on all modes of transport is that of seasonal change. Though the full extent of seasonal changes and the impacts they may have on the transport network are not explored in detail in any of the existing literature, there are a number of sources which hint at the possible impacts which may be experienced as a result in the changes of timing of the seasons. Changes in the length and timing of seasons have already been noticed with the identification that each of the seasons now arrive, on average, 11 days earlier than they did in the 1970s (Thackeray et al., 2010 in Krebs et al., 2010).

One impact of seasonal changes may be an increase in vegetation growth periods and therefore the possibility of more leaves on railway lines has to be considered as this will significantly affect leaf fall timetables, especially the dates when they are enforced and the length of time they are enforced for (Department for Transport, 2005). Leaf fall timetables have to be enforced due to leaves falling on rail tracks and then being compacted into mulch on the rails. There is a possibility that more mulch will be created on rail tracks due to climate change (Clarke et al., 2002a). One such change to the growing season which may impact on leaf production, and hence leaf fall on railway lines, is the number of frost days occurring during the year (Eddowes et al., 2003). Fewer frost days will result in longer growing seasons and therefore greater leaf production. Though it is currently difficult to predict how wind will change in the future, if there are increased autumn winds this may cause further problems in the form of concentrated leaf fall and result in ineffective braking for trains, causing problems with safety (Eddowes et al., 2003).

SEA-LEVEL CHANGE

Any increases in sea level as a result of future climate change will pose a number of challenges particularly for coastal areas (Walsh et al., 2007). The threat of flooding from sea level rise (both as a result of climate change and also continued isostatic change) as well as an increased risk of flooding from more heavy precipitation events means that coastal transport infrastructure is at more risk than inland transport systems. Any rise in sea levels would cause these sites to be inundated meaning that decisions have to made about defence mechanisms or even relocation of the sites. The Rail Safety and Standards Board have identified that a specific research project is needed to determine more clearly how railway locations may be affected by sea level rise across the UK, and it has already been recognised that there are several point in Wales which are likely to be inundated (Eddowes et al., 2003).

Associated with increases in sea level are increased frequency and intensity of storm surges (Koetse & Rietveld, 2009). Ensemble modelling, modelling using multiple numerical predictions generated using slightly different initial conditions, as part of the IPCC suggests that there may be more extreme storm surges under some of the emission scenarios which pose the threat of extensive coastal flooding as well as flooding in estuaries (Alcamo et al., 2007). Whilst it is known that there is potential for storm surges to increase in both frequency and intensity under the predicted climate change scenarios, the actual empirical evidence for Europe is somewhat lacking (Koetse & Rietveld, 2009) and so this needs further investigation to be able to begin to understand the exact implications for coastal transport infrastructure.

Whilst there have been some attempts to determine what the potential impacts of sea level rise may be on surface transportation in the US, little has been done to investigate the possible implications in the United Kingdom. The work of Suarez, Anderson, Mahal, and Lakshmanan (2005) assesses the potential impacts of climate change on Boston's transport systems as a result of increased flooding. It was found that delays may be doubled and journeys cancelled. This study provides information of modelling used to determine the possible outcomes of future flooding on the transport infrastructure and it is a similar approach to this which would be beneficial in determining what problems the UK may face.

EXTREME EVENTS

There are a number of disruptions which will occur for all modes of transport as a result of extreme weather incidents occurring (West & Gawith, 2005). With the identification by the IPCC that transport infrastructure is more sensitive to the effects of changes in local climatic extremes than changes in the mean (Wilbanks et al., 2007), it is crucial that the understanding of how the transport infrastructure in the United Kingdom is likely to respond to an increase in the frequency of extreme weather events is developed. An increase in extreme weather events will significantly affect all modes of transport and the ways in which authorities need to prepare for and deal with such events. For example, Wales has already been identified as being susceptible to disruption to transport infrastructure from extreme weather events in future. There are concerns that the terrain of the country will increase the risk of disruption to transport, with movement along the north-south corridor being especially vulnerable (McKenzie Hedger et al., 2000).

Impacts on the Road Network

There is the possibility with more extreme weather events occurring there may be a reduction in safety for users on the road network, causing more accidents and increasing delays (Koetse & Rietveld, 2009). It is possible that fog and wind may have an increasing effect on the number of road accidents occurring, though this is an area which needs more research to fully understand the potential extent of this relationship under climate change (Edwards 1998; Hermans, Brijs, Stiers, & Offermans, 2006). Fog has been found to increase travel times and so if there are to be increased occurrences of fog, it can be inferred that this will lead to greater journey times as well as an increased number of accidents (Tu et al., 2007 in Koetse & Rietveld, 2009).

If there are more windstorms as a result of climate change (although predictions for wind are uncertain at present) then there will be problems for road transport relating to the amount of debris and vegetation which is blown onto roads and motorways (Clarke et al., 2002a). If this is the case then it will cause problems as more time and resources need to be spent on clearing roads following the occurrence of a windstorm to prevent extensive delays on the network. In the event of increased numbers of windstorms, there will be implications for high sided vehicles using the road network which are more likely to be blown over during strong wind events. The knock-on effects of such accidents include damage to road infrastructure such as bridges and barriers, as well as road closures leading to increased travel times and delays for road network users.

Impacts on the Rail Network

Wind is also an element which can significantly affect transport operations, causing delays on railways due to overhead lines being blown down. Unfortunately, the prediction of future changes in wind speeds as a result of climate change is very difficult and as a result it is hard to know exactly what the impacts of wind will be on the UK transport network. High winds may affect the safety of both air and land transport (Alcamo et al., 2007) and so these impacts need careful consideration in order to incorporate possible adaptation strategies into future plans for the transport network.

Case Study: Windy Thursday

Thursday 18th January 2007 was a day of particularly adverse weather with strong winds across the UK causing problems for both road and rail transport and the day has since been known as 'Windy Thursday' as a result of the disruption caused. The entire Highways Agency road network was affected with disruption causing an estimated £50 million of delay throughout the day (Highways Agency, 2007a). Goods vehicles were most affected with 48 being blown over across the network causing some parts of the network to be closed (Highways Agency, 2007b). It is not just vehicles which are affected, with implications for equipment as well which can affect signalling, reducing the capacity of the network and causing further disruption. It was not only road transport that was disrupted, rail transport was also affected with a number of operators reducing services and ferry crossings were also cancelled (BBC News, 2007b).

As predictions for changes in wind are problematic, it is difficult to determine whether events similar to Windy Thursday will be more likely to occur in the future. However if there are more events like Windy Thursday as a result of climate change, then there need to be measures in place to cope with and mitigate for potential disruption that may occur.

CITY-SPECIFIC IMPACTS

London needs extensive consideration in any research investigating the possible impacts of climate change on UK transport as effects are likely to be exacerbated in London in comparison to other UK cities (Clarke et al., 2002a). At the launch of the London Climate Change Partnership report entitled 'London's Warming', mayor Ken Livingstone identified that the size of London's population means that pressure on resources is already high and therefore, it is essential to ensure that such resources can adapt to and deal with new demands as a result of climate change. There are already a number of reports emerging which consider the potential impacts of climate change on London as a city, with some attention given to transport systems – for example the London Climate Change Partnership reports *Climate change and London's transport systems* (2005) and *Wild weather warning: a London climate impacts profile* (2009). Reports such as this are of great value as they give a detailed insight into how transport in the capital city may be affected in future. Whilst understanding of how individual cities will be impacted is highly beneficial, it is crucial to understand how transport networks across the whole country will be affected in order to determine the resilience of the networks.

IMPACTS OF EXISTING WEATHER CONDITIONS

Much of the current literature on the topic of climate change and its impacts on transport focuses on studies in the United States – for example the Committee on Climate Change and US Transportation special report into the potential impacts of climate change on US Transportation (2008). There are a handful of studies which focus on the effects of weather on transport in the United Kingdom and there are more reports emerging which examine what the potential effects of climate change may be on transport networks in the United Kingdom. However, the majority of these reports are focused on London (e.g. Arkell & Darch, 2006; Clarke et al., 2002a; London Climate Change Partnership, 2005) or the reports focus on a single mode of transport as opposed to examining the impact on all transport modes (e.g. Eddowes et al., 2003). In order to be able to fully understand and prepare for the effects of climate change on the UK's transport network, an integrated approach needs to be taken to further research looking at impacts across the whole of the UK and also on all modes of transport.

To understand how future climate change may affect transport, it is essential to understand how current and past weather conditions impact on transport networks and infrastructure. Literature detailing existing weather conditions gives an indication of the thresholds which currently exist for various modes of transport, for example the temperatures at which speed restrictions have to be imposed on the railways (Eddowes et al., 2003). This knowledge can then be applied to develop ideas of how these thresholds may change in the future and what adaptation will need to occur.

Many existing studies focus on the effects of weather on a specific mode of transport. Road transport has been studied in the most detail with studies focusing on weather impacts in a number of countries (Al-Hassan & Barker, 1999; Andrey, Mills, Leahy, & Suggett, 2003; Edwards, 1998; Golob & Recker, 2003). Rail also has a number of mode specific studies, especially those carried out by the Rail Safety and Standards Board (Eddowes et al., 2003; Rail Safety and Standards Board, 2010). It is these existing studies which provide the background knowledge which can then be applied to develop understanding of how transport may be affected as the climate changes.

Past extreme weather events receive a great deal of attention due to the adverse effects and costs incurred. For example, the summer of 2003 was much hotter than usual and this caused numerous problems in the United Kingdom, with many problems arising for the transport network. There were around 130,000 additional delay minutes on the UK railway network as a result of the hot weather (Dobney et al., 2009) which led to large costs. Events such as this are more likely to occur in the future as a result of climate change and so it is clear that adaptation is key to ensure that the impacts on the transport networks can be reduced and therefore costs associated with these impacts can also be reduced. A report into the costs of the summer of 2003 (Hunt et al., 2006) was commissioned and the report details the implications for transport of the hot weather as well as detailing what the costs were in the summer of 1995 where temperatures were also higher than average. It is important to learn from past occurrences to be able feed our knowledge of how to deal with the occurrence of extreme weather events in the future.

The Rail Safety and Standards Board issued a research brief in August 2010 identifying the need to better understand how extreme weather events affect the operations of the railways. This research brief identifies the current lack of knowledge which the industry has and details its aims to develop this knowledge in order to be able to provide a reliable transport system in the future which is safe, reliable and meets capacity demands (Rail Safety and Standards Board, 2010). Other European countries are also realising that their knowledge and planning for climate change in their transport sectors is lacking – Sweden in particular have identified that '...anticipatory, proactive and planned adaptation strategies for future climate change are lacking' (Lindgren et al., 2009). The sooner more decision makers become aware of the need to integrate climate change adaptation and mitigation strategies then the better plans and development of transport infrastructure will be as the systems' resiliencies are improved.

CONCLUSIONS

The impacts of weather and climate on all transport infrastructure have to be taken into account at the beginning of the planning stage for new transport infrastructure, and these effects need continued consideration throughout the planning process as well as during design, construction, operation and maintenance (Mills & Andrey, 2003). With this in mind, understanding the potential impacts of future climate change on the UK transport network is crucial to ensure that future developments are built with sufficient resilience to the changing climate and also that existing transport networks can be adapted and modified to cope with the predicted changes. For the UK transport network to adapt to climate change, adaptation and mitigation strategies need to be built into the planning stages of new developments and incorporated into upgrades and maintenance of existing infrastructure to minimise future costs and maximise the resilience of the UK transport network over the coming years (Arkell & Darch, 2006). The response to climate change needs to be coherent across the UK to ensure that transport networks are able to continue operating effectively across the entire country. It is not just transport infrastructure which will experience the impacts of climate change in future, there will also be changes in demand for each of the transport modes (Metroeconomica Limited, 2005–2006), and transport systems will have to adapt to both of these impacts to ensure their capability to continue to run as efficiently and effectively as possible.

The impacts of climate change are at the forefront of research at present and there is a range of further research which is needed to develop understanding of how climate change may impact on all elements of society in future. It is only recently that the impacts of climate change on transport have begun to receive more attention, and as a result, future research in this area is of paramount importance to inform decision making, design and planning for future transport networks and infrastructure to ensure that the UK's transport networks are well equipped to cope with a changing climate.

REFERENCES

Al-Hassan, Y., & Barker, D. J. (1999). The impact of unseasonable or extreme weather on traffic activity within Lothian region, Scotland. *Journal of Transport Geography*, *7*(3), 209–213.

Alcamo, J., Moreno, J. M., Nováky, B., Bindi, M., Corobov, R., Devoy, R. J. N., … Shvidenko, A. (2007). Europe. Climate change 2007: Impacts, adaptation and vulnerability. In M. L. Parry, O. F. Canziani, J. P. Palutikof, P. J. van der Linden & C. E. Hanson (Eds.), *Contribution of working group II to the fourth assessment report of the intergovernmental panel on climate change* (pp. 541–580). Cambridge, UK: Cambridge University Press.

Andersson, A. K., & Chapman, L. (2011). The impact of climate change on winter road maintenance and traffic accidents in West Midlands, UK. *Accident Analysis and Prevention*, *43*, 284–289.
Andersson, A. K., Gustavsson, T., Bogren, J., & Holmer, B. (2007). Geographical distribution of road slipperiness in Sweden, on national, regional and county scales. *Meteorological Applications*, *14*, 297–310.
Andrey, J., Mills, B., Leahy, M., & Suggett, J. (2003). Weather as a chronic hazard for road transportation in Canadian cities. *Natural Hazards*, *28*, 319–343.
Arkell, B. P., & Darch, G. J. C. (2006). Impact of climate change on London's transport network. *Municipal Engineer*, *159*, 231–237.
BBC News. (2005). *Railway closes for storm repairs*. Retrieved from http://news.bbc.co.uk/1/hi/wales/north_west/4456298.stm
BBC News. (2007a). *Rescue teams tackle flood chaos*. Retrieved from http://news.bbc.co.uk/1/hi/uk/6909633.stm
BBC News. (2007b). BBC Weather Centre – World Weather – News – 18/01/2007 – Storm batters the UK causing chaos. Retrieved from http://www.bbc.co.uk/weather/world/news/18012007news.shtml
Chapman, L., Thornes, J. E., Huang, Y., Cai, X., Sanderson, V. L., & White, S. P. (2008). Modelling of rail surface temperatures: A preliminary study. *Theoretical and Applied Climatology*, *92*, 121–131.
Clarke, G. R. T., Hughes, D. A. B., Barbour, S. L., & Sivakumar, V. (2006). The implications of predicted climate change on the stability of highway geotechnical infrastructure: A case study of field monitoring of pore water response. *EIC Climate Change Technology*, 1–10. Availalbe at http://ieeexplore.ieee.org/xpl/articleDetails.jsp?tp=&arnumber=4057383&contentType=Conference+Publications&refinements%3D4293579519%26sortType%3Dasc_p_Sequence%26filter%3DAND%28p_IS_Number%3A4057291%29. Accessed on May 21, 2012.
Clarke, S., Kersey, J., Trevorrow, E., Wilby, R., Shackley, S., Turnpenny, J., … Crichton, D. (2002a). *London's warming: the impacts of climate change on London – Technical report, London climate change partnership*. Retrieved from http://www.london.gov.uk/gla/publications/environment/londons_warming_tech_rpt_all.pdf
Clarke, S., Kersey, J., Trevorrow, E., Wilby, R., Shackley, S., Turnpenny, J., … Crichton, D. (2002b). *London's warming: The impacts of climate change on London – Summary report. London climate change partnership*. Retrieved from http://www.london.gov.uk/gla/publications/environment/londons_warming02.pdf
Committee on Climate Change and US Transportation. (2008). *Potential impacts of climate change on U.S. transportation – Special report 290, transportation research board, 296*. Retrieved from http://www.nap.edu/catalog/12179.html
Daily Post. (2011). *Conwy Valley rail line shut for another week after flooding*. Retrieved from http://www.dailypost.co.uk/news/north-wales-news/2011/02/12/conwy-valley-rail-line-shut-for-another-week-after-flooding-55578-28158982/
Department for Transport. (2005). *The changing climate: Its impact on the Department for Transport*. Retrieved from http://www.dft.gov.uk/pgr/scienceresearch/key/thechangingclimateitsimpacto1909
Dijkstra, T. A., & Nixon, N. (2010). Climate change and slope stability in the UK: Challenges and approaches. *Quarterly Journal of Engineering Geology and Hydrogeology*, *43*, 371–385.

Dobney, K., Baker, C. J., Chapman, L., & Quinn, A. (2010). The future cost to the United Kingdom's railway network of heat-related delays and buckles caused by the predicted increase in high summer temperatures due to climate change. *Proceedings of the Institution of Mechanical Engineers, Part F: Journal of Rail and Rapid Transit*, *224*(1), 25–34.

Dobney, K., Baker, C. J., Quinn, A. D., & Chapman, L. (2009). Quantifying the effects of high summer temperatures due to climate change on buckling and rail related delays in south-east United Kingdom. *Meteorological Applications*, *16*, 245–251.

Eddowes, M. J., Waller, D., Taylor, P., Briggs, B., Meade, T., & Ferguson, I. (2003). *Rail Safety and Standards Board: Safety implications of weather, climate and climate change*, 63 pp. Retrieved from http://www.rssb.co.uk/pdf/reports/Research/Safety%20implications%20of%20weather,%20climate%20and%20climate%20change.pdf

Edwards, J. (1998). The relationship between road accident severity and recorded weather. *Journal of Safety Research*, *29*(4), 249–262.

Edwards, J. (2002). Motorway speeds in wet weather: The comparative influence of porous and conventional asphalt surfacing. *Journal of Transport Geography*, *10*, 303–311.

Garnaut, R. (2008). *The Garnaut climate change review – Final report*. Australia: Cambridge University Press.

Golob, T. F., & Recker, W. W. (2003). Relationships among urban freeway accidents, traffic flow, weather and lighting conditions. *Journal of Transportation Engineering*, *129*(4), 342–353.

Hermans, E., Brijs, T., Stiers, T., & Offermans, C. (2006). The impact of weather conditions on road safety investigated on an hourly basis. *Transportation Research Board Annual Meeting, CD-ROM Paper 06-1120*, Washington, 17pp.

Highways Agency. (2007a). *Highways agency – Adverse weather, March 2007*. Retrieved from http://www.highways.gov.uk/business/14717.aspx

Highways Agency. (2007b). *Highways agency – Severe weather alert for goods vehicles*, October 2007. Retrieved from http://www.highways.gov.uk/business/17025.aspx

Hunt, A., Boyd, R., Taylor, T., Kovats, S., Lachowyx, K., Watkiss, P., & Horrocks, L. (2006). *Report on the costs of the hot summer of 2003, DEFRA – Climate change impacts and adaptation: Cross regional research programme*, Metroeconomica Limited, 95pp.

Jaroszweski, D., Chapman, L., & Petts, J. (2010). Assessing the potential impact of climate change on transportation: The need for an interdisciplinary approach. *Journal of Transport Geography*, *18*(2), 331–335.

Jenkins, G. J., Murphy, J. M., Sexton, D. M. H., Lowe, J. A., Jones, P., & Kilsby, C. G. (2009). *UK climate projections: Briefing report*. Met Office Hadley Centre, Exeter, UK, 59pp.

Keay, K., & Simmonds, I. (2006). Road accidents and rainfall in a large Australian City. *Accident Analysis and Prevention*, *38*, 445–454.

Kirshen, P., Ruth, M., Anderson, W., Lakshmanan, T. R., Chapra, S., Chudyk, W., … Vogel, R. (2004). Infrastructure systems, services and climate change: Integrated impacts and response strategies for the Boston metropolitan area – Climate's long-term impacts on metro Boston (CLIMB). *World Resources Review*, 106–122. Available at http://cfpub.epa.gov/ncer_abstracts/index.cfm/fuseaction/display.abstractDetail/abstract/285/report/F. Accessed on May 21, 2012.

Koetse, M. J., & Rietveld, P. (2009). The impact of climate change and weather on transport: An overview of empirical findings. *Transportation Research Part D*, *14*, 205–221.

Krebs, J., Dlugolecki, A., Fankhauser, S., Hall, J., Johnson, A., Palmer, T., ... Young, B. (2010). How well prepared is the UK for climate change? *Committee on Climate Change*, London, 66pp.

Lindgren, J., Jonsson, D. K., & Carlsson-Kanyama, A. (2009). Climate adaptation of railways: Lessons from Sweden. *EJTIR*, *9*(2), 164–181.

London Climate Change Partnership. (2005). *Climate change and London's transport systems – Summary report*. Retrieved from http://www.london.gov.uk/lccp/publications/docs/climatetransportsept05.pdf

Manning, L. J., Hall, J. W., Kilsby, C. G., Glendinning, S., & Anderson, M. G. (2008). Spatial analysis of the reliability of transport networks subject to rainfall-induced landslides. *Hydrological Processes*, *22*, 3349–3360.

McKenzie Hedger, M., Gawith, M., Brown, I., Connell, R., & Downing, T. E. (Eds.) (2000). *Climate change: Assessing the impacts – Identifying responses, The first three years of the UK Climate Impacts Programme, UKCIP Technical Report, UKCIP and DETR*. Oxford, 192pp.

Mearns, L. O., Hulme, M., Carter, T. R., Leemans, R., Lal, M., & Whetton, P. (2001). Climate scenario development. Climate change 2001: Working Group I. The Scientific Basis, Cambridge, pp. 741–768.

Metroeconomica Limited. (2005–2006). Climate change impacts and adaptation: Cross-regional research programme: Project E – Quantifying the cost of impacts and adaptation, Chapter 5 – Transport, DEFRA.

Mills, B., & Andrey, J. (2003). *Climate change and transportation: Potential impacts and interactions*. Retrieved from http://climate.dot.gov/documents/workshop1002/mills.pdf

Murphy, J. M., Sexton, D. M. H., Jenkins, G. J., Boorman, P. M., Booth, B. B. B., Brown, C. C., ... Wood, R. A. (2009). UK Climate Projections Science Report: Climate change projections. Met Office Hadley Centre, Exeter, 193pp.

Peterson, T. C., McGuirk, M., Houston, T. G., Horvitz, A. L., & Wehner, M. F. (2008). *Climate variability and change with implications for transport* (90pp). Washington: Transportation Research Board.

Rail Safety and Standards Board. (2008). *Research brief: Impact of climate change on coastal rail infrastructure*. Retrieved from http://www.rssb.co.uk/sitecollectiondocuments/pdf/reports/research/T643_rb_final.pdf

Rail Safety and Standards Board. (2010). *Research brief: Tomorrow's railway and climate change adaptation*. Retrieved from http://www.rssb.co.uk/sitecollectiondocuments/pdf/reports/research/T925_rb_final.pdf

Standley, S., Miller, K., Okamura, S., Wynn, D., Greenhalgh, S. & Horrocks, L. (2009). *Wild weather warning: A London climate impacts profile*, Greater London Authority, London, 85pp. Retrieved from http://www.london.gov.uk/lccp/publications/docs/wild-weather-warning.pdf

Suarez, P., Anderson, W., Mahal, V., & Lakshmanan, T. R. (2005). Impacts of flooding and climate change on urban transportation: A system wide performance assessment of the Boston Metro Area. *Transportation Research Part D*, *10*, 231–244.

The Royal Academy of Engineering. (2011). *Infrastructure, engineering and climate change adaptation – Ensuring services in an uncertain future*, London, 108pp. Retrieved from http://www.raeng.org.uk/news/publications/list/reports/Engineering_the_future_2011.pdf

Thornes. (1991). Thermal mapping and road weather information systems for highway engineers. In L. J. Symons & A. H. Perry (Eds.), *Highway meteorology* (pp. 39–67). London: E & FN Spon.

Walsh, C. L., Hall, J. W., Street, R. B., Blanksby, J., Cassar, M., Ekins, P., … Watson, S. J. (2007). *Building knowledge for a changing climate- collaborative research to understand and adapt to the impacts of climate change on infrastructure, the built environment and utilities*, 80pp. Available at http://www.ukcip.org.uk/wordpress/wp-content/PDFs/BKCC-Results.pdf. Accessed on May 21, 2012.

West, C. C., & Gawith, M. J. (Eds.). (2005). *Measuring progress: Preparing for climate change through the UK Climate Impacts Programme*. Oxford: UKCIP. 72pp.

Wilbanks, T. J., Romero Lankao, P., Bao, M., Berkhout, F., Cairncross, S., Ceron, J.-P., … Zapata-Marti, R. (2007). Industry, settlement and society. In M. L. Parry, O. F. Canziani, J. P. Palutikof, P. J. van der Linden & C. E. Hanson (Eds.), *Climate change 2007: Impacts, adaptation and vulnerability. Contribution of Working Group II to the Fourth Assessment Report of the Intergovernmental Panel on Climate Change* (pp. 357–390). Cambridge, UK: Cambridge University Press.

CHAPTER 6

THE IMPACTS OF CLIMATE CHANGE ON THE NATIONAL FREIGHT SECTOR

David Jaroszweski

ABSTRACT

Purpose – *The UK road freight sector is an example of an economically important sector that can be viewed from two different perspectives with regards to climate change: firstly, as a significant contributor to the problem in terms of greenhouse gas emissions, and secondly as a likely recipient of significant impacts (both positive and negative) of any resulting change in climate, especially any increase in extreme weather. It is this relatively neglected second perspective, freight as a vulnerable sector, which is explored.*

Methodology/approach – *The way in which the freight sector develops in the future will be explored both in terms of the future type of freight operations and potential changes in vehicle technology.*

Findings – *There is a need to understand how the overlying socio-economic scenario influences the development of the freight sector.*

Transport and Climate Change
Transport and Sustainability, Volume 2, 137–173

ISSN: 2044-9941/doi:10.1108/S2044-9941(2012)0000002009

Originality/value – *The examination of the impact of climate change on the freight sector is a particularly original aspect of this chapter.*

Keywords: Freight; climate change; climate change impact assessment; adaptation; socio-economic scenarios

INTRODUCTION

In terms of climate change, road freight transport is traditionally looked at through the lens of emissions and mitigation. Worldwide, road freight emissions have grown as increased global trade, personal consumption and changes in production and logistics have led to increases in the amount of transport. The exact quantification of this is not clear, although Kahn and Kobayashi (2007) estimate that freight transport accounts for around 8% of worldwide CO_2 emissions, with additional emissions being associated with warehousing and handling. Much of the increase in emissions is due to changes in logistics. Logistics is 'the term now widely used to describe the transport, storage and handling of products as they move from raw material source, through the production system to their final point of sale or consumption' (McKinnon, 2010). The reduction in the cost of transport across all distances has enabled the modern form of global supply chain, where cost barriers to sourcing cheaper supplies from greater distances have been reduced significantly. The 'food miles' issue (e.g. Böge, 1995) where food is sourced and processed over longer distances is an example of this. It has also promoted the use of centralised logistical structures, less efficient practices such as badly laden loads and the use of more polluting modes of transport such as road freight, often as a result of government promotion away from rail freight during the second half of the twentieth century.

As will be shown, the case in the United Kingdom is complex. During the last two decades, there have been several changes in logistics and the economy having the effect of increasing the demand for transport (notwithstanding the economic crisis) and logistic practices adding to a greater amount of transport per unit of GDP. However, the picture in the United Kingdom can also be seen as potentially promising, and perhaps instructive to other countries, as recent changes have pointed towards ways in which freight emissions may be reduced. As will be discussed, this is both through technological changes and changes to the economy.

The view of transport as a vulnerable sector to climate change is a relatively new study area in climate impact studies, and is something that this book as a whole contributes to. However, one issue that has been largely overlooked until recently is that climate change will also impact on the freight and logistics sector through a change in the frequency and intensity of hazardous and disruptive weather events. Over several decades freight has become increasingly time-sensitive with the introduction of low inventory just-in-time deliveries (Wu & Olson, 2008) which, unlike passenger transport have far less capacity to be delayed, rescheduled or cancelled. Furthermore, although improvements to vehicle design, technology and driver training have increased safety, freight vehicles often have inherent vulnerabilities to meteorological hazards compared to their passenger transport equivalents. For example, high sided heavy goods vehicles (HGVs) are more vulnerable to cross winds than passenger vehicles (Young & Liesman, 2007), whereas the opposite is true with rail freight (Baker, Hemida, Iwnicki, Xie, & Ongaro, 2011).

Wind events, in particular extreme events, are arguably the most highly visible hazardous weather type. Such events are especially dangerous for road freight drivers due to the nature of their vehicles. During the 'Great Storm' of 1987 gust speeds of over 80 knots (kn) were recorded across much of the South East England and East Anglia, with higher localised speeds also being recorded (Shine, 1987), causing 65 wind-related road freight accidents across Great Britain. In comparison the similarly powerful 'Burn's Night Storm' of 1990 (Baker & Reynolds, 1992) had a much greater effect on freight than that of the Great Storm. This storm initially tracked across southern Scotland, moving southwards and causing severe gales and storm force winds across large parts of England and Wales, with a maximum gust speed of 93 kn recorded at Aberporth (McCallum, 1990). Overall, 220 wind-related road freight accidents were recorded, compounded by the fact that it occurred on a weekday, thereby increasing exposure to the hazardous driving conditions.

A notable recent example of how weather can impact on transport and how this feeds into the UK economy were the events of 'Windy Thursday' on 18 January 2007. This event resulted in the overturning of almost 50 goods vehicles and was estimated to have caused £50 million in delay costs across the United Kingdom (Highways Agency, 2007). Over this period an intense depression swept over large parts of England, Scotland and Wales with gusts of 65–75 kn (Eden, 2007). In total 111 accidents were recorded. These large-scale wind events can also cause lengthy periods of disruption. For instance, it was reported that roads in some areas took up to 16 days to reopen following the Great Storm, mostly due to uprooted trees (Fishwick, 1990).

This frequency and intensity of such events is likely to change in the future, potentially (although importantly not in all cases) making freight less safe for the drivers as well as less reliable for the customers. So, not only does the logistics industry need a mitigation plan to reduce its contribution to climate change, it also needs to assess the potential impacts climate change will impose on it and to formulate adaptation strategies to counter these.

A relatively new area that is also explored in this chapter is the idea that the sector that will suffer the impacts of climate change will be significantly different to that of today. During the course of this chapter it is suggested that the way in which freight and logistics will develop in the future will be determined both by direct reactions to climate change in terms of mitigation and adaptation, but moreover as a result of changes to the surrounding economy and society of the UK that drive the sector. This will determine both future emissions and resilience to weather and climate-related disruption and accidents. As a result, it is suggested that mitigative and adaptive strategies can and should be linked, and that this task is more viable under certain socio-economic conditions than others.

Jaroszweski, Chapman, and Petts (2010) have previously used the concept of concurrent climatic and socio-economic change, displayed intuitively by Tol (1998) in Fig. 1 to discuss potential climate impacts. The upper left section of the diagram represents the present-day situation, and as will be discussed, many studies have been conducted to assess the relationship

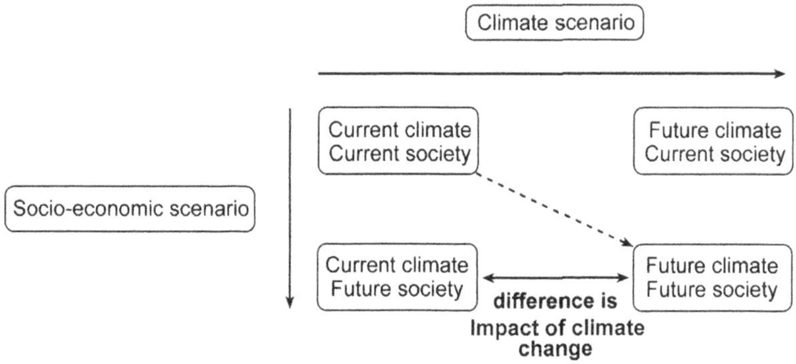

Fig. 1. Climate Change and Socio-Economic Scenarios.
Source: Tol (1998).

between the current climate and the transportation network in its present form. The upper right section represents studies where climate is perturbed, but society remains unchanged, that is non-climatic factors are not taken into consideration and the fundamental relationship between meteorology and transport is unchanged. The lower left represents a theoretical state where climate remains as today but society changes; hence, transportation develops, becoming either more or less vulnerable to the effects of meteorological events. Studies in this area would isolate the impact of socio-economic change. Finally, the bottom right section represents the 'complete' assessment, where both scenarios are incorporated.

International shipping is covered in Chapter 3, hence, this chapter is principally concerned with the movement of freight across shorter distances on a national scale (in this case the United Kingdom). The focus is on road freight transport, the dominant mode of freight transport in the United Kingdom.

FREIGHT AS A VULNERABLE SECTOR TO CLIMATE CHANGE

Until recently the impact climate change may have on freight has remained relatively unexplored. This is an important oversight, due to the aforementioned time-sensitive nature of freight operations and the well-documented observed impacts of rain, wind, high temperatures, ice and snow. Several recent reviews have begun to address this issue, highlighting a range of impacts that predicted climate change may have on transportation, notably by Peterson, McGuirk, Houston, Horvitz, and Wehner (2008) and Koetse and Rieveld (2009).

However, it must be noted that most previous and ongoing research investigates the impact on passenger transport, which is important as road freight transport has a significantly different operational exposure and physical vulnerability to severe weather that needs to be distinguished. For example, heavy goods vehicles are generally more vulnerable to high winds than passenger transportation and often travel in conditions that other road users might choose to avoid. To answer these questions the vulnerability of freight to the most common weather hazards in the UK is discussed below, along with a discussion of how the most recent UK climate projections show the frequency, intensity and spatial distribution of these hazards are likely to change in the future.

Table 1. Atmospheric and Road Surface Conditions Recorded at the Site of Road Freight Accidents in Great Britain between 1999 and 2008.

Hazard	Percentage
Fine without high winds	77.8%
Raining without high winds	13.8%
Snowing without high winds	0.4%
High winds	3%
Fog or mist	1%
Ice	1%
Other	1.7%
Unknown	1.2%

Note: Data obtained by DfT STATS19 (2009), access via ESDS, 2009.

Rain

In terms of safety, the relative importance of rain in the UK is clear; excluding fine conditions, rain is reported in more road freight accidents than all other meteorological parameters combined (Table 1). Rain causes accidents through two main mechanisms; the reduction of friction between the tyre and the road, and the reduction of visibility both through precipitation and spray in the atmosphere and by directly obscuring the windscreen. It is likely that it is the combination of these two mechanisms that make rain such a hazardous weather type, with Andrey and Yagar (1993) demonstrating that the increase in accident rates during rain events is almost completely concentrated into the period when rain is falling, with a sharp decrease when rain recedes even though road surface remains wet.

Although rain has been shown to increase the accident rate in studies across the world (Qiu & Nixon, 2008), it is unclear whether rain increases the rate and severity of injuries resulting from accidents. Edwards's (1998) study seems to suggest that rainfall reduces the proportion of fatal accidents compared to those of a lower severity. An interpretation of these results may be that drivers take more care in these conditions and limit their speed. One of the more curious aspects of the relationship between precipitation and accident rates is that the magnitude (and in some studies, even sign) has been found to change depending on the temporal scale it is being analysed at. This initially puzzling aspect has been highlighted by Eisenberg (2004), who noticed that analysing the number of fatal accidents in a given day against that day's rainfall total gave a strong positive relationship, whereas analysing a month's fatal accident total against that month's precipitation total had a strong negative relationship. Eisenberg explained this with respect to driver behaviour, suggesting that after an initial day of heavy rain, driver's awareness of the difficult conditions increases, allowing them to cope better in rainy conditions.

Flooding can also cause severe disruption that has economic consequences. The wet summers of 2007 and 2009 and the wet autumn of 2000 have shown the disruptive effect of pluvial and fluvial flooding. These have often been cited in the media as examples of the current impact of climate change, which although contentious, has some validity; a highly influential study by has shown that anthropogenic climate change had a two in three chance of increasing the odds of this event.

The most recent UKCP09 projections predict the mean annual precipitation amounts over the United Kingdom to remain roughly unchanged (Murphy et al., 2009). However, the projected seasonal changes, especially for summer and winter, pose several challenges to future road freight operations and safety. For example, winter precipitation across the western UK is projected to increase, although winter precipitation amounts in the Scottish Highlands may see a small decrease. This overall increase in winter precipitation is important, as hazards are usually exacerbated by low lighting conditions. Conversely, summer precipitation sees a strong decreasing trend. Again, Scotland shows the least potential change to its precipitation regime. The daily intensity of precipitation is also set to change, with the wettest day of winter increasing in rainfall amount by up to 56% and the wettest day of summer decreasing by 38%.

Temperature

Temperature is an atmospheric parameter which causes a large number of accidents, and one that features heavily in the literature. Hot weather can cause accidents by creating uncomfortable driving conditions, mainly through heat stress (Stern & Zehavi, 1990; Nofal & Saeed, 1997). In Britain low temperatures and the resultant road ice are currently of more interest, causing around 1,000 road freight accidents a year (DfT STATS19, 2009). Ice generally forms at 0°C, although the freezing point may be lower depending on the composition of the water solution. Ice is most slippery at 0°C (Moore & Copper, 1972), which is significant for Britain because of the marginal climate which results in a high frequency of icing events at this temperature.

In theory, winter road maintenance should make ice a fully preventable road hazard. However, in reality this is not achievable, mainly due to the difficulties in forecasting where and when certain sections of the road network will reach freezing temperatures. The microclimate for road surface temperatures is determined by geographical features of the area. These include altitude, topography, road surface construction, land-use, and most importantly sky-view factor, a measure of overhead exposure (Chapman, Thornes, & Bradley, 2001). In the United Kingdom, the main preventative measure comes through treating roads with rock salt. There are thresholds related to the types of maintenance that can be used, with below −10°C rock salt becoming less effective and below −5°C urea being ineffective (Thornes, 1992). Since the humidity of the United Kingdom is usually above 80%, the hygroscopic properties of rock salt mean that salt remains in solution and makes the roads even more slippery than when they are dry.

Vehicles themselves also play an important role in determining risk. By thermal-mapping different lanes of motorways, Chapman and Thornes (2006) found temperature differentials of 1.5°C. This correlates with the traffic volumes of the lanes, with vehicles warming the road (or preventing it to cool) through the numerous mechanisms mentioned by Prusa, Segal, Temeyer, Gallus, and Takle (2002), including heat from the engine, friction from the tyres and the blocking of long-wave radiation from the road surface. Future traffic volumes and the way they are managed on the roads could have a significant impact on this phenomenon. It is clear that in terms of infrastructure, traffic flows and maintenance, the impact of temperature can be considered to have a more complex relationship with socio-economic development than precipitation.

The problem is infrequent enough to rule out the use of studded tyres on automobiles as has been used in other countries (e.g. Germany, Austria) to confer a slight increase in safety (<5% reduction in accidents, e.g. Elvik, 1999). However, as studded tyres have increasingly been banned in most countries due to the negative impacts on the road surface, winter tyres are now more commonly used. These tyres use softer rubber for better traction at lower temperatures, as well as a tread pattern that allows water to escape, reducing hydroplaning (Hjort & Jansson, 2010).

A significant reduction in frost days is forecast in the United Kingdom. Several regions will see annual frost days approaching single figures by 2080 under medium emissions (Jones, Kilsby, Harpham, Glenis, & Burton, 2009). The obvious impact of the above projections, at least at first glance, is a safer operating environment for road freight transportation. It is highly likely that while the absolute number of annual frost days will reduce, the relative accident rate per frost day may increase, as drivers become less used to these conditions, again as would be predicted by Elvik (2006).

It is also likely that this change in relative accident rate will vary spatially. For instance, freight drivers frequently exposed to conditions in Scotland may be able to retain coping skills for ice, as the number of frost days is projected to reduce less than the rest of the country. In regions seeing a significant reduction in frost days there are several potential problems. The maintenance infrastructure that exists today is designed to cope with a significant number of ice days per year. A significant decrease in road ice events could lead to cuts in the service provided, with the potential for complacency (Andersson & Chapman, 2011). The difficulties of forecasting road icing events have long been established (Thornes & Stephenson, 2001). Due to the topographic influences on ice formation described by Chapman and Thornes (2006), it is foreseeable that under a new climatic environment roads which would have completely frozen before might now only freeze in certain locations. The difficulty of forecasting the surface temperature of stretches of road and pinpointing which specific locations require road-salting is significant.

Wind

Due to its position in the mid-latitude westerlies, Britain is one of the windiest countries in the world (Perry, 1990; Edwards, 1994). The literature regarding wind-related road accidents contains the highest proportion of studies focussing solely on the impact on freight vehicles. This relates to the physical nature of the vehicles themselves, having high and sheer sides and the potential for a far higher centre of gravity than other road vehicles. As well as having strong winds, Great Britain is also vulnerable due to the way its road network is orientated. As traffic generally moves along a northwest to southeast direction, traffic is frequently and disproportionately at right angles to the prevailing westerly wind direction (Perry, 1990; Edwards, 1996). Often high winds lead to major roads and bridge crossings being closed as a precaution, with traffic being diverted to less suitable minor roads, which reduces speed. Baker and Reynolds (1992) suggest that vehicle speed should be limited when gust speeds exceed 38–45 mph.

Summerfield and Kosior (2001) used a dynamic simulation to show that vulnerability is increased further when trailers are lightly loaded or empty, as the reduced downward force allows lower wind speeds to exert a great shear force and increases the likelihood of overturning. The risk of overturning is not purely a function of the weight and shape of the truck and wind speed, as it has been found that a fluctuating wind is more dangerous to truck instability than a steady wind, due to the resonance effect on the trailer (Summerfield & Kosior, 2001). Another significant problem is that of sliding-related accidents, where the force of the wind interacts with a truck driving on wet or icy roads (Saiidi & Maragalas, 1995). An additional indirect hazard is caused through the fallen trees, wall or panelling which are blown onto roads during wind events (Edwards, 1998).

Edwards (1999) shows an increased number of wind-related accidents between October and March associated with winter storms, with injury analysis (Edwards, 1998) showing an interesting spatial pattern that highlights the geographical application of Elvik's (2006) laws of accident causation. The study showed that some UK counties reported an increase in accident severity during high winds, whilst others showed a decrease. It was hypothesised that counties such as Nottinghamshire which saw an increase in severity, are those which are known for infrequent high wind events, reducing the exposure of drivers, limiting their acquisition of the skills to cope under these conditions and hence increasing vulnerability. Wind is also a good example of the disproportionate level of exposure of freight vehicles compared to other modes of road transportation. For example, the United States' road freight fleet makes up 3% of all vehicles, yet accounts for 7% of all vehicle miles travelled and 12% of the total fatality count (Cerrelli, 1998).

The North Atlantic Oscillation (NAO) is a major determining factor of the severity of the winter storm season. The phase of the NAO (either positive or negative) has a large influence on the climate of Europe, in particular the spatial distribution of storm winds and precipitation. This fundamental system has yet to be properly modelled within climate change projections; hence, the results for future wind speeds and direction are inconclusive. Although individual model runs do show a slight variance, when averaged they predict changes in mean wind speed of a few percent over the United Kingdom (Murphy et al., 2009). However, there is probably the potential for extreme outlying scenarios which may not be properly captured with the techniques currently available. Rockel and Woth (2007) argue that one problem with many models is the lack of parametrisation (the representation of subgrid scale processes) of gust speed. Those that included this showed a 20% increase in the frequency of storm peak events greater than 8 on the Beaufort scale over central Europe. Significantly, storms (also classed as gales) of this intensity and above are known to cause vehicles to veer, as well as presenting the additional hazards of damaged trees and wind-blown debris (Royal Meteorological Society, 2010).

Snow

The wintery events of January and December 2010 in the UK have highlighted the widespread problems snow can bring. It has been shown that accident rates can increase by up to 1,000% when snow is present (Qiu & Nixon, 2008), but a more complicated question is whether the injury rate increases or decreases during snowfall. It has been suggested that although the overall accident rate increases, the slower speeds that most drivers travel at during these events means that the severity of the accident may be less than during clear conditions, resulting in less severe injury. Eisenberg and Warner (2005) and Fridstrøm (1999) demonstrate the reduced severity of injuries during snow events, but the low number of crashes makes statistical analysis difficult. Martin, Perrin, Hansen, and Quintana (2000) show a reduction of 25% for wet and slushy conditions, with Knapp, Smithson, and Khattak (2000) showing an even higher figure of 29% for heavy snow.

Like wind, snow is not handled very well by climate models, owing to the complex nature of snow formation and the importance of small-scale topographic features which are not captured in the model parameters (Murphy et al., 2009). The 2009 edition of the UKCP climate scenarios present a less detailed picture of future snowfall than previous iterations, in part due to the acknowledgment of the weaknesses of the methodology. In much the same way as wind, only a small note on the change in snowfall is given, estimating that snowfall will be reduced by between 65% and 80% over mountainous areas of the UK and by 80–95% everywhere else. Similar to temperature, the impact of climate change on snow-related accidents is likely to be positive, apart from potential institutional and driver complacency brought about by a more benign climate. However, it has been theorised that changes in regional climate may lead to an increase in the frequency of heavy snow events such as those seen in early and late 2010.

Fog

The reduction of visibility during fog events is a significant cause of road traffic accidents, and acts in a similar way to the aforementioned visual effects of precipitation. Fog forms either through radiative cooling or humidification, with topography playing a major role in its temporal persistence (Pagowski, Gultepe, & King, 2004). Part of the construction criteria of motorways is a design which limits the build-up of fog

(Musk, 1988). Thornes (1992) points out that although it is always recommended that drivers reduce speeds when visibility is less than 200 m, evidence suggests that this is not done until visibility is less than 100 m.

Hermans, Brijs, Stiers, and Offermans (2006) found a statistically increasing impact of fog on accidents. Although now dated, the figures from Moore and Cooper (1972) are striking, showing that despite traffic volumes being reduced by 20% during fog events, accident numbers increase by 16%. Again, Edwards (1996) provides evidence of the role of exposure and experience to adverse driving conditions. It was shown that the areas known to be the most fog prone, such as the Midlands, report a decrease in accident severity during fog events, compared to those with a low occurrence of fog. Although no statistical studies exist for the Great Britain, it could be argued that Elvik's laws of accident causation (2006) may suggest an increasing sensitivity in the Great Britain to individual fog events, as improvements in air quality have reduced the number of smog events, a condition which was common before the clean air acts of 1956 and 1968.

Fog is not a parameter generally produced by climate models, partly due to the required topographic resolution and partly due to the prioritisation of temperature and precipitation projections. It can be inferred that there will be a reduction in the number of fog days (Murphy et al., 2009). An increase in the occurrence of fog creates hazardous driving conditions and may cause more accidents. In addition there may be a need for better road design to reduce the formation of fog.

DETERMINANTS OF FREIGHT EXPOSURE TO CLIMATE HAZARDS

The volume of freight vehicles on the road, their design and performance, the ways in which they are used and the type of training given to drivers are continuously changing, and together determine the severity of the impact of a given meteorological hazard. Table 2 summarises the factors that influence the exposure and sensitivity of freight vehicles to the main meteorological hazards. The way in which the freight sector develops during the coming decades is therefore a key issue.

For a variety of reasons, freight transport has been observed to be linked to economic growth. Much recent work has looked at more detailed reasons for these trends, as ways of promoting positive decoupling between economic growth and freight activity. Government schemes such as the

Table 2. Summary of Atmospheric Hazards, Their Effect on Road Freight Transportation and Likely Impact of Climate Change.

Hazard	Mechanism of Hazard	Current Temporal Distribution of Accidents	Current Spatial Distribution of Accidents	Climate Projection	Likely Impacts
Precipitation	Reduction of friction between tyre and road surface	Prevalent throughout the year	Affects entire United Kingdom, with similar county figures for percentage of total accidents (For England and Wales – Edwards, 1996)	Large decreases in summer precipitation amounts (central estimate of 40% by 2080 in South-East England)	Potential disproportionate increase in accidents during winter due to combination with other risk factors, for example low daylight hours
	Reduction of atmospheric visibility	Increased accident rates during final three months of the year		Large increases in winter precipitation amounts (central estimate of 33% by 2080 across western United Kingdom).	Strong decrease in summer rain-related road accidents
Ice	Reduction of friction between tyre and road surface	Little research on temporal distribution of accidents Road ice season between October and April	Ice season determined by geography and latitude	Large increase in average mean temperature Increase in minimum daily temperatures Large reduction in frost days through to 2080	Large reduction of ice-related accidents Possible increase in the relative impact of each ice event as drivers become less used to this driving condition
Wind	Vehicle overturning due to cross-winds	Greatest number of accidents between October and March, although accidents occur throughout the year	Roads with a northwest to southeast orientation greatest affected	No consensus of sign of change, although magnitude is limited to around 10%	UKCIP projections suggest little impact
	Skidding Wind-blown debris			Very little scientific certainty on wind projections	However, individual model runs suggest both slight increases and decreases in accident rates

Table 2. *(Continued)*

Hazard	Mechanism of Hazard	Current Temporal Distribution of Accidents	Current Spatial Distribution of Accidents	Climate Projection	Likely Impacts
Snow	Reduction of friction between tyre and road surface	Greatest number between December and February	Relative accidents rates highest for areas with infrequent return periods for snow	Little spatial detail on projections, although large-scale reductions (up to 90% by 2080) are likely	Large reduction in vehicle accidents
	Reduction of atmospheric visibility	Shows accidents through all months apart from August			Possible increase in the relative impact of each ice event as drivers become less used to this driving condition
Fog	Reduction of atmospheric visibility	Greatest number of accidents between November and February	Areas such as the West Midlands show the greatest prevalence of fog	Inferred reduction in fog days	Potential reduction in fog-related accidents
		Accidents reported throughout year	However, areas with infrequent fog events show the greatest increase in accident rates	However, not routinely handle by climate models	

UK modal shift allowance have been implemented to reduce freight transport's contribution to global climate change. This combined with expected increases in fuel efficiency are currently seen as the main ways of reducing emissions. However, when dealing with a sector that can be seen as responding to the demands of the economy, we also need to know how the future trends of the economy itself affects the emissions produced by freight.

The components in the following subsections are those used by McKinnon (2007) to explain the period of positive decoupling between 1997 and 2004. This period provides a potentially useful example of how freight may be decoupled and emissions reduced. This provides a useful analytical framework for discussing the components that determine road freight exposure. The decoupling framework first analyses the factors influencing the amount of goods produced by an economy and then goes on to those which influence the ways in which it is transported. For example, the value, weight and volume of goods produced and consumed is the highest order, and is determined by the composition of GDP, dematerialisation and the erosion of industrial activity to other countries. The amount of this freight taken by road is determined by the modal split, an aspect of transport which is well catered for in recent socio-economic scenarios. Finally, the number of road kilometres used is determined by the number of links in the supply chain, the rate of spatial concentration and the efficiency of freight transport.

Changing Composition of GDP

Changes in the composition of GDP determine the type of goods which are produced and transported. Scenarios that promote a shift towards sectors which produce lighter goods may therefore reduce exposure, especially if the goods themselves experience further dematerialisation. However it must be noted that the volume of goods is perhaps a better indicator for exposure. The UK has seen a long-term shift away from heavy industry and manufacturing towards services, with a resulting reduction in the weight of goods lifted, noted by McKinnon (2007) as a reason for the observed decoupling of freight and GDP.

Sorrell, Lehtonen, and Stapleton (2009) makes the important point that future demand for freight will depend on the future demand for manufactured goods, as opposed to services. The rise of e-commerce is also an important influence on exposure. There is evidence that the impact of e-ecommerce in terms of increasing dematerialisation of certain products into electronic forms discussed by Crowley (1998) is becoming a reality. Although the impact of e-readers on the printed press is still uncertain, emerging technologies such as 3D printers may one day allow production and manufacturing to be shifted to home, eliminating almost all links in the supply chain.

Dematerialisation

Changes in the relative contribution of different economic sectors to overall GDP, as well as changes to the weight of the goods they produce are potentially key determinants of future freight exposure. Furthermore, they are factors that regularly feature in socio-economic toolkits, often diverging quite significantly under different political and economic environments. The reduction of the weight of goods in the economy is termed as dematerialisation. McKinnon (2007) mentions several processes which cause dematerialisation including the miniaturisation of products, replacement of heavier materials with lighter ones and the substitution of physical goods for electronic ones, with the shortening of product life cycles and the growth of packaging having the opposite effect. It is likely that any move towards dematerialisation will promote positive decoupling of freight from GDP and therefore reduce exposure.

Modal Split

Changing the split in modes that carry freight is seen as a promising way of reducing emissions. McKinnon (2007) calculates that the decline of road transport's share of the freight market accounted for 22% of the observed decoupling of road freight from GDP in the United Kingdom between 1997 and 2004. This is partly attributed to the increase in market share of water-based services, from 21% in 1997 to 24% in 2004. The privatisation of rail freight services in 1996 also spurred an increase in rail's market share from 7% to 8%. Although the amount of freight transported by road still rose during this period (DfT, 2009), the effect of the increase in rail and water-based transport's share of the freight market acted to limit this increase, and almost certainly also acted to reduce the exposure of vehicles to meteorological hazards in terms of kilometres travelled. McKinnon (2007) also calculates that a further 12% of the observed decoupling can be explained by the 16.7% increase in the real cost of road freight transport between 1997 and 2004. Although the main effect of this was to increase freight efficiency through better vehicle utilisation, it is plausible that it promoted at least some of the modal shift seen during this time.

The movement towards rail taking a greater share of freight has been predicted (Browne & Allen, 1997), and has been observed over the last decade (DfT, 2009). The Netherlands demonstrate the ability to reduce freight intensity by encouraging modal shift towards rail and inland waterways through greater integration (Haq & Bolhuis, 1998). Recent examples in the UK include Eddie Stobart which now transports freight for Tesco by rail between England and Scotland and has also started rail deliveries of refrigerated goods from Spain, both of which were previously delivered by truck.

Given transport's role as a CO_2 contributor to climate change (Chapman, 2007), it is highly likely that modal split will be an important determinant of future exposure, with a key factor in this importance being the extent to which the British government will promote a shift towards other modes in the coming decades. Much research has been conducted on promoting greater use of rail freight. Forkenbrock (2001) shows the external costs on society involved with rail freight, (including accidents, noise, particulate pollution and greenhouse gases) to be around three times lower than those associated with road freight. For instance, the greenhouse gas externalities for road freight are estimated at 0.15 cents per ton-mile compared to 0.2 cents for rail freight. However, it must be remembered that this is a US study where road freight vehicles and distances travelled are considerably different to those in Great Britain. A British study has demonstrated that freight trains can remove approximately 50–120 HGVs from the road per day (DfT, 2008).

Several studies have attempted to forecast the likely impact of efforts to promote modal shift within the freight sector. For instance, Piecyk and McKinnon's (2008) expert Delphi of the environmental impact of road freight predicts a fall in road's share of the freight market from 64% to 59.9% by 2020. The study also predicts rail's share increasing from 9% to 11.3%, inland waterways and coastal shipping from 23% to 24.4% and pipeline's share from 4% to 4.4%. However, it must be noted that freight traffic volumes were forecast to increase by around 25% from 257 billion tonne-kilometres in 2005 to 322 billion tonne-kilometres in 2020; hence, the total exposure of road freight vehicles to meteorological hazards is still likely to rise.

Akerman (2011) demonstrates the potential of high-speed rail to provide reduction in emissions by shifting both passenger transport and freight transport from more polluting modes. Through lifecycle analysis, it was sound that of the reduction of 550,000 tonnes of carbon projected from the proposed European high-speed rail track in Sweden by 2025/2030 m, 60% would be through a shift from road freight and 40% through a shift from passenger air travel.

In terms of rail, freight trains withstand high winds and are less likely to derail compared to passenger trains. Baker et al. (2011) shows that at mean crosswind speeds up to 20 m/s with gusts of up to 30 m/s derailment is not predicted. Passenger rail and unladen freight derailment is predicted when mean wind speed of 25 m/s, whereas the laden freight vehicle was not predicted to derail. However, rail is more susceptible to other types of hazards, especially that of extreme high temperatures causing rails to buckle, which are predicted to increase in the future. Dobney, Baker, Chapman, and Quinn (2010) calculate the number of buckles may increase, costing the rail industry. Any increase in freight traffic will also increase the loading on the tracks, an important determinant of the lifetime of this infrastructure.

Reduction in Number of Links in the Supply Chain

The number of links in the supply chain is an important determinant of the amount of freight transport. Historically, the move towards larger and more complicated supply chains has been enabled by lower transport costs and faster travelling speeds, termed as a reduction in the 'friction of distance' (Sessa et al., 2009). This is particularly noticeable with 'food miles', the increasing amount of transport that occurs in the farming and processing of food (e.g. Böge, 1995). Although this trend appears intractable, over periods of 20–50 years it is quite conceivable to imagine scenarios where changes in socio-economic values and governance promote changes which are inconceivable in the short-term to medium-term. For instance, the research by (Böge, 1995) suggesting that the total vehicle-km involved in strawberry yoghurt production could be reduced by 67% if the nearest suppliers were used. Any trend in the economy towards regional production would therefore contribute to negative decoupling between freight growth and GDP.

Anderson, Browne, and Allen (1999) stress the importance of the European Commission's (EC) Packaging and Packaging Waste Directive and the subsequent 1997 UK Producer Responsibility Obligation requiring companies to take responsibility for the recycling of their packaging.

Increased home delivery freight driven by e-commerce journeys also act to increase the number of links in the supply chain. It can also be argued that this shift in operational practice increases exposure to winter-related hazards, as a significant proportion of freight transport may increasingly be moved away from the trunk road network and onto minor roads, which are less likely to be gritted. However, McKinnon (2007) mentions that home delivery by vans to some extent merely replaces car journeys that previously delivered the 'last mile' in the supply chain. This development also has significant environmental benefits if deliveries are routed efficiently (Edwards & McKinnon, 2009). Hence, when taking road transport as a whole, this move should act to reduce total exposure.

Spatial Concentration

Spatial concentration refers to the trend towards the observed centralisation of production and distribution, as transport costs fell and average travel speeds increased during the second half of the twentieth century, allowing firms to serve a wide geographic region from fewer locations (Quarmby, 1989; McKinnon, 2007). This had the effect of increasing the average leg length of journeys, therefore increasing exposure. The effect of this process can be seen in the large-scale increase in European freight transport after the liberalisation of freight markets allowed firms to concentrate production and distribution into a small number of sites serving large parts of the continent.

It also has to be remembered that the road network of Great Britain is to a large extent fixed in its current form, with little scope for major reorganisation. As a result, certain logistical changes become difficult to implement, especially further logistical centralisation. For example, the cost savings estimated from a reduction in the number of distribution depots can only be implemented with concurrent improvements to the road network (Quarmby, 1989). However, environmental issues may also emerge as a key driver for changes in the location of distribution centres within the existing road network. For instance, Browne and Allen (1997) concentrate on reducing energy use by freight in cities through a range of measures, including raising vehicle payloads, reducing empty running and the use of urban transhipment centres. This would shift the exposure outside of the city towards the countryside, which as Edwards (1996, 1998) shows, increases accident rates.

Efficiency of Freight Transport

Emissions and exposure may be reduced as freight operations become more efficient. Several components of the 1997–2004 decoupling explored by McKinnon (2007) relate to this. A useful way of introducing these concepts is through the work of Samuelsson and Tilanus (1997) who define efficiency in terms of actual transport output compared to the theoretical maximum output, the case that would exist if goods were moved non-stop from origin to destination along a minimum distance travelling at maximum speed. It is suggested that actual efficiency (E) is determined by four-dimensional efficiencies, time (T), distance (D), speed (S) and capacity (C) so that:

$$E = T \times D \times S \times C$$

Time refers to the percentage of available time that a vehicle is utilised, the theoretical maximum being non-stop 24-hour utilisation. Distance efficiency describes the extent of the deviation away from an idealised minimum radial distance. The speed efficiency is a measure of the travelling speed against the maximum allowable speed and the capacity efficiency, referring to the weight or space used in trailer compared to the maximum weight/space capacity. A future scenario that promotes an improvement to any of these ratios should result in a more efficient freight. Although this may not result in less exposure per tonne kilometre (for instance, a large increase in vehicle capacity without a reciprocal increase in lifted freight would lower the ratio), in practice a decrease is likely.

Better vehicle routing and improvements in vehicle utilisation as a by-product of increased freight cost are the two components mentioned by McKinnon (2007), which relate most to Samuelsson and Tilanus's (1997) concept of efficiency. Improved vehicle routing through computerised methods show significant reductions in distance travelled. As McKinnon (2007) shows, although investment in infrastructure was at a low level during the period of decoupling, the way the existing road network was navigated became more efficient due to the use of computerised vehicle routing and scheduling (CVRS) packages. It is estimated the use of these packages can cut vehicle kilometres by 5–10%, with even greater reductions in travel times achieved by bypassing known congested areas (McKinnon, 2007).

Increases in the real cost of road freight lead to more efficient vehicle utilisation. The empty running of trucks has been reducing in the United Kingdom due to a number of factors including the obligation for operators to collect return loads (McKinnon, 1996). This includes own account (vehicles owned and operated by the company) returning waste products and recyclables as well as stock, and third part logistics companies picking up loads on an ad hoc or agreed arrangement. This third party movement has been especially aided by electronic systems, allowing companies to obtain a suitable backhaul.

However, statistics suggest little change in the percentage of vehicles running empty between the years of 1998 and 2008, with a small increase from 27.5% to 29% (DfT, 2009). McKinnon (2000) shows evidence that just-in-time replenishment (the reduction of buffer stock and lead times in the supply chain, for example Garreau, Lieb, & Millen, 1991) has not encouraged under-utilisation of truck space, with companies taking extra measures to add consolidation to loads at different points in the supply chain. One problem in vehicle utilisation studies is that most countries record load factor by weight rather than volume, which is important as some vehicles are full before their weight limit is reached, leading to an underestimation of vehicle utilisation (McKinnon, Button, & Nijkamp, 2002).

Increasing vehicle weight and size limits can also reduce the number of vehicles on the road (McKinnon, 2007). The use of 'double deck' or 'high cube' trailers is unique to the United Kingdom due to the higher clearance under bridges. This clearance of 4.9 metres compared with the European standard of 4 metres allows taller trailers to be used, having the effect of decreasing exposure, as a greater volume of freight can be carried by the same length of trailer. However, this also has the effect of exacerbating the problems of high-sided vehicles operating in windy conditions previously discussed by Baker and Reynolds (1992).

There are also calls to increase the weight and length and weight limits on UK freight towards those seen on mainland Europe. The current weight limit is 44 tonnes, with the length limit being 16.5. Knight et al. (2008) discuss the impact of raising these limits up to a maximum of 82 tonnes and 34 metres. It was hypothesised that road wear would increase, vehicles may be trapped by road geometry, bridges would need to be strengthened and parking facilities altered. Again this would have the effect of decreasing exposure and increasing sensitivity, due to the greater surface area presented to crosswinds.

Penman (1997) highlights the potential improvements in utilisation simply through using more efficient pallet load systems. For instance, it was shown that by using higher pallets it is possible to utilise more of the available trailer space in grocery vehicles. The potential impact of this if implemented across Europe would be a reduction of 15% of the vehicles used in the sector. McKinnon and Campbell (1998) demonstrate that similar efficiency gains can be made by double-decking loads. Vehicle utilisation also has a bearing on handling under high wind conditions (Summerfield & Kosior, 2001).

ADAPTING TO A NEW CLIMATE

Even whilst benefitting from the effects of increasing efficiency, there is clearly a need to reduce some of the currently observed vulnerabilities. Sensitivity is not a commonly forecasted characteristic, nor is it one which has a relationship with the overlying economy that can be easily modelled. Ongoing improvements in vehicle braking technology, telematics, vehicle body design as well as anticipated future technologies such as automation are all likely to reduce accidents. Furthermore, sensitivity will also be determined by the quality of infrastructure on which freight vehicles operate. Much of any future change in road freight transport safety may come about by as yet unknown 'disruptive technologies', developments which have the potential to fundamentally alter the nature of a given industry. Unlike fuel technology whose take-up is often hastened in response to price shocks (Sessa et al., 2009) the introduction of safety technologies often follow a gentler curve, where economies of scale reduce the price. This is especially true where technologies need to reach a critical mass before their true safety gain is seen, such as automated vehicles. However, this section attempts to draw links with the economy. Although difficult to predict, this section reviews the current understanding how sensitivity changes within the freight sector, and highlights a number of potential developments in both the short-term and long-term.

Technology and Vehicle Design

As shown in Table 3, many the aspects determining sensitivity are concerned with the design of heavy goods vehicles. The UK Foresight Programme (2002) provides an interesting example of how the future vehicle design can be assessed in a structured process. These trends can then be mapped onto various technologies to determine how the overlying economy may promote or impede its development. An interesting example of this is the integration of software into road vehicles. Duff (2006) has previously postulated that

Table 3. Example of Factors Affecting Sensitivity and Exposure to Atmospheric Hazards.

Hazard	Factor
All hazards	Traffic volume (e.g. Qiu & Nixon, 2008)
	Road lighting (e.g. Fridstrøm, Ifver, Ingebrigtsen, Kulmala, & Krosgsgard Thomsen, 1995; Wanvik, 2009)
	Forecasting ability (e.g. Thornes & Stephenson, 2001)
	Weather warning systems (Carson & Mannering, 2001)
	Rural/urban split (e.g. Edwards, 1998)
Precipitation	Road surface construction (e.g. Leden et al., 1998; Sokolovskij, 2007)
	Vehicle technology – visibility (Andrey & Yagar, 1993)
	Vehicle technology – braking (e.g. Edwards, 1996)
Ice	Winter road maintenance infrastructure (e.g. Andersson & Chapman, 2011; Norrman et al., 2000; Thornes & Stephenson, 2001)
	Road surface construction (e.g. Leden et al., 1998; Sokolovskij, 2007)
	Urban heat island (Graham, 1996)
	Winter tyre use (Hjort & Jansson, 2010; Vaiskunaite et al., 2009)
Wind	Vehicle design (e.g. Baker, 1993)
	Empty running/under-loading (e.g. Summerfield & Kosior, 2001)
Snow	Winter road maintenance infrastructure (e.g. Andersson & Chapman, 2011; Thornes & Stephenson, 2001)
	Road surface construction (e.g. Leden et al., 1998; Sokolovskij, 2007)
	Winter tyres use (e.g. Hjort & Jansson, 2010; Vaiskunaite et al., 2009)
Fog	Vehicle lights/distance detectors (e.g. Cavallo et al., 2001)

drive-by-wire and brake-by-wire are possible future developments as well as sensor technology that will determine distances between vehicles and ensure speed limits are kept to. Even aspects like tyre wear could be monitored electronically, which again has impacts on braking distances in adverse weather.

However, the methodology that the Foresight (2002) utilised has several limitations. Firstly, it is based on a single scenario of socio-economic change over a 20-year period. This does not allow for potentially divergent trends in governance and values over this period. Secondly, the scenarios are based on backcasting. This is where a desirable outcome in the future is selected, and the necessary steps to achieve this end-point are envisaged. Lastly, these scenarios provide little scope for clear breaks with the past. Disruptive technologies are only envisaged at the very far end of the timeframe.

The Foresight Programme's later approach (2006) looks at three time horizons, allowing for both innovations in existing and new, as yet unknown technology to be considered. The first horizon describes a continuation of the present, with the introduction of new but understood technology, with fairly predictable innovation. An important point about this second phase is that even when the trend of innovation is clear, the form it is achieved in can be unpredictable. An example they use to illustrate this is the Internet, as researchers and business were predicting a global information network in the 1980s, but did not know the model it would take. They suggest similar broad advancements can be predicted within transport, such as automation, which is highly anticipated even though the technology does not yet exist.

Vehicle Usage

Exposure and sensitivity are often linked. For example, the increase in reverse logistics of packaging waste may have a detrimental effect on the sensitivity of a vehicle to wind. Compared with drivers running empty vehicles who can tie back the side curtains to reduce the chance of overturning during heavy wind events, drivers operating on a backhaul would not have this option, and as a result would potentially be operating partly loaded, leaving the vehicle more vulnerable to crosswinds (Summerfield & Kosior, 2001). Interestingly it has been suggested that an increase in pressure to find backloads usually happens during recessions due to the intense competition in the road haulage industry, potentially adding a cyclical aspect to wind vulnerability. Congestion charges could lead to more night-time deliveries (Duff, 2006) which could increase the effect of adverse weather due to low light conditions.

A significant recent trend in the UK freight and logistics sector has been the increased penetration by foreign-registered operators. For example, in 1997 52% of all freight vehicles arriving at British ports were foreign registered, in 2004 this rose to 75% (Department for Transport [DfT], 2005; in Mckinnon, 2007). Cabotage, the transport of domestic freight by foreign operators has also increased. Differences in driver training, vehicle standards and maintenance as well as the driver's unfamiliarity of driving on the left-hand side of the road could all theoretically contribute to increased sensitivity. Socio-economic scenarios often include levels of EU integration, a potentially useful measure when discussing future rates of cross-border operations.

Certain scenarios promote better expenditure on infrastructure. Some proposed techniques involve bypassing surface transport altogether, such as hydraulic tunnels (Howgego & Roe, 1998; Taniguchi & van der Heijden,

2000), would reduce exposure and sensitivity to meteorological hazards to zero. Similar underground tunnels have been suggested by Browne (2006) as a potential solution to future distribution demands as well as tighter planning and aesthetic restrictions. This would solve many of the problems of city logistics including congestion mentioned by Ogden (1984). Although seemingly far-fetched, several examples of enclosed and subterranean freight systems have already existed, including the Post Office's automated railway which transported mail along a 23-mile network between sorting offices in London until its closure in 2003 (Aldhous et al., 1995; Howgego & Roe, 1998). The involvement of British architect Norman Foster in the planned Masdar City project in Abu Dhabi is an example of the concept of enclosed transport applied to an entire city, with all transport being confined to guided electric vehicles on a subterranean level.

Autonomous Adaptation

In terms of accidents, there are also behavioural effects that may act as automatic stabilisers to reduce the impact of a changed environment, in some senses a form of autonomous adaptation on the individual level. There are associated behavioural and psychological impacts which, although less prevalent in the literature, have arguably equal importance in accident causation. While these less direct impacts are inherently less suited to empirical investigation, Elvik (2006) presents a typology of mechanisms or 'laws' of vehicle accidents through which they can be discussed. The laws can be summarised as follows:

1. *The universal law of learning – This states that the ability to detect and control traffic hazards increases uniformly as the amount of travel increases. This law implies that accident rate per unit of exposure will decline as the amount of exposure increases.*
2. *The law of rare events – This states that the more rarely a certain risk factor is encountered the larger is its effect on the accident rate.*
3. *The law of complexity – This law states that the more units of information per unit of time a road user must attend to, the higher the probability an error will be made.*
4. *The law of cognitive capacity – The more the cognitive capacity of the driver approaches its limits, the higher the accident rate.*

Caution is required when using the term 'law', as they cannot be considered as true scientific laws, as epistemologically there is a distinction between scientific laws and accidental generalisation, where phenomena are observed a great number of times but do not represent an invariant relationship (Elvik, 2006). However, the majority of observed behaviour under meteorological conditions can be explained through these principles. For example, the first two laws can be used to explain why ice is such a dangerous hazard, why its effect is felt with such severity in marginal climates as the United Kingdom, and why freight drivers should theoretically be more able to deal with this hazard compared to other drivers. The law of rare events would explain why a road icing event has a higher relative impact than other driving conditions, as this condition is encountered infrequently and offers drivers fewer opportunities to learn how to respond. It also explains why a road icing event would have a higher relative accident rate in a marginal climate, as a ground frost temperature is reached less frequently than in colder climates, again reducing exposure. The law of learning suggests that as freight drivers spend more time on the road, their exposure to icing events is increased, hence increasing their ability to overcome this hazard compared with other road users who travel less.

This theory has been demonstrated by an observed sharp drop-off in the relative accident rates on snow or ice covered roads as the exposure to these conditions increases (Brüde & Larsson, 1980). However, one potential problem not addressed in the literature is that of complacency, both individual and institutional, towards more common forms of meteorological risk, such as precipitation. The fact that precipitation is often referred to as a 'chronic' hazard (e.g. Andrey, Mills, & Leahy, 2003), is perhaps a telling indication of the effect its prevalence has on risk perception, in that it becomes something which is lived-with and accepted, despite its overwhelming cumulative impact.

The insights by Elvik (2006) help explain the paradox that the severity of a climate does not necessarily determine the level of risk to transportation. In less marginal northern regions, such as the majority of Scandinavia, certain weather conditions are taken as a given at particular times of the year. For instance, snow can be considered as a near constant in northern Sweden during the winter (Eriksson, 2001), which impacts both how drivers behave in these conditions including adaptation of their vehicles and how authorities design the infrastructure. This is mitigated by drivers being permitted to use studded tyres or tyre chains to counter the effect of reduced friction (Elvik, 1999). Conversely, in the UK it is not always cost effective to climate-proof infrastructure to perform safely in events with very low probabilities, or for private companies or the public to purchase vehicles which are suited to these conditions. It is certainly the case that the majority of freight vehicles used in the UK do not perform as well under severe conditions as those designed for the harsher environments of North America or Scandinavia.

THE ROLE OF SCENARIOS

Traditionally, climate change impact assessments determine the current relationship between weather and a vulnerable sector and then extrapolate these onto future climates (Tol, 1998). However, what this approach cannot tell us is what the freight industry will look like in the future and how this will affect the impacts of climate change.

For example, it cannot tell us the extent to which vehicle braking technology will improve, whether better forecasts will inform drivers of hazardous conditions or whether roads will be designed and maintained in such a way as to reduce the influence of meteorology. Neither can it tell us whether the emission scenarios that promote the most extreme climate change will also be those which promote a much larger freight fleet, or whether a greener future would mean fewer vehicles operating in a climate where the worst impacts of climate change have been avoided. Theoretical studies can predict that empty or half-loaded trailers are more sensitive to high winds, but cannot tell us whether in the future this problem will be overcome by advances in 'pallet systems' that intelligently route vehicles and promote efficient loading. These considerations lie at the cutting edge of climate-impacts research, and have grown in their prominence from a recommended yet often neglected component of CIA (UKCIP, 2001), to an essential step for both the identification of impacts and promotion of stakeholder involvement (e.g. Holman et al., 2008).

Climate change impacts are not simply a function of the magnitude and frequency of meteorological events, but also the vulnerability of the network they are impacting upon. This changing vulnerability has been seen during the past 50 years, with increases in population and commerce driving a large expansion of the transport network, increasing the number of vehicles exposed to meteorological hazards and shifting this exposure between the different modes of transportation. Concurrently there have also been increases in the economic costs involved with each disruptive event. For example, the move towards low-inventory 'just-in-time' delivery (Wu & Olson, 2008) in the logistics sector has increased the risk of loses to the economy during disruptive periods. The continuous replenishment necessitated by this logistical approach has lead to an increased frequency of deliveries. The short shelf-lives of the products carried on these trips may also increase the pressure on carriers to travel during hazardous weather. These factors have, however, been offset to some extent by a reduced sensitivity of the vehicles themselves to hazardous weather with innovations such as anti-lock braking systems (ABS) for road transport (Broughton & Baughan, 2002). These types of considerations must be taken into account in a meaningful climate change impact assessment on the transport sector. It is also clear that an understanding of the underlying reasons for changes in the sector must be attained along with an idea of how these may drive development in the future.

In so doing, there is a need to move yet further away from the fields of climatology and meteorology towards that of futurology, and embrace a truly multidisciplinary approach. In some senses this crossing of disciplines is not new within the climate community. Future socio-economic scenarios postulating the way regions and nations may look and act in the future have often been used in climate change models in order to imagine a range of potential future society types and predict the extent to which each one will emit GHGs, thus providing 'emission scenarios' such as those formulated by the IPCC (IPCC, 1998). These 'futures' are often highly descriptive, detailing how a society may develop and how certain sectors may be favoured due to the underlying political and social values which drive the scenario. More recently, these tools have been co-opted by the climate impacts community to provide a realistic range of future scenarios of how vulnerable future societies may be to climate change (for example, the BESEECH scenarios (UKCIP, 2007)). As transport accounts for over a quarter of global CO_2 emissions (International Energy Agency, 2000), its inclusion in these scenarios is already relatively strong. However, to arrive at a more useful set of futures, it is important to elaborate on these basic scenarios to include a greater level of detail on the potential future nature(s) of Britain's freight sector.

This is a concept that has been adopted across a number of sectors. For example, the vulnerability of coastal communities to cliff erosion and flooding is dependent on the in part the level of sea level rise and the type and extent of building that is undertaken. These are both driven by the underlying social values and governance of the country, including attitudes to economic growth and the environment, as shown by Dawson et al. (2009) which imagines a number of different socio-economic scenarios and how these influence coastal settlement and emissions.

An Application of the Scenario Approach to Freight and Logistics

According to UKCIP (2001), the National Enterprise scenario sees a reliance on manufacturing, with bulkier goods potentially decreasing the efficiency of freight and increasing exposure. The relatively low proportion of public transport carried by rail (7%; UKCIP, 2001) would suggest a similar reluctance of the government to promote the movement of freight by rail, again increasing exposure relative to the other scenarios, especially Local Stewardship and Global Sustainability. Economic growth is concentrated around London and the South East in this scenario, which would potentially expose greater traffic volumes to areas identified as among the highest risk of future winter precipitation-related accidents. Sensitivity can be assumed to be relatively high under this scenario, with little investment in infrastructure. It is also suggested that investment in innovation is low, again reducing the implementation of vehicle technology.

The low population and GDP growth under Local Stewardship, along a shift towards rail and a move to more regionalised economies makes this the scenario with the lowest level road freight transportation and hence lowest exposure. This scenario is the most likely to see the type of supply chain restructuring mentioned by (Böge, 1995). The move to regional production of goods and the narrowing of the geographical sourcing of supplies may mean that less long-distance travel is involved in the production and distribution of goods. In this way it also agrees with Waller's (2006) vision of a move away from increasingly lean production cost abroad and towards local production, based on environmental and social concerns. The fact that the even spread of economic growth is promoted may also mean that the exposure of freight vehicles is also more evenly distributed across the country. The low investment in infrastructure may cause an increase in sensitivity, although this scenario may still promote safety through the predicted high rates of investment in innovation. This scenario will benefit from a much lower increase in winter-precipitation accidents than National Enterprise, although it will see a smaller reduction in summer-precipitation and winter ice accidents.

World Markets and Global Sustainability differ from the previous two scenarios by taking a more globalised economic standpoint. However, they differ on social and political values, especially regarding the environment. Hence World Markets sees a similar reliance of road transportation as the National Enterprise scenario. Global Sustainability has higher economic growth than Local Stewardship and hence has a greater demand for transport. However, there is a trend towards eco-efficiency, with low emission vehicles and a move towards rail. There is also increased miniaturisation, which would again act to reduce exposure by allowing the same number of products to be carried by fewer vehicles. Sensitivity under Global Sustainability and World Markets is likely to fall due to high investment in infrastructure and technology. Both seem likely to promote centralisation of distribution centres, although the precise nature of this is likely to differ. For example, World Markets may see the introduction of road trains, whilst the environmental values of the Global Sustainability scenario would be more likely to promote the integrated use of rail to link distribution centres. World Markets will see increases in winter precipitation-related accidents, but also significant reductions in winter ice and summer precipitation-related accidents. Global Sustainability shows a smaller increase in winter precipitation-related accidents, but also a smaller decrease in winter ice-related accidents.

Longer-Term Developments and Sideswipes

Scenarios are often too conservative and do not include sideswipes: unexpected breaks with the past (Berkhout et al., 1999). It has been suggested that incremental reductions in emissions may not be sufficient and that a 'revolution' in the way transport is conducted needs to happen to achieve the necessary reduction in fossil fuel usage (Gilbert & Perl, 2007). Many forecasting and visioning techniques such as the previously mentioned Foresight (2002) study do not allow for the consideration of low probability high impact developments. Part of the reasoning for this is that the process of constructing scenarios often involves building consensus among a group of experts. In reference to the Delphi methodology Linstone & Turoff (1975) argue that consensus-based techniques often produce conservative views of the future. In suppressing radical views such methods have the effect of reinforcing existing paradigms.

However, thought-provoking insights can be found by consulting individual experts in the industry. For example, the Chartered Institute of Logistics and Transport UK (CILTUK, 2006) elicited future transport visions at an 80-year horizon, extending from its implementation in 2006 to an anchor year of 2086. McKinnon (2006) suggests both positive and negative scenarios for this timeframe. The positive scenario sees the introduction of a market-based approach to network allocation, with electronic auctions being used to determine use. This demonstrates the importance of disruptive developments on levels of freight exposure. The positive scenario also sees the automation of most freight, which would have huge and largely unknown impacts in the relationship between weather and freight accidents. Interestingly both the positive and negative scenarios suggest a point where freight growth will cease, either due to miniaturisation and more efficient freight, or through economic collapse brought about by climate change and restrictions put in place on reactionary environmental grounds.

Interestingly, Sessa et al. (2009) mention that as a result of a Delphi survey, safety was reported as one of the most predictable factors, whilst paradoxically technology is perceived as one of the most difficult. This is probably an issue of detail, as respondents will have lived through an era of improving technology, which they are confident of persisting into the future, without necessarily being able to identify the specific technological improvements through which this increased safety will be achieved. This is similar to Moore's Law in computing (1965), where aspects such as processing speed and memory capacity can be accurately predicted to multiply at a standard rate, without knowing the advancements that will lead to this.

CONCLUSION

There is a requirement for change in freight transport driven by two factors; mitigation of climate change through reduced emissions, and adaptation to the impacts climate change will impose. By examining current accident rates in the United Kingdom it is clear that precipitation is the most frequently recorded weather-type at the scene of accidents, even though this may not always infer a causal link. The impact of climate change shows potential for significant shifts in the temporal distribution of weather-related accidents, and in some cases large-scale decreases across the year. For example, ice-related accidents should become far less widespread by 2080 under most emission scenarios. This is also true for snow-related accidents. However, the most interesting and potentially significant change will be in precipitation-related accidents, which should see a reduction in the summer and an increase in the winter. Changes must be promoted in freight and logistics that allow emissions to reduce and exposure to reduce. Without proactive actions, the overlying socio-economic environment described by Global Sustainability seems to offer the best chance of this happening. The best options for adaptation are those soft options that reduce exposure or move exposure to less vulnerable modes.

REFERENCES

Akerman, J. (2011). The role of high-speed rail in mitigating climate change: The Swedish case Europabanan from a life cycle perspective. *Transportation Research Part D – Transport and Environment*, *16*(3), 208–217.

Aldhous, P., Anderson, A., Coghlan, A., Mullins, J., O'Neill, B., & Spinney, L. (1995). Beneath your feet. *New Scientist*, *146*, 5.

Anderson, S., Browne, M., & Allen, J. (1999). Logistics implications of the UK packaging waste regulations. *International Journal of Logistics: Research and Applications*, *2*(2), 129–145.

Andersson, A. K., & Chapman, L. (2011). The impact of climate change on winter road maintenance and traffic accidents in West Midlands, UK. *Accident Analysis and Prevention*, *43*(1), 284–289.

Andrey, J., Mills, B., & Leahy, M. (2003). Weather as a chronic hazard for road transportation in Canadian cities. *Natural Hazards*, *28*(2-3), 319–343.

Andrey, J., & Yagar, S. (1993). A temporal analysis of rain-related crash risk. *Accident Analysis Prevention*, *25*(4), 465–472.

Baker, C. J. (1993). The behaviours of road vehicles in unsteady cross winds. *Journal of Wind Engineering and Industrial Aerodynamics*, *49*(1-3), 439–448.

Baker, C. J., Hemida, H., Iwnicki, S., Xie, G., & Ongaro, D. (2011). Integration of crosswind forces into train dynamic modelling. *Proceedings of the Institution of Mechanical Engineers Part F – Journal of Rail and Rapid Transit*, *225*(2), 154–164.

Baker, C. J., & Reynolds, S. (1992). Wind induced accidents of road vehicles. *Accident Analysis and Prevention*, *24*(6), 559–575.

Berkhout, F., Hertin, J., Lorenzoni, I., Jordan, A., Turner, K., O'Riordan, T., ... Skea, J. (1999). Non-climate futures study: Socio-economic futures scenarios for climate impact assessment. Final report. SPRU, Brighton, Sussex, UK.

Böge, S. (1995). The well-travelled yogurt pot: Lessons for new freight transport policies and regional production, world transport policies and regional production. *World Transport Policy and Practice*, *1*(1), 7–11.

Broughton, J., & Baughan, C. (2002). The effectiveness of antilock braking systems in reducing accidents in Great Britain. *Accident Analysis and Prevention*, *34*, 347–355.

Browne, M. (2006). Low carb future – Carbon trading and surface transport. *Back to the future? The next 80 years of joined up thinking and joined up journeys*. CILTUK. Retrieved from http://www.ciltuk.org.uk/pages/downloadfile?d=1D695F83-2EEB-4933

Browne, M., & Allen, J. (1997). *Forecasting the Future of Freight Transport and Distribution in Britain*. London: University of Westminster Transport Studies Group.

Brüde, U., & Larsson, J. (1980). Samand vintertid mellan vaderlek-vaglag-traffikolyckor. Statistisk bearbentning och analys. VTI-rapport 210. Statens vag – och trafikinstitut (VTI), Linkoping, in: Elvik, R. 2006. Laws of accident causation. *Accident Analysis and Prevention*, *38*(4), 742–747.

Carson, J., & Mannering, F. (2001). The effect of ice warning signs on ice-accident frequencies and severities. *Accident Analysis and Prevention*, *33*(1), 99–109.

Cavallo, V., Colomb, M., & Dore, J. (2001). Distance perception of vehicle rear lights in fog. *Human Factors*, *43*(3), 442–451.

Cerrelli, E. C. (1998). *Trends in large truck crashes* (NHTSA Technical Report, DOT HS 808 690). National Highway Traffic Safety Administration, Washington, DC.

Chapman, L. (2007). Climate change and transport: A review. *Journal of Transport Geography*, *15*, 354–367.

Chapman, L., & Thornes, J. E. (2006). A geomatics based road surface temperature prediction model. *Science of the Total Environment*, *360*, 68–80.

Chapman, L., Thornes, J. E., & Bradley, A. V. (2001). Modelling of road surface temperature from a geographical parameter database. *Part 1: Statistical, Meteorological Applications*, *8*(4), 409–419.

CILTUK. (2006). *Back to the future? The next 80 years of joined up thinking and joined up journeys*. Retrieved from http://www.ciltuk.org.uk/pages/downloadfile?d=1D695F83-2EEB-4933

Crowley, J. A. (1998). Virtual logistics: Transport in the marketplace. *International Journal of Physical Distribution and Logistics Management*, *28*(7), 547–574.

Dawson, R. J., Dickson, M. E., Nicholls, J. W., Hall, M. J. A., Walkden, P. K., Mokrech, M., … Watkinson, A. R. (2009). Integrated analysis of risks of coastal flooding and cliff erosion under scenarios of long term change. *Climate Change*, *95*, 249–288.

DfT. (2008). Carbon pathways analysis: Informing development of a carbon reduction strategy for the transport sector, 111pp.

DfT. (2009). *Road freight statistics 2008*, 186pp.

DfT STATS19. (2009). *Data access via ESDS access and preservation*. UK Data Archive. Accessed on March 3, 2009.

Dobney, K., Baker, C. J., Chapman, L., & Quinn, A. D. (2010). The future cost to the UK's railway network of heat related delays and buckles caused by the predicted increase in high summer temperatures due to climate change, Proceedings of the Institution of Mechanical Engineers Part F. *Journal of Rail and Rapid Transit*, *224*(1), 25–34.

Duff, A. (2006). Aviation – Are the skies really the limit? *Back to the future? The next 80 years of joined up thinking and joined up journeys*. CILTUK. Retrieved from http://www.ciltuk.org.uk/pages/downloadfile?d=1D695F83-2EEB-4933

Eden, P. (2007). Weather Log January 2007: Wet, wind and exceptionally mild until the 20th; much quieter last ten days. *Weather*, *62*(3), i–iv.

Edwards, J. B. (1994). Wind-related road accidents in England and Wales 1980–1990. *Journal of Wind Engineering and Industrial Aerodynamics*, *52*(1-3), 293–303.

Edwards, J. B. (1996). Weather-related road accidents in England and Wales: A spatial analysis. *Journal of Transport Geography*, *4*(3), 201–212.

Edwards, J. B. (1998). The relationship between road accident severity and recorded weather. *Journal of Safety Research*, *29*(4), 249–262.

Edwards, J. B. (1999). The temporal distribution of road accidents in adverse weather. *Meteorological Applications*, *6*(1), 59–68.

Edwards, J. B., & McKinnon, A. C. (2009). Shopping trip or home delivery: Which has the smaller carbon footprint? *Logistics and Transport Focus*, *11*(7), 20–24. The Chartered Institute of Logistics and Transport (UK), Corby.

Eisenberg, D. (2004). The mixed effects of precipitation on traffic crashes. *Accident Analysis and Prevention*, *36*(4), 637–647.

Eisenberg, D., & Warner, K. E. (2005). Effects of snowfalls on motor vehicle collisions, injuries and fatalities. *American Journal of Public Health*, *95*(1), 120–124.

Elvik, R. (1999). The effects on accidents of studded tires and laws banning their use: A meta-analysis of evaluation studies. *Accident Analysis and Prevention*, 125–134.

Elvik, R. (2006). Laws of accident causation. *Accident Analysis and Prevention, 38*(4), 742–747.

Eriksson, M. (2001). Regional influence on the occurrence of road slipperiness during winter precipitation events. *Meteorological Applications, 8*(4), 449–460.

Fishwick, P. M. (1990). The October 1987 storm: The role of a highways authority. *Weather, 45*, 34–35.

Foresight. (2002). Foresight vehicle technology roadmap: Technology and research directions for future road vehicles, 67pp. London: Department of Trade and Industry.

Forkenbrock, D. J. (2001). Comparisons of external costs of rail and truck freight transportation. *Transportation Research A, 35*(4), 321–337.

Fridstrøm, L. (1999). *Econometric models of road use, accidents, and road investment decisions* (Vol. 2, TOI Report 457). Oslo: Institute of Transport Economics.

Fridstrøm, L., Ifver, J., Ingebrigtsen, S., Kulmala, R., & Krosgsgard Thomsen, L. (1995). Measuring the contribution of randomness, exposure, weather and daylight to the variation in road accident counts. *Accident Analysis and Prevention, 27*(1), 1–20.

Garreau, A., Lieb, R., & Millen, R. (1991). JIT and corporate transport: An international comparison. *Logistics Management, 21*(1), 42–47.

Gilbert, R., & Perl, A. (2007). *Transport revolutions: Moving people and freight without oil.* London: Earthscan.

Graham, E. (1996). Urban heat island of Dublin city during the summer months. *Irish Geography, 26*, 45–57.

Haq, G., & Bolhuis, M. (1998). Dutch transport policy: From rhetoric to reality. *World Transport Policy and Practice, 4*(1), 4–8.

Hermans, E., Brijs, T., Stiers, T., & Offermans, C. (2006). The impact of weather conditions on road safety investigated on an hourly basis. Transportation Research Board Annual Meeting. Hasselt University, Belgium and Ministry of Transport, Netherlands.

Highways Agency. (2007). *Severe weather alert for goods vehicles.* Retrieved from http://www.highways.gov.uk/business/17025.aspx

Hjort, M., & Jansson, J. (2010). Handling of buses on slippery roads during the influence of side wind: A study of the effects of different tyres. *Accident Analysis and Prevention, 42*(3), 972–977.

Holman, I. P., Rounsevell, M. D. A., Cojacaru, G., Shackley, S., McLachlan, C., Audsley, E., … Richards, J. A. (2008). The concepts and development of a participatory regional integrated assessment tool. *Climatic Change, 90*, 5–30.

Howgego, T., & Roe, M. (1998). The use of pipelines for the urban distribution of goods. *Transport Policy, 5*, 61–72.

International Energy Agency (IEA). (2000). *CO_2 emissions from fuel combustions 1971–1998* (2000 Edn). Paris: OECD.

IPCC. (1998). Special report on emission scenarios, intergovernmental panel on climate change, Cambridge University Press, Cambridge.

Jaroszweski, D., Chapman, L., & Petts, J. (2010). Assessing the potential impact climate change on transportation: The need for an interdisciplinary approach. *Journal of Transport Geography, 18*(2), 331–335.

Jones, P. D., Kilsby, C. G., Harpham, C., Glenis, V., & Burton, A. (2009). *UK climate projections science report: Projections of future daily climate for the UK from the Weather Generator.* Newcastle upon Tyne, UK: University of Newcastle.

Kahn, R. S., & Kobayashi, S. (2007). Transport and its infrastructure. Fourth assessment report: Climate change 2007 – Mitigation of climate change, Inter-governmental Panel on Climate Change (IPCC), Geneva.

Knapp, K., Smithson, L. D., & Khattak, A. J. (2000). The mobility and safety of winter storm events in a freeway environment. *Mid-Continent Transportation Research Symposium Proceedings 2000* (pp. 67–71). Center for Transport Research and Education, Iowa State University, Ames.

Knight, I., Newton, W., McKinnon, A., Palmer, A., Barlow, T., McCrae, I., ... Taylor, N. (2008). *Longer and/or longer and heavier goods vehicles (LHVS): A study of the effects if permitted in the UK*. Final report. Transport Research Laboratory, Wokingham.

Koetse, M. J., & Rieveld, P. (2009). The impact of climate change and weather on transport: An overview of empirical findings. *Transportation Research Part D, 14*, 205–221.

Leden, L., Hamalainen, O., & Manninen, E. (1998). The effect of resurfacing on friction, speeds and safety on main roads in Finland. *Accident Analysis and Prevention, 30*(1), 75–85.

Linstone, H., & Turoff, M. (1975). *The Delphi method*. London: Addison-Wessley.

Martin, P. T., Perrin, J., Hansen, B., Quintana, I. (2000). *Inclement weather signal timings* (Research Report MPC01-120). Salt Lake City, UT: Utah Traffic Lab (UTL), University of Utah.

McCallum, E. (1990). The burn's day storm, 25 January 1990. *Weather, 45*, 166–173.

McKinnon, A. C. (1996). The empty running and return loading of road goods vehicles. *Transport Logistics, 1*(1), 1–19.

McKinnon, A. C. (2000). Sustainable distribution: Opportunities to improve vehicle loading. *Industry and Environment, 23*(4), 26–30.

McKinnon, A. C. (2006). Logistics in the future, two opposing scenarios. Back to the future? The next 80 years of joined up thinking and joined up journeys. CILTUK. Retrieved from http://www.ciltuk.org.uk/pages/downloadfile?d=1D695F83-2EEB-4933

McKinnon, A. C. (2007). Decoupling of road freight transport and economic growth trends in the UK: An exploratory analysis. *Transport Reviews, 27*(1), 37–64.

McKinnon, A. C. (2010). Environmental sustainability: A new priority for logistics managers. In A. McKinnon, S. Culinane, A. Whiteing & M. Browne (Eds.), *Green logistics: Improving the environmental sustainability of logistics*. London: Kogan.

McKinnon, A. C., Button, K., & Nijkamp, P. (Eds.). (2002). *Transport logistics*. Cheltenham: Edward Elgar Publishing.

McKinnon, A. C., & Campbell, J. B. (1998). The double decking of road goods vehicles: An assessment of the opportunities and constraints, World Conference on Transport Research, Antwerp.

Moore, G. E. (1965). Cramming more components onto integrated circuits. *Electronics, 38*(8).

Moore, R. L., & Cooper, L. (1972). *Fog and road traffic* (TRRL Report LR 446). Crowthorne, UK: Transport and Road Research Laboratory.

Murphy, J. M., Sexton, D. M. H., Jenkins, G. J., Boorman, P. M., Booth, B. B. B., Brown, C. C., ... Wood, R. A. (2009). UK Climate Projections Science Report: Climate change projections. Exeter: Met Office Hadley Centre.

Musk, L. (1988). The assessment of local fog climatology for new motorway and major road schemes. Proceedings of the IVth International Conference on Weather and Road Safety, Florence (pp. 777–797).

Nofal, F. H., & Saeed, A. A. (1997). Seasonal variation and weather effects on road traffic accidents in Riyadh City. *Public Health, 111*, 51–55.

Norrman, J., Eriksson, M., & Lindqvist. (2000). Relationships between road slipperiness, traffic accident risk and winter road maintenance activity. *Climate Research*, *15*(3), 185–193.

Ogden, K. W. (1984). A framework for urban freight policy analysis. *Transportation Planning and Technology*, *8*(4), 253–265.

Pagowski, M., Gultepe, I., & King, G. (2004). Analysis and modelling of an extremely dense fog event in Southern Ontario. *Journal of Applied Meteorology*, *43*, 3–16.

Penman, I. (1997). Efficient unit loads. *Logistics Focus*, *5*(5), 4–6.

Perry, A. H. (1990, November). *Highwinds on highways*. Paper given to Royal Meteorological Observing Systems, Meteorological data in support of road and rail transport. London: Institute of Mechanical Engineers.

Peterson, T. C., McGuirk, M., Houston, T. G., Horvitz, A. H., & Wehner, M. F. (2008). *Climate variability and change with implications for transportation* (90pp.). Washington, DC: Transportation Research Board.

Piecyk, M., & McKinnon, A. (2008). A survey of expert opinion on the environmental impact of road freight transport in the UK in 2020, Logistics Research Centre, Heriot-Watt University, Edinburgh, 6pp.

Prusa, J. M., Segal, M., Temeyer, B. R., Gallus, W. A., & Takle, E. S. (2002). Conceptual and scaling evaluation of vehicle traffic thermal effects on snow/ice covered roads. *Journal of Applied Meteorology*, *41*, 1225–1240.

Qiu, L., & Nixon, W. A. (2008). Effects of adverse weather on traffic crashes systematic review and meta-analysis. *Transportation Research Record*, (2055), 139–146.

Quarmby, D. A. (1989). Developments in the retail market and their effects of freight distribution. *Journal of Transport Economics and Policy*, *23*(1), 75–87.

Rockel, B., & Woth, K. (2007). Extremes of near-surface wind speed over Europe and their future changes as estimated from an ensemble of RCM simulations. *Climatic Change*, *81*, 267–280.

Royal Meteorological Society. (2010). Beaufort Scale for land areas. Retrieved from http://www.rmets.org/weather/observing/beaufort.php

Saiidi, M., & Maragalas, E. (1995). *Identification of trigger wind velocities to cause vehicle instability*. Carson City: Nevada Department of Transportation.

Samuelsson, A., & Tilanus, B. (1997). A framework efficiency model for goods transportation, with an application to regional less-than-truckload distribution. *Transport Logistics*, *1*(2), 139–151.

Sessa, C., Anderson, P. B., Enei, R., Fiedler, R., Fischer, D., Larrea, E., Timms, P.M., & Ulied, A. (2009). *Report on transport scenarios with a 20 and 40 year horizon* (Final Report), Funded by DG TREN, Rome, Italy.

Shine, K. P. (1987). News: October storm. *Weather*, *42*(11), 394.

Sokolovskij, E. (2007). Automobile braking and traction characteristics on the different road surfaces. *Transport*, *22*(4), 275–278.

Sorrell, S., Lehtonen, M., & Stapleton, L. (2009). Decomposing road freight energy use in the United Kingdom. *Energy Policy*, *37*(8), 3115–3129.

Stern, E., & Zehavi, Y. (1990). Road safety and hot weather: A study in applied transport geography. *Transactions of the Institute of British Geographers*, *15*(1), 102–111.

Summerfield, S., & Kosior, J. (2001). *Simulation of heavy trucks in inclement weather*. Canadian Transportation Research Forum, 209-15 Innovation Boulevard, Saskatoon, Saskatchewan Canada.

Taniguchi, E., & van der Heijden, R. E. C. M. (2000). An evaluation methodology for city logistics. *Transport Reviews*, *20*(1), 65–90.

Thornes, J. E. (1992). The impact of weather and climate on transport in the UK. *Progress in Physical Geography*, *16*(2), 187–208.

Thornes, J. E., & Stephenson, D. B. (2001). How to judge the quality and value of weather forecast products. *Meteorological Applications*, *8*(3), 307–314.

Tol, R. S. J. (1998). Socio-economic scenarios. In J. F. Feenstra, I. Burton, J. B. Smith & R. S. J. Tol (Eds.), *Handbook on methods for climate change impact assessment and adaptation strategies* (Chapter 2). New York: United Nations Environment Programme.

United Kingdom Climate Impacts Programme. (2001). Socio-economic scenarios for climate change impact assessment: A guide to their use in the UK Climate Impacts Programme, UKCIP, Oxford.

United Kingdom Climate Impacts Programme. (2007). Building knowledge for a changing climate, collaborative research to understand and adapt to the impacts of climate change and infrastructure, The Built Environment and Utilities, UKCIP, Oxford.

Vaiskunaite, R., Laurinavicius, A., & Miskinis, D. (2009). Analysis and evaluation of the effect of studied tyres on road pavement and the environment (II). *Baltic Journal of Road and Bridge Engineering*, *4*(4), 203–211.

Waller, A. (2006). Apocalypse now but there is hope for supply-chains of the future. *Back to the future? The next 80 years of joined up thinking and joined up journeys*. CILTUK. Retrieved from http://www.ciltuk.org.uk/pages/downloadfile?d=1D695F83-2EEB-4933

Wanvik, P. O. (2009). Effects of road lighting: An analysis based on Dutch accident statistics 1987-2006. *Accident Analysis and Prevention*, *41*(1), 123–128.

Wu, D. S., & Olson, D. I. (2008). Supply chain risk, simulation and vendor selection. *International Journal of Production Economics*, *114*, 646–655.

Young, R. K., & Liesman, J. (2007). Estimating the relationship between measured wind speed and overturning crashes using a binary logit model. *Accident Analysis and Prevention*, *39*, 574–580.

CHAPTER 7

THE ROLE OF WALKING AND CYCLING IN REDUCING THE IMPACTS OF CLIMATE CHANGE

Colin G. Pooley, Dave Horton, Griet Scheldeman, Miles Tight, Helen Harwatt, Ann Jopson, Tim Jones, Alison Chisholm and Caroline Mullen

ABSTRACT

Purpose – *To examine the potential for switching short trips in urban areas from cars to walking and cycling, and the possible contribution, this could make to a reduction in transport-related greenhouse gas emissions.*

Methods – *Case studies in four urban areas combining a questionnaire survey, interviews with households and during journeys and in-depth ethnographies of everyday travel.*

Findings – *The barriers to an increase in walking and cycling in British urban areas are emphasised. It demonstrates that motivations for walking and cycling are mostly personal (health and local environment) and that the complexities and contingencies of everyday travel for many households, combined with inadequate infrastructure, safety concerns and the fact that walking and cycling are seen by many as abnormal modes of travel, mean that increasing rates of walking and cycling will be hard. Given that the contribution of trips less than 2 miles to transport-related*

Transport and Climate Change
Transport and Sustainability, Volume 2, 175–195

ISSN: 2044-9941/doi:10.1108/S2044-9941(2012)0000002010

greenhouse gas emissions is relatively small, it is argued that any gains from increased walking and cycling would mostly accrue to personal health and the local environment rather than to the UK's carbon reduction target.

Social implications – *Positive attitudes towards walking and cycling are motivated mainly by personal concerns rather than global environmental issues.*

Originality – *Use of detailed ethnographic material in policy-related transport research.*

Keywords: Walking; cycling; greenhouse gas emissions; ethnographies; complexity

INTRODUCTION

This chapter argues that whilst there are many benefits to be derived from increased walking and cycling, including improved personal health, better environmental quality and reduced congestion, the reductions in greenhouse gas emissions that would accrue would make only a small contribution to the UK's carbon reduction target. Furthermore, it is argued that this is due not only to the nature of the trips themselves (and their existing contribution to greenhouse gas emissions) but also to the difficulty of effecting a large-scale switch from cars to more sustainable forms of transport.

CONTEXT

In Britain, walking and cycling have been increasingly promoted as healthy and environmental-friendly means of everyday travel. This can be seen not only in the material produced by organisations explicitly committed to promoting sustainable travel (e.g. Sustrans, Living Streets, Walk 21 and Cycling England) but also in policy documents and advice for local authorities produced by the national government. In 2004, the Department for Transport produced 'Walking and Cycling: An Action Plan' (DfT, 2004) and this has been followed up in a range of publications. The message in all these documents is similar. For instance, the 2009 publication 'Delivering Sustainable Low Carbon Travel: An Essential Guide for Local Authorities' states that walking and cycling are 'the most sustainable forms of local transport', and that the 'potential for greater take up is enormous'. Positive benefits from increased walking and cycling are listed as follows: reduced congestion, improved health, low-cost and reduced air pollution and carbon emissions (DfT, 2009). A similar message is included in the DfT's (2007) advice to 'Act on CO_2' (http://actonco2.direct.gov.uk/home.html, accessed on 10 December 2011) where walking and cycling are promoted as a key strategy to (amongst other benefits) reduce carbon emissions. Current government strategy (DfT, 2011) is focused mainly on strategies to encourage more sustainable travel (including walking and cycling) rather than on direct intervention and regulation, though a recent House of Lords Science and Technology Select Committee report (2011) has raised questions about the likely success of such policies.

At the same time, there has been increased academic interest in the study of walking and cycling, mainly from the perspective of health benefits, safety and the ways in which urban design can encourage more sustainable travel (Cevero, Sarmiento, Jacoby, Gomez, & Neiman, 2009; Elvik, 2009; Hume et al., 2009; Jacobsen, 2003; Jacobsen, Racioppi, & Rutter, 2009; Lorenc, Brunton, Oliver, Oliver, & Oakley, 2008; McMillan, 2007; Ogilvie, Egan, Hamilton, & Petticrew, 2004; Pikora, Giles-Cirti, Bull, Jamrozik, & Donovan, 2003; Pucher & Dijkastra, 2003; Pucher, Dill, & Handy, 2010; Saelens, Sallis, & Frank, 2003; Timperio, Crawford, Telford, & Salmon, 2003). Such studies emphasise the importance of making walking and cycling an easy choice through good urban design (footpaths, cycle routes etc.), restrictions on car use and land-use planning that enables services to be accessed without motorised transport. In particular, the relatively high rates of walking and cycling in cities such as Copenhagen and Amsterdam are contrasted with the very low rates in Britain and North America. For instance, Pucher and Buehler (2008, p. 495) argue that

> The key to achieving high levels of cycling appears to be the provision of separate cycling facilities along heavily travelled roads and at intersections, combined with traffic calming of most residential neighbourhoods. Extensive cycling rights of way in the Netherlands, Denmark and Germany are complemented by ample bike parking, full integration with public transport, comprehensive traffic education and training of both cyclists and motorists, and a wide range of promotional events intended to generate enthusiasm and wide public support for cycling. In addition to their many pro-bike policies and programmes, the Netherlands, Denmark and Germany make driving expensive as well as inconvenient in central cities through a host of taxes and restrictions on car ownership, use and parking. Moreover, strict land-use policies foster compact, mixed-use developments that generate shorter and thus more bikeable trips. It is the coordinated

implementation of this multi-faceted, mutually reinforcing set of policies that best explains the success of these three countries in promoting cycling.

A smaller number of studies directly relate the promotion of walking and cycling to carbon reduction and climate change. These mostly make the (quite reasonable) assumption that increased walking and cycling will contribute to a reduction in carbon emissions and then focus on strategies for and barriers to increasing walking and cycling (Chapman, 2007; Maibach, Steg, & Anable, 2009). Higgins and Higgins (2005) are unusual as (in a US context) they attempt to estimate the oil and carbon emissions saved by increased walking and cycling (together with other forms of sustainable transport). Their study attempts to calculate the reductions in carbon emissions and gasoline use that would follow if levels of walking and cycling in the USA were increased to meet recommended daily levels of walking and cycling and, in consequence, eliminate obesity from the adult population. They argue that 'Assuming substitution of cycling for driving, the reduction in gasoline demand is equivalent to 34.9% of current domestic oil consumption. This constitutes considerably more oil than is recoverable from the Arctic National Wildlife Refuge. The concomitant reduction in US CO_2 emissions would constitute approximately 10.9% relative to 1990 net US emissions and would be a substantial step toward satisfying the Kyoto Protocol' (Higgins & Higgins, 2005, p. 1). However, they deal only briefly with the barriers to achieving such increases whilst recognising that 'Willingness to walk or cycle for transportation is rare and large obstacles such as weather, logistics, poor health, disability and time constraints must all be overcome. At the same time, the unequal distribution of distances driven within the population and on a daily basis reduce the capacity to substitute all walking or cycling for car travel' (Higgins & Higgins, 2005, p. 4).

Recent papers have developed a scenario-modelling approach to the energy and carbon implications of different UK transport futures (Anable, Brand, Tran, & Eyre, 2010; Brand, Tran, & Anable, 2010). Using sophisticated modelling techniques, they argue for an integrated approach to transport modelling which combines assumptions of both lifestyle and technological change and conclude that 'lifestyle change can make a significant contribution to delivering UK carbon emission goals, and assist early action, but that alone it is insufficient to deliver an 80% reduction goal, as this requires a wider transformation of the energy system' (Anable et al., 2010, p. 12). This is based on a model of change that sees energy demands in the transport sector halved by 2050, and with lifestyle changes reducing the need for rapid electrification of the motor transport system. They then argue that there is need for UK energy policy to pay much more attention to demand-side changes rather than to technological developments in supply, though the modelling approach adopted in these studies does not go on to explore the difficulties of changing everyday travel practices amongst the general population.

This chapter attempts to test some of the above assumptions by examining, through a series of case studies, the extent to which environmental factors are significant in encouraging people to walk or cycle, and the household and other barriers to increased walking and cycling. It then discusses the potential for increased walking and cycling to make a significant contribution to a reduction in the UK's carbon footprint given the resistance to changes in everyday travel expressed by respondents in our surveys. Whilst agreeing that changes in demand are highly desirable, it concludes that achieving the level of change assumed in the models discussed above will be difficult. It does not attempt to measure precise carbon reduction scenarios as these are dependent on a range of factors well beyond the scope of the research project on which this chapter is based.

SOURCES AND METHODS

The research effort was focused on four English case study towns (Leeds, Leicester, Worcester and Lancaster) chosen to reflect a range of urban environments with average levels of walking and cycling, and different existing interventions to promote sustainable travel (for a full analysis of the Lancaster data, see Pooley et al., 2011). For instance, Worcester has been designated a Sustainable Travel Town and Lancaster a Cycle Demonstration Town, whilst Leicester, with no official designation, has pursued markedly green transport and environmental policies through local initiatives. In contrast, interventions in Leeds have been minimal. In each town four main research strategies were adopted. Firstly, a large-scale questionnaire on attitudes to and experiences of walking and cycling was carried out in a stratified random sample of households across the four case study towns (see below); secondly, analysis of the spatial structure of the built environment was completed to examine the permeability of the urban structure with respect to walking and cycling (not reported here); thirdly, a series of household and 'go-along' interviews were conducted with respondents selected from across the four towns; and, fourthly, in-depth ethnographic research was conducted with approximately five households in specific neighbourhoods of each urban area. Data drawn from the interviews and ethnographies are used to develop the argument below where more information is given about the precise methods employed. Quotes cited are selected to reflect a range of views identified after extensive reading of 262 separate transcripts covering over 1.5 million words of text.

ENVIRONMENTAL ATTITUDES

There is an extensive literature on public attitudes to environmental issues, and this has been reviewed and critiqued extensively in Anable (2006). In the report to the UK Department for Transport, Anable examines literature on public attitudes to climate change and towards the links between transport and climate change; the various models of behaviour change are critiqued and avenues for future research are suggested (see also Shove, 2010 and House of Lords, 2011). This chapter does not explore environmental attitudes in detail, but uses a series of questions (embedded in a much larger survey of walking and cycling) to provide evidence on attitudes to walking and cycling in the four case study cities. This is used as context for the in-depth qualitative evidence that follows.

The influence of environmental attitudes on walking and cycling behaviours was measured using a questionnaire survey carried out in 2009. Some 15,000 survey schedules were distributed in four study towns (Leeds, Leicester, Worcester, Lancaster) with the sample stratified according to the index of multiple deprivation for lower super output areas (LSOA) in the four urban areas. The aim of the survey was to examine both practices of, and attitudes towards, walking and cycling in a representative population sample with questions designed to be analysed within the framework of the Theory of Planned Behaviour. Separate schedules were used for walking and cycling and a response was sought from one adult (age 18+) in each household. Only questions relating to environmental attitudes with regard to walking and cycling are analysed in this chapter. In total 1,417 responses were received, giving a response rate of 8.3% from the cycling survey and 9.3% from the walking survey. Although low, this level of response is not unusual in unsolicited postal surveys (Moser & Kalton, 1979). The sample also produced some familiar biases. Responses were highest for females, adults in the middle and older age ranges and from those with more education and higher status occupations. Non-car owning households were also substantially under-represented in the survey.

Respondents were asked to rank their level of agreement with a series of statements on a scale of 1–5 (where 1 = strong agreement, 3 = neutral and 5 = strong disagreement). Tables 1 and 2 summarise the mean scores for selected statements for the four study towns. In this chapter, the focus is on statements that relate to people's engagement with the local environment, their awareness of transport's contribution to environmental pollution and climate change and personal health and well-being. Responses are remarkably similar for all four study areas suggesting that attitudes vary little across the country and between large cities (such as Leeds) and smaller country towns (such as Worcester and Lancaster). In general, most scores were close to neutral with a mild agreement with most statements. Statements that respondents associated most clearly with were that walking and cycling would benefit health and reduce contribution to local air pollution. Respondents to the walking questionnaire quite strongly associated walking with enjoyment but this was less strong for cyclists. In general, most cyclists were less positive than walkers about the enjoyment and well-being associated with the activity, though cyclists and walkers

Table 1. Environmental Attitudes Towards Walking (Mean Scores).

Statement	Leicester	Lancaster	Leeds	Worcester
If I make or were to make journeys on foot, I would find walking enjoyable.	1.9	1.9	1.8	1.9
If I make or were to make journeys on foot, I would get a sense of freedom.	2.0	2.1	2.1	2.0
If I make or were to make journeys on foot, I would feel part of my community.	2.7	2.7	2.7	2.7
If I make or were to make journeys on foot, I would find it relaxing.	2.1	2.1	2.0	2.0
If I make or were to make journeys on foot, it would benefit my health.	1.5	1.5	1.5	1.5
If I make or were to make journeys on foot, it would mean I contribute less to climate change.	2.1	2.1	2.1	2.0
If I make or were to make journeys on foot, it would increase my exposure to air pollution.	3.0	3.1	3.0	3.0
If I make or were to make journeys on foot, it would contribute less to local air pollution.	1.9	2.1	2.0	1.9

Note: 1: strongly agree; 3: neutral; 5: strongly disagree.
Source: Questionnaire survey (2009).

Table 2. Environmental Attitudes Towards Cycling (Mean Scores).

Statement	Leicester	Lancaster	Leeds	Worcester
If I make or were to make journeys by bicycle, I would find cycling enjoyable.	2.3	2.3	2.5	2.3
If I make or were to make journeys by bicycle, I would get a sense of freedom.	2.3	2.3	2.4	2.2
If I make or were to make journeys by bicycle, I would feel part of my community.	3.1	3.1	3.2	3.0
If I make or were to make journeys by bicycle, I would find it relaxing.	2.6	2.5	2.7	2.6
If I make or were to make journeys by bicycle, it would benefit my health.	1.6	1.6	1.6	1.7
If I make or were to make journeys by bicycle, it would mean I contribute less to climate change.	2.0	2.1	2.0	2.0
If I make or were to make journeys by bicycle, it would increase my exposure to air pollution.	2.7	2.9	2.6	2.7
If I make or were to make journeys by bicycle, it would contribute less to local air pollution.	1.8	1.9	1.9	1.9

Note: 1: strongly agree; 3: neutral; 5: strongly disagree.
Source: Questionnaire survey (2009).

equally recognised the association with health, reduced air pollution and lower carbon emissions. It can be suggested from these data that the main 'environmental' motivations for walking and cycling relate to personal health and concern about local air pollution. Whilst there is some awareness of broader factors, these are more muted.

These results reinforce the findings of other studies on public attitudes to environmental issues (Anable, 2006) and have implications for policies that seek to encourage sustainable transport to mitigate the impacts of transport-related greenhouse gas emissions on climate change. It can be suggested that reductions in levels of car use may be achieved most readily by stressing those benefits that would be felt most personally by individuals. Improvements in health together with a more attractive local environment resonate with respondents in a way that global concerns about climate change do not. Such attitudes are not surprising, but suggest that there needs to be effective integration of policies in health promotion, local environmental improvement and carbon reduction to maximise likely effects. It also suggests that policies to encourage more sustainable travel to reduce carbon emissions may meet with limited success.

HOUSEHOLD CONSTRAINTS AND OPPORTUNITIES

Household constraints on using walking and cycling for everyday journeys in the four urban areas were investigated using a range of in-depth qualitative research techniques. In each town, approximately 20 interviews and an in-depth ethnographic study of at least 5 households were undertaken. This gave a total sample size of approximately 80 interviews and 20 ethnographies. Interviews were divided equally between household-based interviews (often with more than one person present) and accompanied journeys or 'go-alongs'. Respondents were selected from the original questionnaire survey and reflect a range of characteristics. The go-along is now a reasonably well-established research tool in mobility research (Carpiano, 2009; Kusenbach, 2003; Ricketts, Evans, & Jones, 2008) and such interviews can provide real insights into the experience of travelling and the everyday practical barriers to movement that are often neglected in a static interview. Conducting interviews whilst cycling presents particular logistical problems (Fincham, McGuiness, & Murray, 2009) and in this study data collection was adjusted to suit the chosen route and prevailing traffic conditions. Where it was not possible to talk whilst cycling, data were collected via in-depth interviews both before and after the ride, and through a commentary by the researcher. The aim of the investigation was to probe below the (inevitably) superficial responses that are gained though one-off interviews (either static or mobile) and to gain a deeper understanding of how everyday travel decisions are made in a household setting. The ethnographic investigations utilised a number of different techniques which were deployed in ways that suited individual households. These included repeated in-depth interviews, accompanied journeys, mobility inventories within the household (e.g. storage of bicycles, ownership of suitable outdoor clothes and footwear), mapping exercises, travel diaries and observations. Although not constituting full ethnographies (this would have required a much longer period of interaction with each household and would almost certainly have been deemed too intrusive by respondents), a range of ethnographic methods that have been found to be effective in other studies were utilised (Descartes, Kottak, & Kelly, 2007; Silverstone, Hirsch, & Morley, 1991; Wallman, 1984). Ethnographic respondents were not meant to be statistically representative but were selected to reflect a range of household types clustered within a single neighbourhood.

Many studies have sought to identify the reasons why people do not walk or cycle more frequently. These can be grouped broadly into three groups: issues relating to infrastructure and the physical environment, issues relating to perceptions of risk and personal safety and issues relating to family and household constraints (Anable, 2005; Ogilvie et al., 2007; Mackett, 2003). It is argued that what is especially important is the complex ways in which many of these factors interact to create barriers to walking and cycling in particular circumstances that may make it difficult for even families and individuals committed to sustainable travel to commit fully to walking and cycling for short trips in urban areas. In particular, it is posited that it is the normality of car use within the British urban environment that makes it difficult for many households – even those with a desire to travel more sustainably – to change their travel behaviour on a regular basis. Car use, even for short trips, is habitual and expected, irrespective of other factors. The detailed ethnographies and accompanied interviews allow the relationships between such factors to be explored in detail. Selected examples are used in this chapter, focusing especially on household and family constraints.

Walking and cycling, especially with a young family, are activities that require substantial levels of organisation and commitment. They require quick and easy access to the right equipment (bikes, helmets, outdoor clothes etc.) and the ability to organise all family members to leave in sufficient time to walk or cycle for an everyday journey. For many it is perceived to be easier to jump in a car and drive even over short distances. One male participant emphasised this point with respect to access to a bike: if it requires a substantial amount of time and effort to get a bike out from a back yard or shed, then the chances are that it will be rarely used, yet many households do not have the space (or have not thought about creating the space) to allow cycles to be stored in a convenient location.

> Yes. One of the important things about bikes is having ready access to them I find. I've just been fixing up a bike for a friend and I said you have to make it somewhere where you can get at it quickly otherwise you won't use it. It has to be somewhere where a couple of seconds and it's ready rather than having to go in the shed and have to do it and have to do this and have to do that, so it's there. (Male participant, cycling go-along, Lancaster)

In a similar vein another male participant commented on the lack of secure cycle parking in the town centre which thus acted as a disincentive for people to take their bikes into the city centre for shopping:

> I think that if W town centre, where the shops are and there's a nice pedestrianised area now but if that also had a secure undercover bike locked area, excuse me, that would be very helpful indeed and that would encourage far more people to do likewise. (Male participant, cycling go-along, Leicester)

This participant also went on to comment that parking for cars near the shops was free and plentiful, thus emphasising the degree to which cities are structured to accommodate motorised transport which in turn makes it harder for individuals to adopt more sustainable forms of mobility.

Although cities are planned for cars, much less attention is usually paid to planning cities for people to walk or cycle; but as a female participant succinctly put it, at an individual level even walking needs planning: there is 'need for a fair degree of planning for walking, especially if you are intending to carry something' (female participant, walking go-along, Leeds). What is interesting is that this respondent was a regular walker, she frequently walked distances of 2 miles to the shops and for other purposes and was quite happy to admit that others might see her travel behaviour as abnormal (field notes, Leeds). This female participant was single, with adult or teenage children, and has a reasonable degree of freedom to travel as she wishes. Levels of planning required are much greater with young children. For instance, walking to and from school with young children requires a substantial amount of organisation to get them all ready at the same time with the right clothes and shoes. Such organisation can, for some, be an enjoyable part of everyday travel, but for many families it creates barriers to walking or, at least, breaks down when time is short or other factors intervene. There are many quotes from respondents emphasising the ways in which the presence of young children makes travelling on foot a challenge for many households. These are summarised in Box 1.

Some respondents had thought very carefully about how they organised their travel arrangements to enable them to travel sustainably under almost any conditions. But this required very careful planning and a strong commitment. Thus, one male respondent's family had a carefully planned routine, with necessary clothes and equipment to hand, but even so weather conditions or other commitments might disrupt such intentions and lead to car use, and one female participant also considered herself well organised so that she could cycle to work on a regular basis.

> J goes to work in Morecambe and endeavours to go on her bike or on the train whenever possible. And that's often down to weather, or whether there are any jobs to do on the way back or places to go where public transport and the like is not possible ... We both try and walk, J cycles whenever we possibly can, I obviously walk and use public transport, then this again would apply to both of us and the boys of course as well. This is if you are walking or using public transport we need to be equipped so I have set up waterproofs, coat, trousers, waterproof trousers, hats of various varieties depending on cold, sun, rain; shoes. (Male participant, Mobility Inventory, Lancaster)

Box. 1. Family and Household Constraints on Walking and Cycling.

We'd like to [cycle more] but when children are smaller it's actually, there's very few places where they can safely cycle, in terms of roads ... [Cycling as a family] ... that's very difficult while my youngest is still on stabilizers and not confident. I don't want her to put her on a road with much traffic. (male participant, Leicester).

Children influence walking routes (both through wanting (insisting) to go a particular way and through parents wanting to take them a nicer/safer/less polluted way). (female participant, Leeds).

As the kids got older they liked it [walking] less and less. (male participant, Leicester).

[When you have children] *You don't have any sleep and you just can't do it* [3/4 of an hour journey each way]. *You can't get up at half six every day and go to work.* (female participant, Leicester).

With the demands of family and work and everything there's not much time or energy [for walking and cycling]. (male participant, Worcester).

I think time between activities makes a difference, even though it may be a short journey, we probably wouldn't walk or cycle because we've only got so much time to get from one place to another. (male participant, Worcester).

> Weather doesn't really bother me when cycling, I've got waterproofs. There are changing facilities and showers at work. I have a locker at work and there are good cycle parking facilities. (Female participant, Worcester)

In general, young children were perceived to make walking (and cycling) more difficult either because they were too tired, or the journey was too complex or it was just too much bother to try to walk or cycle. The following two quotes are typical:

> If (3 year old child) has only just woken up from his afternoon nap, he can't walk straightaway, it takes a while for him to 'come around', you know? So then I'd drive. If he's really tired, too, I sometimes drive. Perhaps I should expect him to walk, but it's hard isn't it? ... And then if we're going to swimming class at the University, obviously we're in the car for that. (Reconstructed quote from ethnographic field notes, Lancaster)

> Oh we have a daughter, she's six and she goes to D school so yeah she goes to D so the only sort of journeys that we need to do are err taking Isabel to school and we do drive

> (coughs) mainly, because it's, well you've seen it's a long way. So it's quite far ... yeah. Sometimes in summer, we'll walk with her on the way home, on the way to school it's always a rush. But on the way home we'll, we'll maybe walk into the school and then, um, walk back. (husband) sometimes um has cycled with her. (Female participant, Lancaster)

However, not all families are deterred by such factors, and some find that once they find ways to deal with the inherent complexities of the journey, then they obtain real pleasure from the time spent walking with their children. The comments of the following female participant are fairly typical of such respondents:

> Well, I've always enjoyed walking anyway but I think it's basically just the fact that I feel much more refreshed when I get to work when I've walked and don't feel as sluggish, you know, you just feel like it kind of wakes you up on the way. Also the fact that when you've got the kids, you know, you've got the chance to talk to them on the way to school and that type of thing it's er, I mean you know, from here going to Lancaster, going to where I work in town I would have to go all the way round the one way system, it would probably take me just as long and then you'd get there feeling really stressed and harassed. (Female participant, Lancaster)

OTHER FACTORS INFLUENCING WALKING AND CYCLING

Respondents identified many other barriers to walking and cycling in addition to household constraints. These included poor cycling infrastructure, obstructed or uneven pavements, traffic speed and volume, safety concerns, topography and the weather. Boxes 2 and 3 give a selection of quotes highlighting the range of concerns that were expressed by respondents. Box 2 highlights selected general and infrastructural issues, whilst Box 3 focuses more specifically on safety concerns. The full implications of the views expressed in these quotes cannot be explored in this chapter, but they give an indication of the range of concerns expressed by respondents. Whilst some barriers to walking and cycling (such as weather and topography) are unavoidable, many others can be tackled through infrastructural interventions and regulation (and increasingly are being improved, especially in towns which have recognised the need to improve infrastructure through specific interventions and through the efforts of organisations such as Sustrans and their Connect2 project (http://www.sustrans.org.uk/what-we-do/connect2). However, despite such interventions, the fact remains that for most people road conditions in British towns and cities are currently a major deterrent to on-road cycling for all

Box. 2. Views on Cycling and Walking in Four Case-Study Towns.

If there were cycle lanes … I certainly would cycle more (male participant, Worcester)

Cycle paths are a joke as most cyclists know cycle paths I think in some ways increase rather than decrease the danger to cyclists by giving the appearance of safety so yeah. … [we need] *designated cycle routes, that are separate to pedestrian paths as well because pedestrians and cyclists don't get on very well* (male participant, Leeds)

Cycling [to work] wouldn't be very practical that's the main thing. Because we've not got any showers there and you don't want to arrive at work all hot and sweaty. (male participant, Leicester).

As a cyclist you're not really recognised as a road user but you can't use the footpaths so its bit frustrating. (male participant, Leicester).

If I want to stop at the shops on my way home from work, sometimes it's quite difficult because there isn't sufficient places to safely lock the bike … I'm not comfortable locking my bike … So most places you can't stop. … When there's a lot of lane changes if you've got to get across a couple of lanes then it's pretty hard … Sometimes I cannot get across the road, it's impossible. There are not enough breaks in the traffic for me to actually get across three lanes of traffic. So then I have to go on to the pavement and cross the pedestrian crossing to get back across the road. (male participant, Leicester).

I don't get scared cycling on the road I just find it easier on a pavement … On the road you have to watch out for cars. I can't ride with one hand so even if I want to signal I have to stop the bike and then put my hand out so it's better if we just like ride on the pavement I think. (female participant, Leicester).

I find cycle lanes put you on the pavement and then you have to come out again into the traffic and it's usually slower. I don't see why I should be marginalized, I can see why people like them but I'm comfortable on the road so I prefer to do it where possible. (male participant, Leeds).

The streets that are busy and haven't got bus lanes put me off cycling, so I would tend to choose routes that avoid those sort of roads unless its early in the morning or later at night. (male participant, Leicester).

[Attention should be paid to] *the absolutely disgusting state of our pavement slabs.* (female participant, Leicester).

Where the paving stones are very uneven and slippery, I don't like the unevenness so I don't go that way. (female participant, Lancaster).

> *People do just walk off the pavement and onto the road because the pedestrian crossings are not where they want them to be.* (male participant, Leeds).
>
> *That road is awful, the pavement is very narrow, and in autumn it's covered in leaves so you slip over half the time, it's terrifying. But by car the road is fine.* (female participant, Lancaster).
>
> *I don't like walking on busy roads. I hate the big lorries. It's the big vehicles that frighten me whether I'm on foot or on the bike.* (female participant, Leeds).

but the most committed cyclists. By far the most frequent request from respondents was for the provision of fully segregated cycle paths, separate from both motor vehicles and pedestrians. Although most urban areas do have pavements, pavement condition was also a disincentive to walking together with the current urban structure that meant that many everyday trips required a journey that was longer than reasonable walking distance. It could thus be argued that the greatest barriers to increased walking and cycling are the combined circumstances of contingency and complexity focused on particular household situations which require substantial organisation and commitment to overcome, together with the fact that walking or cycling requires individuals and families to go against the perceived norms of everyday travel in most British urban areas, and the physical barriers and concerns about safety produced by inadequate walking and cycling infrastructure. Such factors cannot readily be addressed solely through interventions in the built environment or through cycle training. They require much more fundamental adjustments to everyday family routines that can only be generated by individuals themselves, together with changes in the attitudes towards sustainable travel within wider society. Whilst some families do recognise and tackle such issues effectively, others are, not surprisingly, deterred.

CONCLUSIONS

Research reported in this chapter does not attempt to quantitatively model potential reductions in carbon emissions from increased walking and cycling. Indeed, it can be argued that the factors involved are so complex that any figures produced could be misleading. Instead, a range of factors that

Box. 3. Safety Issues When Walking and Cycling.

I find cycling round here incredibly dangerous. (male participant, Leeds).

There's some places I wouldn't go on my own. And there are some places I perhaps would go if I was in a car. (male participant, Leicester).

I wouldn't dream of cycling on the road without a helmet. I'd be too terrified of obviously falling off and hitting my head. (male participant, Leicester).

I feel that the cyclists who are safest are the ones who make themselves big and 'I'm here and I'm keeping up with the traffic and I know what I'm doing' and you make yourself obvious. (female participant, Lancaster).

I feel safer going through that street where there's a lot more people around, rather than that road where you've got the cars but you don't really have many people walking it. (male participant, Leeds).

I don't feel I have to be aware of anything at all when I am walking around. I pretty much know everyone in the area. It's just being familiar with faces, we're not a close knit community but I think people would look out for each other if there was a problem. (female participants, Leicester).

It's quite scary on the ring road on a bike – people come down at 60,70,80 mph. (female participant, Leeds).

I am not comfortable at all with cycling. I am always scared of the traffic around me. (female participant, Leicester).

I wouldn't tend to go walking at night generally. (female participant, Worcester).

People appear to become or are becoming so aggressive when they're on the street, on the road, in a vehicle they become angry almost when they get behind the wheel because they know or they perceive that something's going to stop them going home, getting there. (male participant, Worcester).

Walking through alleyways can be very claustrophobic and you feel a bit scared so I tend to walk the long way round on the road which adds journey time, but it's worth it. (female participant, Worcester).

Cycling, from a woman's point of view, is much safer than walking home. (female participant, Worcester).

influence potential gains are examined, and the potential impacts of current policies to increase walking and cycling for short everyday trips are discussed. It is obvious that any increase in walking and cycling brings benefits, not just in terms of reducing carbon emissions but also in relation to personal health and reduced congestion. It can be argued that the extent to which increases in walking and cycling could make a substantial impact on UK carbon emissions is limited. This is due to the inherent difficulties in increasing walking and cycling to a level that would have a significant impact on total carbon emissions (as demonstrated above) and to the relatively small proportion of total transport-related carbon emissions generated by trips that could reasonably be transferred to walking and cycling (see below).

Even though road transport makes a substantial contribution to the total UK greenhouse emissions (European Commission, 2009), the potential for walking and cycling to reduce this total is limited. The mean trip length for travel on foot in the UK, as recorded in the National Travel Survey, is just 0.7 miles, whilst the mean trip length for bicycle trips is 2.9 miles, with 68% of all cycle trips under 2 miles (DfT, 2010). Clearly there will be variations around this mean, with some individuals prepared to walk or cycle over much longer distances, but it seems reasonable to assume that most people switching travel to more sustainable modes will, at least initially, make journeys at the lower end of the distance distribution. It should also be noted that the NTS means relate to all trips on foot or by bike, and that walking or cycling for leisure purposes is likely to generate longer trips than utility travel (which is the focus of this chapter). It thus seems reasonable to suggest that the greatest opportunities for switching from car use to walking or cycling would be for trips of less than 2 miles. Some 22% of all car journeys are under 2 miles in length (DfT, 2010), but such journeys account for only about 5% of all carbon emissions generated by household car journeys (DfT, 2009). Car journeys of over 2 miles account for the vast majority of car-related CO_2 emissions and few of these could realistically be transferred to walking and cycling. Switching to more sustainable public transport is, of course, an option and there may be the potential to extend cycling to some longer journeys and thus further reduce carbon emissions.

It can thus be suggested that to generate significant carbon reduction from increased walking and cycling it would require either a very large proportion of short trips currently undertaken by car to switch to walking and cycling, or for a significant proportion of the population to be prepared to walk or cycle over much longer distances than they do at present. Whilst such a level of change is not inconceivable, it can be suggested that the complex set of barriers outlined above, together with limited environmental awareness in relation to walking and cycling as revealed in the attitudinal survey, will make large-scale change difficult. In particular, current schemes to promote walking and cycling, focused mainly on improved infrastructure, information and education, though important, are unlikely to be sufficient on their own. It would require much more fundamental change in individual attitudes, in the organisation of society (e.g. more flexible and family-friendly work arrangements) and the physical structure of urban areas (with more facilities closer to residential areas) to achieve large-scale change. The Government campaign to encourage everyone to drive 5 miles less a week (http://actonco2.direct.gov.uk/home/campaigns/drive-5-miles-less-a-week.html, accessed 10 December 2011) is very worthwhile, but the projected reduction in CO_2 emissions if this was achieved accounts for just 1.5% of all UK transport-related emissions. It can be suggested that there are much greater potential gains to be had in the transport sector by encouraging people to switch to public transport for car trips of between 2 and 25 miles (many of which are commuting journeys) and which collectively contribute some 60% of car-related CO_2 emissions.

It is argued that the benefits to be gained from increases in walking and cycling that can realistically be achieved by current policies relate more to the health benefits for individuals and the reduction in congestion and air pollution in urban areas. Even a small contribution to a reduced carbon footprint is important, but increased walking and cycling alone is unlikely to have a major impact on predicted climate change scenarios. Paradoxically, campaigns to increase walking and cycling for health reasons can also have a negative impact on carbon emissions. Many people find it more convenient and comfortable to cycle or walk for leisure purposes rather than for everyday utility travel and, in particular, to cycle off-road. This is because such activity is undertaken without the usual time constraints of journeys to work or school and can be undertaken in what are perceived to be safe environments. This view was summarised by one Lancaster respondent:

> I'm a complete coward when it comes to cycling on roads, everyone I know who has does a considerable amount of cycling has been knocked off and injured in some way, cyclists on the road seem so vulnerable that I don't join them. (Female participant, Lancaster)

However, leisure-related walking and cycling also often generates car journeys as people travel by car to walk or cycle in scenic areas. There has been a significant investment by organisations such as the Forestry

Commission in developing mountain biking facilities such as the 19 km Altura Trail in Whinlatter Forest Park, Cumbria, which attracts large numbers of dedicated cyclists who arrive each weekend by car. In these circumstances increased walking and cycling for leisure purposes, whilst delivering real health benefits, may increase car travel and associated greenhouse gas emissions, rather than lead to a reduction in transport-related carbon emissions.

ACKNOWLEDGEMENTS

Thanks to all the people in Leeds, Leicester, Worcester and Lancaster who willingly gave up their time to be interviewed, to be accompanied and observed during their daily journeys. Research for this United Kingdom-based project was funded by the EPSRC (Engineering and Physical Sciences Research Council) under grant number EP/G00045X/1.

REFERENCES

Anable, J. (2005). Complacent car addicts' or 'aspiring environmentalists'? Identifying travel behaviour segments using attitude theory. *Transport Policy*, *12*, 65–78.

Anable, J. (2006). *An evidence base review of public attitudes to climate change and transport behaviour*. London: DfT. Retrieved from http://www.dft.gov.uk/pgr/sustainable/review-transportbehaviourclimatechange/pdf/iewofpublicattitudestocl5730.pdf

Anable, J., Brand, C., Tran, M., & Eyre, N. (2010). Modelling transport energy demand: A socio-technical approach. *Energy Policy*. doi:10.1016/j.enpol.2010.08.020

Brand, C., Tran, M., & Anable, J. (2010). The UK transport carbon model: An integrated life cycle approach to explore low carbon futures. *Energy Policy*. doi:10.1016/j.enpol.2010.08.019

Carpiano, R. (2009). Come take a walk with me: The 'Go-Along' interview as a novel method for studying the implications of place for health and well-being. *Health and Place*, *15*, 263–272.

Cevero, R., Sarmiento, O., Jacoby, E., Gomez, L., & Neiman, A. S. (2009). Influences of built environments on walking and cycling: Lessons from Bogotá. *International Journal of Sustainable Transportation*, *3*, 203–226.

Chapman, L. (2007). Transport and climate change: A review. *Journal of Transport Geography*, *15*, 354–367.

Department for Transport. (2004). *Walking and cycling: An action plan*. London: DfT. Retrieved from http://webarchive.nationalarchives.gov.uk/+/http://www.dft.gov.uk/pgr/sustainable/walking/actionplan/ingandcyclingdocumentinp5802.pdf

Department for Transport. (2007). *Act on CO_2*. Retrieved from http://actonco2.direct.gov.uk/home.html

Department for Transport. (2009). *Delivering sustainable low carbon travel: An essential guide for local authorities*. London: DfT. Retrieved from http://www.dft.gov.uk/pgr/sustainable/guidelocalauth/pdf/lowcarbontravel.pdf

Department for Transport. (2010). *National Travel Survey*. London: DfT. Retrieved from http://www.dft.gov.uk/pgr/statistics/datatablespublications/nts/

Department for Transport. (2011). *Creating growth, cutting carbon. Making sustainable local transport happen*. London: DfT. Retrieved from http://www2.dft.gov.uk/pgr/regional/sustainabletransport/pdf/whitepaper.pdf

Descartes, L., Kottak, C. P., & Kelly, A. (2007). Chauffeuring and commuting. *Community, Work & Family, 10*, 161–178.

Elvik, R. (2009). The non-linearity of risk and the promotion of environmentally sustainable transport. *Accident Analysis and Prevention, 41*, 849–855.

European Commission. (2009). *EU energy and transport figures: Statistical pocket book*. Brussels: European Communities.

Fincham, B., McGuiness, M., & Murray, L. (2009). *Mobile methodologies*. Basingstoke: Palgrave.

Higgins, P., & Higgins, M. (2005). A healthy reduction in oil consumption and carbon emissions. *Energy Policy, 33*, 1–4.

House of Lords Science and Technology Select Committee. (2011). *Behaviour Change Report*. London: TSO. Retrieved from http://www.publications.parliament.uk/pa/ld201012/ldselect/ldsctech/179/179.pdf

Hume, C., Timperio, A., Salmon, J., Carver, A., Giles-Corti, B., & Crawford, D. (2009). Walking and cycling to school: Predictors of increases among children and adolescents. *American Journal of Preventive Medicine, 36*, 195–200.

Jacobsen, P. (2003). Safety in numbers: More walkers and bicyclists, safer walking and cycling. *Injury Prevention, 9*, 205–209.

Jacobsen, P., Racioppi, F., & Rutter, F. (2009). Who owns the roads? How motorised transport discourages walking and cycling. *Injury Prevention, 15*, 369–373.

Kusenbach, M. (2003). Street phenomenology: The go-along as ethnographic research tool. *Ethnography, 4*, 455–485.

Lorenc, T., Brunton, G., Oliver, S., Oliver, K., & Oakley, A. (2008). Attitudes to walking and cycling among children, young people and parents: A systematic review. *Journal of Epidemiology and Community Health, 62*, 852–857.

Mackett, R. (2003). Why do people use their cars for short trips? *Transportation, 30*, 329–349.

Maibach, E., Steg, L., & Anable, J. (2009). Promoting physical activity and reducing climate change: Opportunities to replace short car trips with active transportation. *Preventive Medicine, 49*, 326–327.

McMillan, T. (2007). The relative influence of urban form on a child's travel mode to school. *Transportation Research, Part A: Policy and Practice, 41*, 69–79.

Moser, C., & Kalton, G. (1979). *Survey methods in social investigation*. London: Heinemann.

Ogilvie, D., Egan, M., Hamilton, V., & Petticrew, M. (2004). Promoting walking and cycling as an alternative to using cars: Systematic review. *British Medical Journal, 329*, 763–766.

Ogilvie, D., Foster, C., Rothnie, H., Cavill, N., Hamilton, V., Fitzsimons, C. F., & Mutrie, N. (2007). Interventions to promote walking: Systematic review. *British Medical Journal, 334*, 1204–1207.

Pikora, T., Giles-Cirti, B., Bull, F., Jamrozik, K., & Donovan, R. (2003). Developing a framework for the assessment of the environmental determinants of walking and cycling. *Social Science and Medicine, 56*, 1693–1703.

Pooley, C., Horton, D., Scheldeman, G., Tight, M., Harwatt, H., Jopson, A., Jones, T., & Chisholm, A. (2011). Household decision-making for everyday travel: A case study of walking and cycling in Lancaster (UK). *Journal of Transport Geography*, *19*, 1601–1607.

Pucher, J., & Buehler, R. (2008). Making cycling irresistible: Lessons from The Netherlands, Denmark and Germany. *Transport Reviews*, *28*, 495–528.

Pucher, J., & Dijkastra, L. (2003). Promoting safe walking and cycling to improve public health: Lessons from the Netherlands and Germany. *American Journal of Public Health*, *93*, 1509–1516.

Pucher, J., Dill, J., & Handy, S. (2010). Infrastructure, programs and policies to increase bicycling: An international review. *Preventive Medicine*, *50*(Suppl. 1), S106–S125.

Ricketts, J., Evans, J., & Jones, P. (2008). Mobile methodologies: Theory, technology and practice. *Geography Compass*, *2*, 1266–1285.

Saelens, B., Sallis, J., & Frank, L. (2003). Environmental correlates of walking and cycling: Findings from the transportation, urban design and planning literatures. *Annals of Behavioural Medicine*, *25*, 80–91.

Shove, E. (2010). Beyond the ABC: Climate change policy and theories of social change. *Environment and Planning A*, *42*, 1273–1285.

Silverstone, R., Hirsch, E., & Morley, D. (1991). Listening to a long conversation: An ethnographic approach to the study of information and communication technologies in the home. *Cultural Studies*, *5*, 204–227.

Sustrans, Connect2 project. Retrieved from http://www.sustrans.org.uk/what-we-do/connect2

Timperio, A., Crawford, D., Telford, A., & Salmon, J. (2003). Perceptions about the local neighbourhood and walking and cycling among children. *Preventive Medicine*, *38*, 39–47.

Wallman, S. (1984). *Eight London households*. London: Tavistock Publications.

Whinlatter Forest Park. Retrieved from http://www.forestry.gov.uk/forestry/infd-79yfhn

CHAPTER 8

A POLICY PERSPECTIVE ON TRANSPORT AND CLIMATE CHANGE ISSUES

Greg Marsden, Ian Bache and Charlotte Kelly

ABSTRACT

Purpose – *To provide a policy perspective on the relationship between transport and climate change.*

Methodology/approach – *Two key themes are identified and discussed: the meaning of a major change in a policy perspective, covering the Climate Change Act and the development of a Low Carbon Transition Plan. A theoretically informed framework applies and highlights the importance of understanding policy change from a historical perspective.*

Originality/value – *The largely incremental nature of the policy change is considered in terms of whether there are real prospects of a radical change in transport policy that will deliver a low carbon transport future, whilst also allowing transport to fulfil its many other roles.*

Findings – *The chapter demonstrates that the current approach to climate change policy has seen only minor adjustments to existing policy tools which are not consistent with the more radical shift in policy targets. This incremental approach may reflect the significant uncertainty over technological change or a reluctance to tackle the difficult issue of travel*

Transport and Climate Change
Transport and Sustainability, Volume 2, 197–223

ISSN: 2044-9941/doi:10.1108/S2044-9941(2012)0000002011

behaviour. Whilst oil prices remain high this may not be problematic but more proactive steering will be necessary in the coming years.

Keywords: Climate change; governance; transport policy

INTRODUCTION

This chapter explores how UK policy has changed in response to the new transport and climate change agenda. It looks ahead also to the potential for more radical future change and discusses the conditions under which more radical change might be enacted.

Giddens (2009) suggests that whilst there is now a wave of awareness of the need to act to tackle climate change, there is a need to embed climate reduction policies 'in our institutions and in the everyday concerns of citizens, and here, ... there is a great deal of work to do' (p. 4). It is argued that the state is an 'all important actor' (p. 5) from a local to an international level, in setting treaties and in enacting their delivery, in supporting embryonic technologies and in working with markets and the private sector to ensure that the true costs of climate change are reflected in prices. The market, as argued by Giddens (2009), can produce results that no other agency or framework is able to – but only if steered to do so. On this issue, Marsden and Rye (2010) consider the interplay between state and non-state actors in the transport sector with respect to climate change. Whilst they agree that the state plays a critical role, they argue that much greater attention needs to be given to the policy structures and processes of policy development if more significant progress is to be made in carbon reduction in the transport sector.

The UK is an interesting case study in climate change policy. It has taken the lead on developing a framework for action. A key actor in the development of the Kyoto Protocol (1997) the UK has since gone on to become the first nation to commit itself, through the Climate Change Act 2008, to a legally binding target of at least an 80% cut in greenhouse gas emissions by 2050 (relative to 1990 levels). Progress is to be tracked by an interim emission reduction target of 34% by 2020, with interim rolling five-year budgets.

To achieve such dramatic decarbonisation of the UK economy will require action across all sectors. The action does not have to be equal across all sectors or to proceed at similar rates. However, as emissions from transport represent 21% of the UK total for domestic emissions, transport must play a significant role in moving to a low carbon economy (Department for Transport [DfT], 2008). In 2009 the DfT released its Carbon Reduction Strategy for Transport where it described decarbonising transport as:

> part of the solution. This will be a major change, but moving to a low carbon economy and transport system also presents huge opportunities; not just for climate change but for our prosperity, health, and the wider environment. (2009a, p. 5)

THE MEANING OF POLICY – AN ANALYTICAL FRAMEWORK

The purpose of this section is to clarify the meaning of a 'policy perspective'. This is done by unpacking the notion of 'policy', drawing primarily on Hall's (1993) approach to social learning. In doing so, the argument advanced by a number of scholars is followed that such disaggregation is valuable for understanding different dynamics at play in different aspects of policy-making.

In his oft-cited analysis of social learning in economic policy-making in Britain, Hall (1993, p. 278) identified policy-making as a process usually involving three variables: overarching policy goals that guide policy, the instruments or techniques used in seeking to attain these goals and the precise setting of the instruments. Hall's distinction between the means and end of policy, and between abstract and concrete features, was an important contribution to the study of policy change. It challenged the then dominant view that tended to reduce all elements of policy into a single variable and to view all change as incremental in nature (Howlett & Cashmore, 2009, p. 36).

Building on this work, Liefferink and Jordan (2004) placed Hall's three variables within the category of 'policy content': a category said to exist alongside the 'subtly interrelated' variables of policy structures and policy style (shown in Table 1). Like the category of policy content, policy structures is a broad category, seeking to encompass both formal and informal notions of structure. Liefferink and Jordan (2004, p. 40) take it to include 'both the formal structure or architecture of the state as well as the norms and rules (both formal and informal) that govern the operation of its constituent parts'. Policy style refers to the cultural dimension or 'standard operating procedures' of national policy and is understood best in comparative perspective. This category has two aspects: (1) government's approach to problem solving, ranging from anticipatory/active to reactive;

and (2) government's approach to other actors involved in policy-making, ranging from seeking consensus to seeking to impose (Liefferink & Jordan, 2004, p. 42).

Also building on Hall (1993), Howlett and Cashmore identified six elements of policy that can undergo change, suggesting that each of Hall's three variables should recognise his own distinctions between policy 'ends' and policy 'means'. The implication of this taxonomy is that 'every 'policy' is in fact a more complex regime of ends and means related goals (more abstract), objectives (less abstract) and settings (least abstract)' (Howlett & Cashmore, 2009, p. 38), and provides a more nuanced framework than Hall's original.

There is something useful in each of these contributions. Hall's seminal contribution feeds into the later pieces and provides insights into the notion of policy change (below). Liefferink and Jordan's taxonomy highlights policy structures and policy style as features that interact with and influence policy content and thus need to be categorised and understood. However, the distinction between structures and content is clearer than that between structures and style: the cultural/sociological aspects of policy structures elide with the cultural aspects of policy style, which may have most value for cross-national comparisons where distinctly national styles can be identified. As Liefferink and Jordan (2004, p. 43) suggest, it is of less use even where national comparisons are drawn from a similar pool (e.g. Europe). As this is a study of transport and climate policy in England, the notion of policy style is kept in view but this is not a major theme in the chapter.

The chapter focuses on changes in policy content as a first step to unpicking the question of how governance in the transport sector is changing in response to the climate change agenda. Howlett and Cashmore's finer taxonomy is adopted as a starting point, which is illustrated in Table 2. Key

Table 1. Liefferink and Jordan's Policy Variables.

Policy Variables	Features
Policy content	Three levels (following Hall): overarching goal; instruments or techniques used; precise setting of instruments
Policy structures	From formal bureaucratic organisations to cultural phenomena (e.g. structures formed and rules and norms of governance)
Policy style	Two dimensions: government's approach to problem solving (ranging from anticipatory/active to reactive); and government's approach to other actors in policy-making (either seeking to reach consensus or to impose decisions)

Table 2. A Taxonomy of Policy Components.

	High-Level Abstraction	Programme-Level Operationalisation	Specific-On-the-Ground Measures
Policy ends or aims	***Goals*** *What general types of ideas govern policy development?* (e.g. environmental protection)	***Objectives*** *What does policy formally aim to address?* (e.g. control of carbon emissions)	***Settings*** *What are the specific on-the-ground requirements of policy?* (e.g. % reduction in carbon for the transport sector, a mode or a city)
Policy means or tools	***Instrument logic*** *What general norms guide implementation preferences?* (e.g. welfare maximising rational economic paradigm)	***Mechanisms*** *What specific types of instruments are utilised?* (e.g. vehicle ownership taxation)	***Calibrations*** *What are the specific ways in which the instrument is used?* (e.g. levels of charges to target particular types of vehicle)

Source: Based on Howlett and Cashmore (2009, p. 39).

changes in policy structures are observed, but it is not yet possible to make firm associations between changes in structure and content.

As well as providing a taxonomy and framework to describe policy change, it is also important to understand how policy changes. Hall's point of departure for explaining policy change through social learning (Hall, 1993) was to set out the position of state theorists on this, which had three features. First, this position emphasised the importance of previous policies in shaping current policies, rather than prevailing social and economic conditions the idea of 'policy legacies' (or what might now in historical institutional terms be described as 'path dependence'). Second, it identified 'experts' as the key agents in promoting policy learning, rather than politicians – experts either working for the state or located at the interface of bureaucracy and the 'intellectual enclaves of society'. Third, it stressed the capacity of the state to act autonomously from societal pressure in shaping policy (Hall, 1993, pp. 277–278).

In response to this position, Hall argued for a clearer definition of social learning, described as 'a deliberate attempt to adjust the goals or techniques of policy in response to past experience or new information (1993, p. 278) and for the concept to be disaggregated. Here the distinction of policy-making as a process that usually involves three different variables (overarching goals, instruments and instrument settings) is important in allowing analysis of policy change of different orders and thus may be subject to very different processes. Specifically, he identified three orders of change in policy-making.

First order change refers to changes in the settings of policy instruments in light of experience and new knowledge, while keeping the overall goals and instruments unchanged. Second order change refers to changes in the instruments themselves as well as their settings in response to experience, while the overall policy goals remain unchanged. Third order change – the least likely – refers to wholesale changes in policy instrument settings, the instruments themselves and the overarching hierarchy of goals behind them. While first and second order changes might be considered normal policy adjustments within a stable paradigm, third order change is generally associated with paradigm change (Hall, 1993, pp. 278–279). Moreover, while first and second order changes are likely to be characterised by incrementalism – analysis consistent with the state theorists' approach to policy learning – third order change is unlikely to be incremental, but more radical. Moreover, first and second order changes do not necessarily lead to third order change: each paradigm is shaped by a different ontology in relation to how the (policy) world operates and thus judgments about how best to proceed are not simply technical. Hall identified three important aspects of paradigm change:

- The replacement of one policy paradigm by another is likely to be more sociological than scientific. That is, although expert opinion is a factor, the choice between paradigms ultimately rests on 'more political judgments' and the outcome dependent 'not only on the arguments of competing factions, but on their positional advantages within a broader institutional framework, on the ancillary resources they can command in the relevant conflicts, and on exogenous factors affecting the power of one set of actors to impose its paradigm over others' (Hall, 1993, p. 280, see also Dudley & Richardson, 2000 for evidence in a transport context).
- Authority over policy is central to paradigm change. That is, particularly where complex issues are at stake, the shift from one paradigm to another 'is likely to be preceded by significant shifts in the locus of authority over policy' (Hall, 1993, p. 280).

- Experimentation and policy failure are likely to play a key role in paradigm change. That is, attempts to respond to anomalies and challenges to the existing paradigm will lead to it being stretched in response, but this is likely to eventually undermine the coherence and precision of the paradigm: 'if the paradigm is genuinely incapable of dealing with anomalous developments, these experiments will result in policy failures that gradually undermine the authority of the existing paradigm and its advocates even further' (Hall, 1993, p. 280).

Hall's challenge to the orthodox incrementalist approach to policy change led to a long scholarly debate, which Howlett and Cashmore (2009, p. 34) suggest has been replaced by a new 'postincremental' orthodoxy as scholars have accepted the idea that 'periods of marginal adaptation and revolutionary transformation are typically linked in a "punctuated equilibrium" pattern of policy change'. This orthodoxy is characterised by an acceptance of the need for historical analysis of policy development; an understanding that political institutions and their policy subsystems are the primary mechanisms of policy reproduction; recognition that paradigmatic change tends to take place only when the policy institutions themselves are transformed; and agreement that paradigmatic change or 'punctuations' are usually the result of 'external perturbations' that disrupt existing ideas, institutions and practices (Howlett & Cashmore, 2009, pp. 35–36). This orthodoxy generates three expectations:

- that policy processes are typically stable (explained by path dependent institutionalisation);
- that policy change ('punctuated equilibrium') results from the breakdown of an institutionalised 'policy monopoly';
- this typically occurs through an alteration in policy subsystem beliefs and membership due to some external (societal) 'perturbation' (Howlett & Cashmore, 2009, p. 36).

Howlett and Cashmore's critique of the new orthodoxy is twofold: they suggest that it is based on a taxonomy that conflates very different elements of policy (hence their sixfold categorisation) and, related to this, that it has an under-developed classification of the different types of changes that different policies undergo. This leads to 'erroneous conclusions being drawn by Hall and others about the factors underlying policy dynamics and their appropriate modelling' (Howlett & Cashmore, 2009, p. 37). They suggest that the reality is far more complex than the binary distinction promulgated by the incremental vs. paradigmatic debate.

In summary, this section has sought to flesh out the dependent variable in our study of policy. Doing this requires us to both outline what should be measured in empirical investigations and to identify the most appropriate data for empirical investigation (Green-Pedersen, 2004). This second task is returned to below. In relation to the former, Howlett and Cashmore's framework highlights the need for greater precision in disentangling the policies that are being studied; to distinguish between changes that move in different directions at different times but do not deviate significantly from the policy equilibrium from those that move in the same direction over time (cumulative change); and to take care in attributing change to either endogenous or exogenous sources, when the reality might be a complex interaction of the two. The main theoretical claim is that policy change takes place through a process of social learning based on past experience or new information.

TRANSPORT POLICY IN ENGLAND (1998–2009)

The previous section identified the need to review change in terms of policy content and policy structures to better understand it. This review could be undertaken at many levels (from local to international). Climate change is clearly a problem that does not respect administrative boundaries and binds state actions with those of non-state actors and institutions. This points to the analysis of climate change policy through the lens of multi-level governance (Bache & Flinders, 2004; Marsden & Rye, 2010). In selecting the geographical boundaries for this work, there are difficulties resulting from the model of asymmetric devolution in the UK characterised by multi-level governance processes in the different constituent parts (England, Northern Ireland, Scotland and Wales) that produce divergent public policies. In this context, therefore, our analysis (below) is developed around climate change and transport policy in England at a national level. The reasons for selecting England are that 86% of vehicle kilometres travelled in the British Isles are in England and its actions on climate change are therefore pivotal in attaining the United Kingdom's commitments on climate change. The focus is on the national level as this is where the Climate Change Act was developed. The DfT also holds a pivotal place in the governance system, exerting stronger control than national transport departments in many other European countries (Marsden, Frick, May, & Deakin, 2011). Moreover, the national level provides a space in which policy across all of the different transport modes can, in theory, be integrated Such an approach makes this analysis tractable but provides clear limitations:

1. In the UK, the impacts of devolution to Northern Ireland, Scotland and Wales has led to some divergence in policy compared to England (MacKinnon, Shaw, & Docherty, 2008). This may influence policy in England although recent analysis suggests that this has only been in a very limited manner with respect to climate change (Marsden & Rye, 2010).
2. Cities are important sources of policy and innovation in their own right and there is evidence of city level policy-making that seeks to bypass aspects of the nation state as part of wider Europeanisation (Kern & Bulkeley, 2009).

The period from 1998 to 2009 was selected for study to enable an understanding of change over time. This period captures the main policy documents from the period of the Labour administration beginning with the 1998 Integrated Transport White Paper and ending in the publication of the 2009 Carbon Reduction Strategy for Transport. Dudley and Richardson (2000) suggest that longer periods of time might be necessary to fully capture the dynamics of policy change in the transport sector, even if third order change processes can be observed over periods of two to three years. However, 1998 represented an important change in the nature of transport policy as it was the first transport White Paper for twenty-one years and confirmed the departure from the large scale roads programmes that preceded it (Roads for Prosperity was published in 1989 setting out a plan for £23 billion investment in 500 road schemes). 2009 was selected as the final date for analysis as the incumbent Conservative–Liberal coalition government has yet to publish a national White Paper, although some changing signals from more recent policy announcements are highlighted. A timeline of key events and policy documents can be found in Table 3.

This period has been divided into three broad epochs for the analysis in the next section of the chapter.

1. Early (1998–2000) – An integrated vision

The Labour government entered office on 2nd May 1997. Shortly after, in June of the same year, the Department for Environment, Transport and the Regions (DETR) was formed (amalgamating the Departments of Transport and Environment). The Ministry was under the purview of the Deputy Prime Minister, John Prescott, who came into power committed to introducing an integrated transport policy. The White Paper that was published

Table 3. Timeline of Key Events.

1998	*Transport White Paper* – A New Deal for Transport Better For Everyone (DETR, 1998) *Introduction of the New Approach to Transport Appraisal* (NATA)
1999	*Greater London Authority Act 1999* – The control of London's Buses, Trains, Underground System, Traffic Lights, Taxis and River Transport moved to TfL *SACTRA Report* – Transport and the Economy (SACTRA, 1999)
2000	*Local Transport Act 2000* – Included the requirement for the production of local transport plans in England and legislation to allow road user charging and workplace parking charges *Transport 10 Year Plan* – (DETR, 2000a) *Guidance on the First Local Transport Plan issued* (DETR, 2000b) *Creation of the Sustainable Development Commission*
2001	*Creation of the Department for Transport Local Government and the Regions (DTLR)* (previously DETR) – The remit for the environment moved to DEFRA
2002	*DTLR was reformed as the Department for Transport (DfT)*
2003	*Start of the London Congestion Charging Scheme* *Aviation White Paper* – The Future of Air Transport (DfT, 2003) *EU Biofuels Directive* (2003/30/EC) *Energy White Paper* – Our Energy Future – Creating a Low Carbon Economy (DTI, 2003)
2004	*Rail White Paper* – The Future of Rail (DfT, 2004a) *Transport White Paper* – The Future of Transport (DfT, 2004b) *Guidance on the 2nd Local Transport Plans issued* (DfT, 2004c) *Smarter Choices* – Changing the way we travel – Cairns et al. (2004)
2005	*EU Emission Trading Scheme (ETS) starts (phase 1)* *UK Government Sustainable Development Strategy* – Securing the Future DEFRA (2005)
2006	*NATA – updated to include monetising carbon dioxide emissions* *Climate Change and Sustainable Energy Act 2006* *Stern Review on the Economics of Climate Change* (Stern et al., 2006) *The Eddington Transport Review* (Eddington, 2006)
2007	*Rail White Paper* – Delivering a sustainable Railway CM7176 (DfT, 2007a) *King Review* – The King Review of Low Carbon Cars (King, 2007) *Transport White Paper* – Towards a sustainable transport system: supporting economic growth in a low carbon world (TASTS), (DfT, 2007b)
2008	*Local Transport Act 2008* *Climate Change Act 2008* *Carbon Pathways Analysis* – Informing Development of a carbon reduction strategy for the transport sector (DfT, 2008)

Table 3. (*Continued*)

	EU Emission Trading Scheme (ETS) Phase 2 starts *Creation of the Department of Energy and Climate Change (DECC)* – bringing together Climate Change Mitigation Policy and Energy Policy (previously covered by DEFRA and BERR respectively)
2009	*Low Carbon Transport Plan* – Low Carbon: A Greener Future, (DfT, 2009a) *Guidance on the 3rd Local Transport Plans Issued* (DfT, 2009b)

in July 1998 was to signal 'a radical change in transport policy' (DETR, 1998, Foreword). The thinking behind the paper was heavily influenced by the government's advisor Phil Goodwin, who had, through his work on the new realism, established widespread acceptance of the proposition that it was no longer affordable, environmentally acceptable or practically feasible to build roads to cut congestion. Therefore a new approach to travel demand management was required (Docherty & Shaw, 2011; Goodwin, Hallett, Kenny, & Stokes, 1991).

The creation of a 'super ministry' was ultimately symbolic more than practically important as it was disbanded in 2001 (Beecroft, 2002). Nonetheless, the White Paper offered an early window on the underlying ideology for transport policy under Labour. The White Paper followed soon after the landmark Kyoto protocol negotiations (in which Prescott was the lead UK Secretary of State) where many developed nations agreed to commit to reduce their climate change emissions. The UK agreed, as part of a broader European commitment, to a 20% reduction by 2020 compared with 1990 levels. The White Paper established, for the first time, the importance of climate change and transport's role in that:

> with our new obligations on climate change, the need for a new approach [to transport] is urgent. (DETR, 1998, Foreword).

The White Paper contained some important provisions. It established a requirement on local authorities in the UK to produce five year Local Transport Plan strategies; it signalled the introduction of legislation for local road user charging schemes; and, whilst not attempting to unpick the privatisation of the railways, sought to bring greater governmental influence to rail policy through the establishment of an arm's length planning body (the Strategic Rail Authority). However, by 2000 the government was being attacked for being strong on ideology and weak on delivery. It launched a 10 Year Plan for Transport that set out what it hoped could be achieved and provided a long-term significantly enhanced funding envelope to achieve it

(DETR, 2000a). By this stage, the Royal Commission on Environmental Pollution had concluded that the UK should put itself on a pathway to a 60% CO_2 emission reduction by 2050 (Royal Commission on Environmental Pollution [RCEP], 2000). This epoch concluded with the fuel duty protests of September 2000 (where farmers and hauliers blockaded oil refineries in protest to the fuel duty escalator), which were widely identified as a major reason for the Labour government stepping back from a more radical approach to tackling the rising demand for travel (Marsden, 2002).

2. Mid (2001–2005) – Events and a crowded agenda

Only one month after the fuel duty protests, the Hatfield rail crash and the ensuing collapse of Railtrack was to change the direction of policy making. In May 2002, the Transport Select Committee published a highly critical review of the likelihood of the government achieving the goals set out in the 10 Year Plan, in particular the congestion reduction target (HOC, 2002). Alistair Darling replaced Stephen Byers as Secretary of State and set about the task of reorganising the rail industry and dulling the media focus on the failure to reduce congestion.

Four important policy publications were produced in this period with a bearing on transport policy. The Social Exclusion Unit, working out of the Cabinet Office, produced its report 'Making the Connections' (SEU, 2003). This identified transport as a potentially important part of the broader social inclusion agenda and led to a requirement for all local authorities to prepare an accessibility plan for their area that tried to address these issues (whilst climate was not seen to be a local priority). The Aviation White Paper set out the framework for airport expansion in the United Kingdom (the first since 1986), anticipating a doubling or trebling of passenger numbers by 2030, an increase that would bring acknowledged increases in climate change emissions (DfT, 2003). A Rail White Paper (DfT, 2004a) set out the new structure for the rail industry with the new not-for-profit Network Rail being directed by central government once again with the abolition of the Strategic Rail Authority. In addition, a new overarching Transport White Paper was published (DfT, 2004b) alongside a report into the feasibility of a national road pricing system and a report into Smarter Choices (Cairns et al., 2004). The Smarter Choices report presented the case for an enhanced investment in measures such as commuter travel plans, walking and cycling and personalised travel planning as a cost effective means of reducing congestion and emissions as well as promoting healthier travel.

3. Late (2006–2009) – External advice

The policy direction was subsequently influenced by a series of influential expert reports, all commissioned in part by the Treasury. The Stern review on the economics of climate change has had a global significance, setting out the potential range of economic costs that could be incurred as a result of climate change. Critically it advanced the argument that the costs of early action would be lower than the costs of inaction and identified benefits to the competitiveness of the UK national economy in taking a leading role in the new green economy (Stern et al., 2006).

The Eddington review (joint with the DfT) looked at the competiveness and connectivity of the UK transport system and made recommendations for how best to prioritise spending to improve the economy. It concluded that small scale network improvements, investment in cities and at key congestion hot spots were critical and that road pricing would be beneficial (Eddington, 2006). The report said little about the environment even though it was produced concurrently with the Stern Review.

Finally, the King Review of low carbon cars (King, 2007 – joint with DfT, DEFRA and BERR) examined the potential to decarbonise car transport. The King Review came down heavily in favour of hybrid electric and electric vehicles being a part of the transition path to a decarbonised car fleet. Whilst at this stage it has not been possible to understand the motivation for Treasury involvement, it is noted that there is a substantial history of Treasury involvement in influencing the policy frame and long-term funding programme for the sector (Dudley & Richardson, 2000).

In response to these expert reports a new Transport White Paper, 'Towards a Sustainable Transport System (TASTS)' was launched. The White Paper anticipated the introduction of the Climate Change Act (2008), which makes the United Kingdom the only country to have a legally binding framework to cut climate change emissions. 'The Act requires that emissions are reduced by at least 80% by 2050, compared to 1990 levels' (Committee on Climate Change [CCC], 2011). Five-yearly budgets are also set for a fifteen-year period to provide a trajectory for progress. The budgets set for the UK as a whole are shown in Table 4.

The timing of the TASTS White Paper meant that further details of how transport was to contribute to the reductions required was still to be provided. This came in the form of the 2009 Low Carbon Transport plan (DfT, 2009a), published as part of a whole government Low Carbon Transition Plan. The change from the Blair to the Brown premiership also brought some further changes in approach with the coming to the fore

Table 4. UK Carbon Budget Targets.

	Budget 1 (2008–2012)	Budget 2 (2013–2017)	Budget 3 (2018–2022)
Carbon budgets ($MtCO_2e$)	3018	2782	2544
Percentage reduction below 1990 levels	22	28	34

Source: DECC (2011a).

of High Speed Rail to the North of England which, under Eddington's thinking, had been put to one side.

This section has provided a brief overview of some of the key events that have shaped the transport debate. It illustrates that climate change is one of many competing policy agendas that influence transport and which, over time, seem to exert different levels of influence. These contextual factors will be drawn on, where relevant, to explain why climate policy in transport has not followed a smooth growth in importance over the period. The analysis is based on four key documents: the transport White Papers of 1998, 2004 and 2007 and the Low Carbon Transport Strategy of 2009.

POLICY CONTENT

Goals

Framework: What General Types of Ideas Govern Policy Development?
There has been a significant shift in the importance of climate change as a goal of transport policy. The 1998 White Paper was the first to signal the relationship between transport and climate change, describing the importance of climate change as a reason to adopt a new approach to transport, quoting the Kyoto protocol as a motivation for tackling the environmental impacts of transport. Climate change is however, treated as a sub-objective of 'environment', one of five overarching goals of the 1998 White Paper. An analysis of the number of references to climate change (or related terms) within the 1998 White Paper relative to other important goals (Fig. 1) shows that it received the same attention as local air quality and less than one quarter of those of road safety.

The analysis from Fig. 1 appears to suggest a growth in the importance of climate change in the 2004 White Paper. Whilst it is trailed in the Foreword

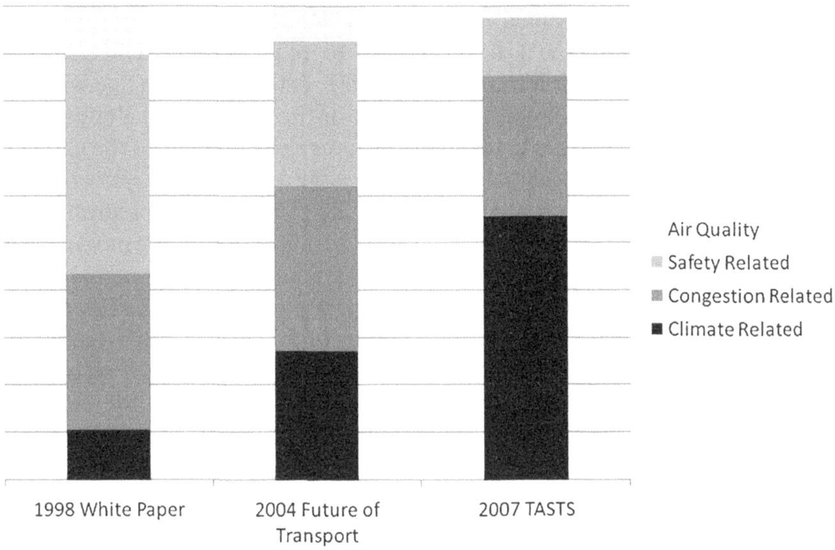

Fig. 1. Prevalence of Policy Goals Featured in the Transport White Papers.

by Prime Minister Blair, the strategy is described as being built around the three key themes of sustained investment, improvements in transport management and planning ahead. Climate remains part of the environment goal. There is a more significant change in 2007 where climate change was elevated to one of five goals: 'Goal 2 is to address climate change, by cutting emissions of carbon dioxide (CO_2) and other greenhouse gases' (DfT, 2007b, p. 9). The first goal was to maximise competitiveness and productivity with the remaining three being 'safety, security and health', 'quality of life' and 'equality of opportunity' (DfT, 2007b, p. 10). The analysis in Fig. 1 underlines the importance of the first two goals with congestion (and its links to productivity) and climate dominating the discourse on goals.

Objectives

Framework: What Does Policy Formally Aim to Address?

As an overarching government objective, cutting CO_2 emissions has remained a consistent theme since the Kyoto protocol was formulated in 1997. This began with a formal commitment to the EU of a 12.5% reduction by 2010 compared with 1990 levels and a national goal of a 20% reduction by 2010. From 2003 (signalled in the Energy White Paper) the Government committed to 'putting the UK on a path to reducing CO_2 emissions by around 60% from current levels by 2050' (DfT, 2004b, p. 23) in line with the recommendation of the Royal Commission on Environmental Pollution.

The objective was further crystallised with the development of the Climate Change Act (2008) and the subsequent decision to adopt an 80% carbon reduction goal across the economy in 2008. This changed the panorama for transport as it is responsible for over 20% of all domestic greenhouse gas emissions. A 60% reduction would allow for a relatively moderate reduction in transport related greenhouse gas emissions if other sectors such as energy production effectively decarbonised. The adoption of an 80% reduction commitment meant that the transport sector had to become a significant part of the solution. The DfT's 2009 strategy stated that 'the effective decarbonisation of the transport sector will play a big part of achieving this goal' (DfT, 2009a, p. 20).

Settings

Framework: What Are the Specific On-the-Ground Requirements of Policy?

The specific requirements for the transport sector with respect to carbon reduction have changed over the period in line with an increasing necessity to make a contribution to the overarching goal as outlined above.

The 1998 White Paper talks in terms of quite significant reductions in climate change emissions, for example 'even without a major change in behaviour, with the key measures in the New Deal for transport, there is the potential to reduce forecast 2010 road traffic CO_2 emissions by 22–27%' (DETR, 1998, p. 27). However, the subsequent 10 Year Plan for transport was acknowledged as focussed on congestion reduction and not climate emissions reduction (Department of Trade and Industry [DTI], 2003). It did not propose a formal target for carbon reduction despite targets being set for many other aspects of transport (Docherty & Shaw, 2011; Marsden & Bonsall, 2006). Even were the 10 Year Plan to have delivered in full on its promises it would have achieved only a 1.6MtC reduction in CO_2 emissions compared with doing nothing, which equates to an absolute reduction of 0.9MtC or less than 1% of surface transport emissions in the year 2000 and an increase on 1990 levels (DETR, 2000b). The approach to carbon emissions management could be characterised as being an approach to stabilising or limiting the absolute growth in emissions rather than cutting them.

In 2004, the DfT was made jointly accountable for the overall carbon reduction goal through a Public Service Agreement to 'Reduce greenhouse gas emissions to 12.5 per cent below 1990 levels in line with our Kyoto commitment and move towards a 20 per cent reduction in carbon dioxide emissions below 1990 levels by 2010, through measures including energy efficiency and renewables, together with the Department for Environment, Food and Rural Affairs and the Department of Trade and Industry' (DfT, 2004b, p. 136). No formal attribution of responsibility for goal achievement was set, however, and the 2004 White Paper forecasted an increase in emissions from transport by 2010 with the prospect of the achievement only of meeting 1990 levels by 2020 (see Fig. 2).

By 2007 the TASTS White Paper was explicitly acknowledging that transport had to play a more significant role in the proposed aim of achieving a 60% reduction of CO_2 emissions by 2050. By 2009 an acceptance that transport would have to significantly decarbonise was evident (see previous section). The setting moved from a cross-departmental Public Service Agreement to a formal legislative commitment across government within the Climate Change Act. Indicative budgets for each Department were set in 2009 although these are not formally binding. The Coalition government has however abandoned an outcome budget approach in favour of an action plan reporting framework where the progress of specific policies and initiatives are assessed (Department of Energy and Climate Change [DECC], 2011b).

The settings have swung from a loosely-specified aspiration to limit the increase in climate change emissions to a more formalised cross-government commitment to achieve specific carbon reduction targets. Whilst it is clear that transport must contribute to these cross-governmental goals it still remains unclear by how much and over what timescale. That is not to say that (as in Fig. 2) there are not expectations of what might be achieved but that the expectation for transport appears to be driven by what it possible not what is necessary.

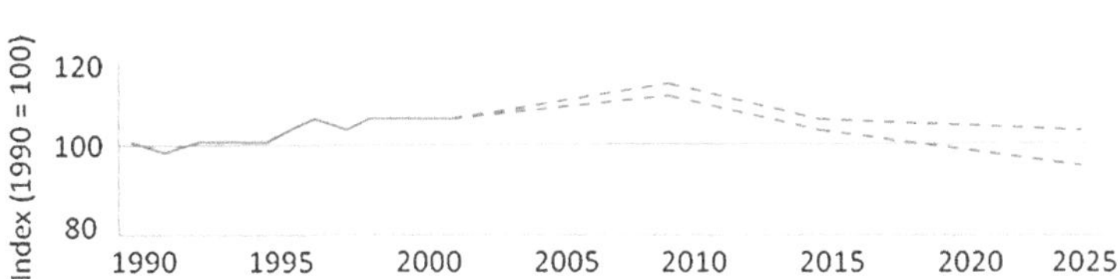

Fig. 2. 2004 Future of Transport White Paper Estimates of Road Traffic Based CO_2. *Source:* DfT (2004b).

Instrument Logic

Framework: What General Norms Guide Implementation Preferences?

The transport sector in the UK has a strongly market-led approach with governmental decision-making processes driven by welfare maximising economic principles. This is evidenced by the privatisation of the bus and rail industries and the adoption of a large-scale public–private partnership (PPP) for the London Underground, for example. The decision-making processes adopted for prioritising transport decisions are based largely around cost-benefit analysis. The initial approach developed and applied to early projects such as the M1 motorway utilised on pure economic costs (construction and operation) and benefits (time and operating cost savings). 1998 signalled the broader evolution of practice into a more open multi-criteria assessment of a range of factors (including environmental degradation, largely in response to major environmental protests (Dudley & Richardson, 2000)). Whilst the cost:benefit ratio is still critical to the decision-making process the benefits have increasingly incorporated a broader range of social benefits (including a shadow price for carbon). A review of the 1998 roads programme showed that other factors such as qualitative environmental assessments were also influencing the decision-making process (Nellthorp & Mackie, 2000).

The underlying instrument logic has remained largely unchanged with a central approach of the various White Papers to be to ensure that market prices reflect, where possible, the full social costs of journeys made (including the environmental externalities) and that these are also reflected in the prioritisation processes of central government. Over the period, the detail of how this has been articulated and implemented has evolved. The emerging EU Emissions Trading System (ETS), which the European aviation sector will join in 2012, is developing a tradable market price for CO_2 whereby the values are based on expected EU ETS allowance prices. Aside from aviation, transport is in the non-traded sector and relies on the incorporation of a shadow price for carbon in appraisal as part of government investment decisions. The 2007 White Paper stated that 'Fundamentally, we need to get the prices right to cover environmental and congestion costs of transport' (DfT, 2007b, p. 5), whilst the 2009 Low Carbon Transport Strategy takes this further: 'Factoring carbon costs into the prices we pay for transport provides incentives for us to be either more energy-efficient or to opt for lower carbon alternatives. It also sends the right long-term signals for investment' (DfT, 2009a, p. 86). This suggests a fairly strong adherence from the DfT to the quasi-market based logic of the Treasury. The DfT acknowledges though that 'carbon pricing is a complex matter. The price itself can be subject to a range of uncertainties' (DfT, 2009a, p. 87).

This allows climate change to be incorporated directly into the existing decision-making processes in transport, although challenges still remain. Whereas in some sectors (e.g. electricity generation), the costs of technology improvement to cut climate emissions can be given a direct cost, for most transport policies (i.e. not solely technology) there are co-benefits in congestion reduction and health impacts that makes identifying the attributable costs of carbon reduction challenging. This may contribute to the transport sector appearing expensive as a source of emissions reductions compared with other sectors (DfT, 2004b). It also allows for schemes which provide more immediate time savings (which provide monetary benefits linked to economic growth) to be constructed whilst doing little to reduce CO_2 emissions.

There is also a lack of trust and credibility in the underlying logic for pricing, exposed by the incoherent messages produced at the time of the Fuel Duty protests (Marsden, 2002). There is no direct link between fuel taxation and transport spending. Fuel duty is a reasonably good proxy for climate externality costs and these are largely covered by current duty levels. The main externality not accounted for is congestion and whilst the language of the White Papers suggests setting in place the right long-term signals for climate emissions 'the [Labour] Government had no plans to introduce a national system of road pricing' (DfT, 2009a, p. 92), a position maintained by Coalition government.

Mechanisms

Framework: What Specific Types of Instruments Are Utilised

There has been a fluctuation in the types of instruments that have been ascribed to tackling climate change as shown in Fig. 3.

Technological development dominates with around 40% of all references to carbon reduction relating to vehicle technology or fuel technology improvements up to 2007. This has increased significantly in 2009, which is a reflection of the influence of the King Review and the prevailing necessity to decarbonise the transport sector. Whilst the earlier White Papers discussed biofuels and the potential in the longer term for hydrogen fuel cell vehicles, the 2009 strategy firmly underlined the role of electric vehicles and plug-in hybrids as part of a transition pathway. This has signalled substantial subsidy for new vehicle purchase and for the establishment of new infrastructures for recharging vehicles.

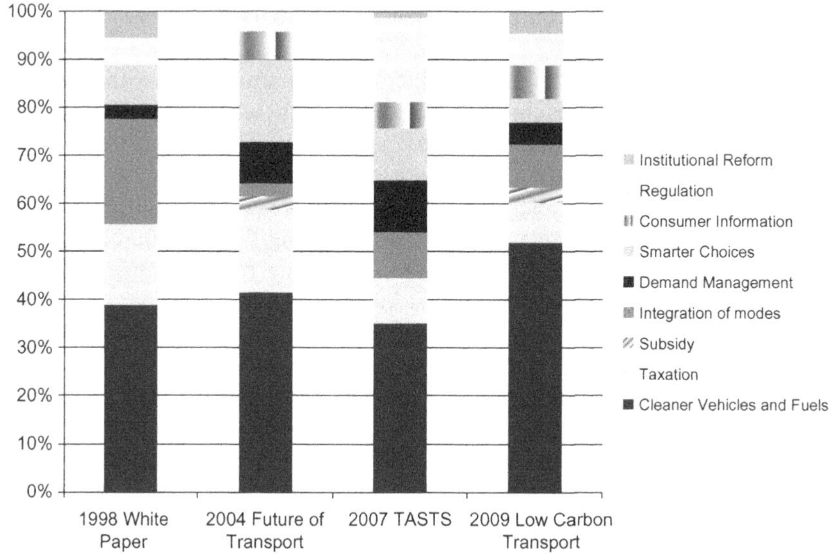

Fig. 3. Types of Instruments Proposed to Tackle Climate Change Emissions.

Taxation has been a consistent theme across the various documents with major changes being made to fuel duty, Vehicle Excise Duty and company car taxation over the period, to provide incentives that are more aligned with lower carbon choices. Subsidy is generally limited, focussing on initiatives such as low carbon buses, new ultra-low carbon car purchases and grants for intermodal freight transfer (to rail and shipping). This implies integration with the Treasury, as it has responsibility for setting these taxes and agreeing spending on subsidies.

The integration of transport modes was unsurprisingly a major feature of the 1998 White Paper. Despite the potential offered by the Smarter Choices report for carbon emissions reduction, this was largely seen as important to congestion, and although frequently cited in the 2004 White Paper, was not identified as having a major impact on climate policy. The 2007 and 2009 analyses put more emphasis on the importance of local and regional solutions as 64% of CO_2 emissions from transport are identified as coming from journeys of under twenty-five miles in length (DfT, 2009a, p. 28).

Regulation has grown in importance, relative to other mechanisms, and also in terms of specificity over the period. There are some common themes. International aviation and maritime emissions fall outside the purview of the UK and European authorities to some extent, requiring agreement by the International Civil Aviation Authority and the International Maritime Organization respectively and these decision-making bodies have generally been slow to react. This has led to the EU seeking to take action where it can, for example through the inclusion of flights within the EU in the EU Emissions Trading Scheme. This was not a feature of the 1998 White Paper, was a possibility in 2004 and was in process by 2009. The car manufacturing industry is also global in its nature (although with a strong European base) and the UK government works within a European framework to influence the efficiency of new cars. This has included a voluntary and now mandatory commitment to the EU to reduce the CO_2 emissions of new cars. Nationally, there has been a shift in emphasis with all forms of transport related regulation incorporating climate objectives. The 2009 Low Carbon Strategy signalled for example, that the rail sector will include environmental performance objectives in the next round of franchise renewals.

Overall, the development of a specific Low Carbon Transport Strategy underlines an increase in the number and sophistication of the mechanisms that can and are being applied to tackle climate change. A note of caution is necessary however, as not all of the policies reported are necessarily effective and the strategy has yet to be reintegrated into a broader national approach. For example, the increase to 100% concessionary fares for over-60s in England (introduced in 2006) is branded as a climate reduction policy when in reality it serves a social inclusion function and may encourage additional bus based trips rather than reducing car travel per se. The decision to pursue high speed rail is also included as a pro-environmental policy when the environmental benefits form only a small fraction of the overall scheme benefits (Gonzalez-Gonzalez, Marsden, & Smith, 2010). There may be strong economic arguments for high speed rail in England but as a climate reduction policy there are very limited co-benefits compared to the opportunity cost of an estimated £16 bn investment.

Calibrations

Framework: What Are the Specific Ways in Which the Instrument Is used?
It is not possible here to review all of the mechanisms set out above so a small number of key exemplars are used:

- Vehicle Excise Duty was reformed in 2005 to incentivise the purchase of lower CO_2 vehicles. Nine bands were established which related to a combination of CO_2 emissions and other toxic air pollutants. In 2010, this was expanded to 13 bands to provide greater differentiation and incentives for purchasing lower carbon vehicles. A specially enhanced first year VED rate was also included which was set at £0 for the lowest carbon vehicles and ranges up to £950 for the highest carbon vehicles.

- The UK was part of a voluntary agreement within the EU for manufacturers to reduce the fleet average new car emissions rates to an average of 140 g/km of CO_2 by 2008 (a 25% reduction from the 1995 level). As a result of slightly disappointing progress (on average) and because of the need to make more ambitious improvements in the coming decades a new agreement to 2020 has been formalised with a target of 95 g/km with manufacturers facing fines for every vehicle sold for each g/km above the threshold.
- An operating subsidy is provided to all bus companies, now referred to as the Bus Service Operators Grant (BSOG). BSOG was, in 1998, a simple Fuel Duty Rebate with operators refunded on the basis of the amount of fuel used. Such an incentive is clearly not well aligned with fuel use reduction but reform to a per passenger subsidy has proven difficult to implement (HoC, 2011). In 2009, an additional rebate was given to low carbon buses, which provides some correction to this.

In general, the calibration of climate related instruments has become more sophisticated over time, as has the evidence base that underpins it. Light goods vehicles and the freight sector more generally appear relatively poorly understood and are not currently treated with the same sophistication as surface passenger transport and efforts here are focussed proportionately more on gathering the evidence base.

The approach to tackling the efficiency of vehicles is far clearer than the approach to reducing the amount of travel and transport activity that takes place. The 2007 White Paper looked ahead to a 2.25% increase in transport spending in real terms per annum up to 2018–19, which would have more than doubled government expenditure on transport from 1997–98. However, little headway was made in curbing the growth in road transport and providing the radical change promised in 1998 with the funds available to 2010 (Docherty & Shaw, 2011). There seems little prospect of a more coherent and ambitious approach to this strand of the strategy in the current environment of reduced funding.

CONCLUSIONS

Returning to the analytical framework, most evidence of first order change can be shown as relating to the settings of policy instruments, as might be expected. Broadly speaking, the same types of instruments and market-based logics have remained in operation over the period, but there have been subtle recalibrations to adjust to new conditions and in response to experience, consistent with Hall's expectations on social learning (Hall, 1993). However, this is only part of the story and social learning alone by no means captures all developments. Alongside established instruments such as fuel taxation, new ones have emerged (e.g. EU Emissions Trading and Air Passenger Duty) suggesting elements of second order change. It would be difficult to argue for paradigmatic change though: wholesale changes in policy are not evident and do not appear to be an immediate prospect. While it might be argued that expert opinion points in the direction of more radical change, this would require a political shift that appears unlikely even with a new government ostensibly committed to becoming the 'greenest ever'.

The framework has proven a useful tool in separating the dimensions of policy change and describing what policy is supposed to achieve (the ends) and what is put in place to achieve it (the means). Further interpretive work is required to understand the appropriateness of the settings and how those in one sector link with those in other (e.g. transport with energy and planning) to understand consistency across policies. In addition, the framework invites but does not formalise an evaluative tier that would consider the degree to which the policies are implemented and how, in a real-world setting, their effectiveness might be impacted by other conflicting or synergistic policies within (e.g. rising bus fares) or outside transport (e.g. the opening up of school choice). In other words, providing a greater focus on the question of 'what difference does this make when policy is put into practice'?

Might current conditions lead to a radical shift in policy? The Climate Change Act creates some of the conditions for such a proposition to emerge in that it provides a framework against which progress will be assessed. This, in turn, might create pressure from outside of the transport sector for accelerated change. The uncertainty and upward trajectory of oil prices may drive greater industrial and individual action more than government policies are able to do and might provide a rare window of opportunity for radical initiatives that would lock in the benefits of change. However, the combination of high oil prices and economic recession also dampen travel demand and allow the government to achieve its short-term ends with limited additional action, perhaps instead creating a window to watch and wait while relevant technology advances.

Although the policy goals of cutting carbon emissions seems relatively uncontested, the politics of developing the solutions are not. The current policy discourse is framed around the potential technological revolution, in the form of a zero emission vehicle fleet, which could also support a green economy and expanding choice. Whilst the prospect of tackling climate change largely through technology exists, it is difficult to see the political need to risk more aggressive behaviour change policies. More restrictive policies are likely to be difficult to implement and hold strong equity implications. Thus, while there are plans for a transition to lower carbon technology, other measures that might reduce the growth of traffic are marked by less commitment. Paradoxically, the transition to lower carbon technology will also, over time, lower the per-mile costs of travel which could create significant adverse impacts on congestion and other aspects of quality of life through the rebound effect.

This chapter has used the case of England to explore how climate change policy is being conceptualised and operated and has focussed on a national-level analysis. Within this, important influences have been identified from a range of actors. These include the European Union, other national level government departments and external agencies and interests including the private sector. The difficulties in adopting an effective long-term strategy under the uncertainty of the future pace of climate change and technological progress and the complex actor networks that define the implementation field are not unique to England. The analytical framework presented here reinforces the importance of understanding the evolution of climate change policy in the broader context of the norms, practices and traditions of area of study and should inform future cross-comparative work. It also suggests that a broad range of analytical tools may be required to fully understand why policy changes.

The analytical framework requires us to resist a snapshot analysis of events and take a longer view. Periods of marginal adaptation can suddenly shift to transformational change, particularly where circumstances external to the policy subsystem change. Such change is less predictable than incremental change and can be prompted by external crisis (war, financial meltdown, natural disasters). The unfortunate irony in this case is that the gradual and incremental nature of change increases the chances of 'external' shock in the form of major environmental change.

ACKNOWLEDGEMENTS

This research was funded by the Economic and Social Research Council (ESRC) as part of research grant ES/J007439/1 'Multi-Level Governance, Transport Policy and Carbon Emissions Management'.

REFERENCES

Bache, I., & Flinders, M. (2004). *Multi-level governance*. Oxford: Oxford University Press.

Beecroft, M. (2002). *From DETR to DfT via DTLR, what are the potential implications for transport planning of these changes in departmental organisation*. Transport Planning Society Position Paper. Retrieved from www.tps.org.uk

Cairns, S., Sloman, L., Newson, C., Anable, J., Kirkbride, A., & Goodwin, P. (2004). *Smarter choices – Changing the way we travel*. Final Report to Department for Transport. Retrieved from http://www.dft.gov.uk/pgr/sustainable/smarterchoices/ctwwt/

CCC. (2011). *Carbon budgets*. The Committee on Climate Change, London. Retrieved from http://www.theccc.org.uk/carbon-budgets

Climate Change Act 2008 c.27. Retrieved from http://www.legislation.gov.uk

DECC. (2011a). *Carbon budgets*. Department for Energy and Climate Change. Retrieved from http://www.decc.gov.uk/en/content/cms/what_we_do/lc_uk/carbon_budgets/carbon_budgets.aspx

DECC. (2011b). *Carbon plan*. March 2011. Retrieved from http://www.decc.gov.uk/en/content/cms/what_we_do/lc_uk/carbon_plan/carbon_plan.aspx

DETR. (1998). *A new deal for transport: Better for everyone*. London: Department of the Environment, Transport and the Regions.

DETR. (2000a). *Transport 2010: The 10 year plan for transport*. London: Department of the Environment, Transport and the Regions.

DETR. (2000b). *Transport 2010: The 10 year plan for transport background analysis*. London: Department of the Environment, Transport and the Regions.

Department for Environment, Food and Rural Affairs (DEFRA). (2005, March). *Securing the future delivering the UK sustainability strategy* (CM6467). London: Department for Environment Food and Rural Affairs. Retrieved from http://www.Defra.gov.uk

DfT. (2003). *The future of air transport* (White Paper, CM6046). London: Department of Transport. Retrived from http://www.dft.gov.uk

DfT. (2004a, July). *The future of rail* (White Paper). London: Department for Transport. Retrieved from http://www.dft.gov.uk

DfT. (2004b, July). *The future of transport a network for 2030* (White Paper, CM6234). London: Department for Transport. Retrieved from http://www.dft.gov.uk

DfT. (2004c). *Full guidance on local transport plans* (2nd edn.). London: Department for Transport. Retrieved from http://www.dft.gov.uk

DfT. (2007a, July) *Delivering a sustainable railway* (CM7176). London: Department for Transport. Retrieved from http://www.dft.gov.uk

DfT. (2007b, October). *Towards a sustainable transport system, supporting economic growth in a low carbon world* (CM7226). London: Department for Transport. Retrieved from http://www.dft.gov.uk/

DfT. (2008, July). *Carbon pathways analysis – Informing development of a carbon reduction strategy for the transport sector*. London: Department for Transport. Retrieved from http://www.dft.gov.uk

DfT. (2009a, July). *Low carbon transport: A greener future. A carbon reduction strategy for transport* (CM7682). London: Department for Transport. Retrieved from http://www.dft.gov.uk/. Accessed on May 12, 2011.

DfT. (2009b, July). *Guidance on local transport plans*. London: Department for Transport. Retrieved from http://www.dft.gov.uk

DIRECTIVE 2003/30/EC of the European Parliament and of the Council. The promotion of the use of biofuels or other renewable fuels for transport. *Official Journal of the European Union*. Retrieved from http://eur-lex.europa.eu

Docherty, I., & Shaw, J. (2011). The transformation of transport policy in Great Britain? 'New Realism' and New Labour's decade of displacement activity. *Environment and Planning Part A*, *43*, 224–251.

DTI. (2003). *Our energy future – Creating a low carbon future* (Energy White Paper). London: Department of Trade and Industry. Retrieved from http://www.berr.gov.uk

Dudley, G., & Richardson, J. (2000). *Why does policy change: Lessons from British transport policy 1945–99*. London: Routledge.

Eddington. (2006). *The case for action: Sir Rod Eddington's advice to government*. London: HM Treasury. Retrieved from http://www.hm-treasury.gov.uk

Giddens, A. (2009). *The politics of climate change*. Cambridge: Polity Press.

Gonzalez-Gonzalez, E., Marsden, G., & Smith, A. E. (2010, July 11–14). *How important are environmental factors in the case for high speed rail? A comparison of the United Kingdom and Spain*. Presented at the 12th World Conference on Transport Research, Lisbon.

Goodwin, P., Hallett, S., Kenny, F., & Stokes, G. (1991). *Transport: The new realism transport studies unit*. Oxford: Oxford University.

Greater London Authority Act. (1999). Retrieved from http://www.legislation.gov.uk

Green-Pedersen, C. (2004). The dependent variable problem within the study of welfare state retrenchment: Defining the problem and looking for solutions. *Journal of Comparative Policy Analysis: Research and Practice*, *6*(1), 3–14.

Hall, P. (1993). Policy paradigms, social learning, and the state: The case of economic policymaking in Britain. *Comparative Politics*, *25*, 275–296.

House of Comments (HOC). (2002). *10 year plan for transport*. Eighth Report of the Select Committee on Transport, Local Government and the Regions, HC 558-I, London, ISBN 0 215 00348 9.

House of Comments (HOC). (2011). *Buses: Grants and subsidies* (House of Commons Library Standard Note SN/BT/1522). Retrieved from http://www.parliament.uk/topics/Buses Archive.htm

Howlett, M., & Cashmore, B. (2009). The dependent variable problem in the study of policy change: Understanding policy change as a methodological problem. *Journal of Comparative Policy Analysis: Research and Practice*, *11*(1), 33–46.

Kern, K., & Bulkeley, H. (2009). Cities, Europeanization and multi-level governance: Governing climate change through transnational municipal networks. *Journal of Common Market Studies*, *47*(2), 309–332.

King Review. (2007, October). *The King review of low-carbon cars. Part I: The potential for CO_2 reduction*. Retrieved from http://www.hm-treasury.gov.uk

Liefferink, D., & Jordan, A. (2004). Measuring Europeanization and policy convergence: National baseline conditions. In A. Jordan & D. Liefferink (Eds.), *Environmental policy in Europe: The Europeanization of national environmental policy* (pp. 32–46). London: Routledge.

Local Transport Act 2000 c.38. Retrieved from http://www.legislation.gov.uk

Local Transport Act 2008 c.28. Retrieved from http://www.legislation.gov.uk

MacKinnon, D., Shaw, J., & Docherty, I. (2008). *Diverging mobilities? Devolution, transport and policy innovation, current research in urban and regional studies* (Vol. 4). Oxford: Elsevier.

Marsden, G. (2002). Fuel taxes and the environment – economy trade off. In G. Lyons & K. Chatterjee (Eds.), *Transport lessons from the fuel tax protests of 2000* (pp. 63–84). Aldershot: Ashgate (ISBN 0 7546 1844 7).

Marsden, G., & Bonsall, P. (2006). Understanding the role of performance targets in transport policy. *Transport Policy*, *13*(3), 191–203.

Marsden, G., Frick, K. T., May, A. D., & Deakin, E. (2011). How do cities approach policy innovation and policy learning? A study of 30 policies in Northern Europe and North America. *Transport Policy*, *18*, 501–512.

Marsden, G., & Rye, T. (2010). The governance of transport and climate change. *Journal of Transport Geography*, *18*, 669–678.

Nellthorp, J., & Mackie, P. J. (2000). The UK roads review: A hedonic model of decision-making. *Transport Policy*, *7*(2), 127–138.

RCEP. (2000). *Energy: The changing climate*. Twenty-second Report of the Royal Commission on Environmental Pollution, London.

Social Exclusion Unit-SEU. (2003). *Making the connections: Final report on transport and social exclusion*. London: Social Exclusion Unit.

Standing Advisory Committee on Trunk Road Assessment (SACTRA). (1999). *Transport and the economy*. The Standing Advisory Committee on Trunk Road Assessment, DETR, London.

Stern, N., Peters, S., Bakhshi, V., Bowen, A., Cameron, C., Catovsky, S., … Zenghelis, D. (2006). *Stern review: The economics of climate change*. London: HM Treasury.

CHAPTER 9

AN ECONOMIC PERSPECTIVE: THE COST-EFFECTIVENESS OF CARBON REDUCTION MEASURES IN TRANSPORT

Abigail L. Bristow and Alberto M. Zanni

ABSTRACT

Purpose – *To examine the cost-effectiveness of UK government policy with respect to the mitigation of carbon emissions from the transport sector.*

Methodology/approach – *Existing policy as set out by the Department for Transport in* Low Carbon Transport: A Greener Future *is examined. This document elaborates a Low Carbon Transport Strategy intended to achieve annual emissions savings of 17.7 $MtCO_2$ by 2020. A wide range of policy areas where further action could be taken to reduce carbon emissions are examined and their cost-effectiveness considered.*

Findings – *Measures that influence behaviour including smarter choices, eco-driving across modes, freight best practice and modest price increases are highly cost-effective. More cost-effective routes to saving 17.7 $MtCO_2$ are identified, as are further cost-effective savings.*

Transport and Climate Change
Transport and Sustainability, Volume 2, 225–259

ISSN: 2044-9941/doi:10.1108/S2044-9941(2012)0000002012

Originality/value – *It appears that government targets could be delivered and indeed exceeded at lower cost than the Low Carbon Transport Strategy. However, policy development is influenced by a wide range of factors which help to explain why cost-effective measures are not always fully exploited.*

Keywords: Transport; carbon mitigation; cost-effectiveness

INTRODUCTION

Economic analysis of climate change can take either a macro- or a micro-perspective. At the macro level, the focus is on either the costs to governments (in terms of gross domestic product (GDP) loss), and society (in terms of welfare), of enhanced global warming particularly in terms of the impacts of rising temperatures and sea levels and the increased frequency of disruptive weather events; or on the cost-effectiveness of particular policies aimed at reducing emissions and hence avoiding some consequences (e.g. Anthoff, Hepburn, & Toll, 2009). Studies taking a micro-perspective have examined, for example, the behavioural impact of carbon reduction policies, both at the individual and firm levels and sometimes covering the acceptability of such measures (e.g. Bristow, Wardman, Zanni, & Chintakayala, 2010; Malueg & Yates, 2006).

In the UK, the economic case for investing in carbon reduction measures was made in 2006 with the publication of the *Stern Review on the Economics of Climate Change* (Stern, 2006). The review made the case for necessity of implementing cost-effective carbon reduction measures.[1]

Here the cost-effectiveness of UK government policy is examined with respect to the mitigation of carbon emissions from the transport sector. In order to address this question, existing policy is examined as set out by the Department for Transport (DfT) in *Low Carbon Transport: A Greener Future* (DfT, 2009a). This document sets out a Low Carbon Transport Strategy (LCTS) intended to achieve annual emissions savings of 17.7 $MtCO_2$ by 2020. A wide range of policy areas are examined where further action could be taken to reduce carbon emissions. It is possible to identify both more cost effective routes to saving 17.7 $MtCO_2$ and additional cost effective savings.

The next section examines the cost-effectiveness of existing policy measures and defines terms used throughout the chapter. This is followed by an exploration of policy areas where additional measures may be developed and where some estimates of cost effectiveness are available. Cost-effective policy packages are then developed and conclusions drawn.

COST-EFFECTIVENESS OF CURRENT POLICY

Before discussing the cost-effectiveness of current policies, a few methodological notes and definitions are provided. In this chapter, greenhouse gas emissions are expressed throughout in terms of CO_2. In order to be consistent with UK policy documents, source emissions are used and international aviation and shipping are excluded. The 1990 baseline for transport emissions is 33.3 MtC or 122.1 $MtCO_2$ (carbon is converted to CO_2 by multiplying by 44/12 throughout). Analysis is based on 2007 when transport emissions were 133.1 $MtCO_2$ (IPCC source category, which excludes electricity consumption for rail operation (DfT 2009b) or 21.8% of total UK CO_2.[2]

The cost-effectiveness of measures is assessed in terms of the cost per tonne saved:

$$\text{Cost per tonne } CO_2 = \frac{(\text{present value of costs}) - (\text{present value of benefits excluding the value of } CO_2 \text{ saving})}{\text{tonnes } CO_2 \text{ saved in the non-traded sector}}$$

The measures in the LCTS and their cost per tonne of CO_2 saved are shown in Table 1. In the table, a negative cost per tonne indicates that the benefits of the measure exceed the costs *before* any carbon benefits are considered. The overall average cost per tonne is £32.81. To place this in context, the central value placed on carbon in the non-traded sector by the Department of Energy and Climate Change (DECC) (DECC, 2009) is £50 (for 2008, the first year for which an estimate is provided).[3] This value is derived from mitigation cost estimates commensurate with the achievement of carbon reduction targets. Compared to the non-traded sector carbon price, the transport sector targets appear to be cost-effective in the context of the long-run mitigation targets. However, approximately 41% of the total savings are to be delivered by ensuring that 10% of fuel is from renewable sources by 2020, a measure that costs £80 a tonne, well in excess of the DECC value of CO_2. The strategy is also very dependent on the delivery of low carbon cars. It is, therefore, worth noting that with respect to the

Table 1. Low Carbon Transport Strategy (LCTS) Carbon Savings and Cost-Effectiveness.

Measures	$MtCO_2$ Per Year in 2020	Cost Per Tonne CO_2 (£)
Revised Baseline		
Fuel duty increases (2p in 2009 and 1p from 2010 to 2013)	2	−916.66[a]
Rail efficiency	0.36	−1[b]
LCTS		
New car emissions 130 g CO_2/km by 2012–2015	3.3	−9
New car emissions 95 g CO_2/km by 2020[c]	3.7	55
Complementary measures cars:		
• Gear shift indicators (GSI)	0.18	−114
• Tyre pressure monitoring systems (TPMS)	0.10	225
• Low viscosity lubricants	0.20	−5
• Low rolling resistance tyres (LRRT)	0.15	−36
• More efficient air conditioning	0.18	173
10% transport fuel from renewable sources by 2020	7.3	80
Possible new EU van emissions regulation	1.9	11
Safe and Fuel-Efficient Driving (SAFED) for bus drivers	0.2	−82
Low carbon emission buses	0.2	−7
Low rolling resistance tyres for HGVs	0.2	−77
Electrification of 750 single track km of rail line	0.2	−40
LCTS overall	17.7[d]	32.81[e]

Source: Adapted from DfT (2009a).
[a]Based on the estimate of cost-effectiveness of the Fuel Duty Escalator.
[b]Own estimate – conservative.
[c]Assumes new car emissions and achieve 98 g/km by 2020.
[d]DfT estimate – does not sum due to rounding.
[e]DfT estimate excluding rail cost-effectiveness.

European Voluntary Agreement to reduce new car emissions, the original forecasts indicated a saving of 14.67 $MtCO_2$, progressively downgraded to 8.43 $MtCO_2$ (Commission for Integrated Transport (CfIT), 2007). In the case of the Renewable Transport Fuel Obligation (RTFO), the 10% requirement is included despite remaining reservations with respect to sustainability (Gallagher, 2008).

There are few measures to encourage behavioural change with respect to personal or commercial travel. The LCTS includes Safe and Fuel-Efficient

Driving (SAFED) for bus drivers, and arguably gear shift indicators and tyre pressure monitoring systems may be classified as behavioural measures as they depend on drivers taking action on information provided to secure any carbon saving. Even including these two technology measures less than 3% of the LCTS savings accrue from behavioural change. Whilst recognising that the adoption of new lower carbon cars also requires behavioural change in the form of purchasing decisions, this is essentially related to the uptake of technology. The LCTS is effectively more dependent on technology to deliver carbon savings in transport than the earlier 2006 Climate Change Programme (Department for the Environment, Food and Rural Affairs [DEFRA], 2006).

Overall, UK policy with respect to reducing emissions from transport, in common with that of other countries, is largely dependent on technological developments both to reduce the carbon intensity of transport fuels and improve the energy efficiency of vehicles. There is, however, increasing recognition both in the academic literature (e.g. Bristow, Tight, Pridmore, & May, 2008; Hickman & Banister, 2007; Stanley, Hensher, & Loader, 2011) and in advice to government from the Committee on Climate Change (CCC, 2008, 2009), of the need for additional behavioural change if a stringent 80% reduction target is to be approached.

However, there appears to be a disconnection, especially in the transport sector, between very ambitious policy targets for carbon reduction and policy reality. With target 80% reductions across the economy, transport cannot rely on other sectors to over-deliver. The LCTS should deliver a reduction of around 5% on 1990 levels by 2020 if all the elements are implemented and perform as anticipated. At present not all are firm and funded, and experience to date with, for example moving to lower carbon cars suggests that progress is often slower than anticipated. Clearly the sector will significantly under-deliver relative to other sectors in the economy. This is partly because the transport sector is 'difficult' in practical terms with millions of individual decisions made every day that influence transport emissions and high levels of car dependency and also politically as the sector is already perceived to be highly taxed and many measures, especially those relating to new technologies, are expensive and uncertain (Bristow et al., 2008).

In estimating the savings from improvements in vehicle efficiency, the DfT analysis allows for a 'rebound effect'. The rebound effect describes the situation in which higher efficiency stimulates an increase rather than a decrease in consumption, and this increase may, in extreme cases, be sufficient to cancel out the original efficiency savings (van den Bergh, 2011).

As improvements in vehicle efficiency reduce the per kilometre cost of motoring, some of the potential savings are taken by consumers in the form of increasing the distance driven and increased in-vehicle comfort. In the case of the 130 g/km target, the losses to the rebound effect are estimated to be 1.3 $MtCO_2$ and for the 95 g/km target 0.9 $MtCO_2$ (DfT, 2009c). Thus, in the absence of complementary pricing measures, 24% of the potential savings from vehicle efficiency are lost and the savings that are secured come at a higher price.

Emissions reductions associated with technological developments are somehow viewed by policymakers as more certain in delivery. However, historically delivery has been slower and less effective with respect to reducing new passenger car emissions than originally forecast. The target reduction in the Voluntary Agreement was 140 g/CO_2 per km to be delivered by 2008. The actual figure in 2008 in the UK was 158 g/CO_2 per km; however, progress since 2007 has been much more rapid than in the earlier years of the agreement reaching 144.2 g/CO_2 per km in 2010. This represents a 24% reduction with respect to 1997, the year in which this data was first reported. 50% of the market, in terms of volume, now consists of vehicles below 140 g/CO_2 per km. Various reasons have been identified for this recent sharp reduction: the recession, the scrappage incentive scheme, very low emissions models appearing in the market, the availability of a greater range of vehicles with low emissions, increases in fuel duty, steeper differentials in Vehicle Excise Duty (VED) including a zero rate for very low emission vehicles, increased oil prices, increased proportion of diesel models and increased environmental awareness. Average emissions for cars currently in use were 172.8 g/km in 2009 (Society of Motor Manufacturers and Traders [SMMT], 2011).

The next section explores whether the planned reductions of 17.7 $MtCO_2$ could be delivered at a lower overall cost and whether greater total savings could be delivered at a lower or similar cost.

COST-EFFECTIVENESS OF ADDITIONAL MEASURES IN THE TRANSPORT SECTOR

This section examines potential measures not yet embedded in adopted policy. The international aviation and shipping sectors are also discussed here, although they are not yet included in carbon budgets. The focus is on measures where some indication of cost-effectiveness is available.

Rail

Government has already adopted a range of measures designed to save energy from rail transport operations and these are now included in the baseline. The launch of two electrification projects at a cost of £1.1 billion builds on this. These are the Great Western Mainline from London to Cardiff and services from Manchester to Liverpool, Preston and Blackpool respectively (Baker, 2011; Hammond, 2011). These schemes are being undertaken on a business case 'Over the medium term this £1.1 billion investment in electrification will be self-financing, paying for itself through lower train maintenance, leasing and operating costs' (DfT, 2009d, p.26).

Network Rail analysis indicates that electric trains are cheaper than diesel for all cost categories on a per vehicle mile basis: 33.3% for maintenance costs, 44.7% for fuel cost and 13.3% for track wear and tear. The lease cost per vehicle is also cheaper by 27.3%. Additional benefits accrue from increased reliability. Carbon emissions per vehicle mile are estimated to be 20.7% lower and will clearly fall further as the energy sector decarbonises. Network Rail, ATOC and Rail Freight Operator's Association (2009) suggests that economically viable electrification could eventually cover 80% of customers.

Clearly electrification should be supported in the context of climate change reduction. As the operating cost savings are high relative to diesel operation, the cost per tonne of carbon as estimated by DfT is negative at −£40 per tonne (for an illustrative example of the electrification of 750 single track kilometres of rail line) (DfT, 2009a). Two specific concerns are firstly, the ability to deliver large-scale public projects to time and budget (Cantarelli, Flyvbjerg, Molin, & van Wee, 2010; Flyvbjerg, Skamris Holm, & Buhl, 2003) and secondly, vulnerability of overhead lines to more frequent extreme weather events in the future. A broader concern is the increasing demand for electricity in a context of aging power stations, which may limit the sectors ability to deliver sufficient clean electricity.

Recently the strategic case has been made for a new high-speed line in the UK, High Speed 2 (HS2) which would run from London initially to Birmingham and then on two lines to Leeds and Manchester respectively. The government argues that it would be instrumental in 'promoting valuable strategic change in the economic geography of Britain, supporting long-term growth and reducing regional disparities' (DfT, 2011b, p. 12). The economic case indicates a cost–benefit ratio of 2:1 including wider economic benefits and 1.6:1 excluding such benefits. The case is not particularly strong as the benefits are highly dependent on price increase assumptions that are lower than those currently being imposed (DfT, 2011a). Here the focus is on any carbon reduction that might be achieved.

At present, emissions per seat kilometre are higher for high-speed rail than for conventional rail; however, this may be offset where higher load factors are achieved leading to lower emissions per passenger kilometre and this will improve further as the electricity sector decarbonises (Network Rail, 2009). The forecast benefits depend on a number of arguable assumptions. Generally, the case for High Speed Rail (HSR) in the UK has been thoroughly reviewed by John Preston for the RAC Foundation (Preston, 2009) who concluded that any carbon savings would be relatively small overall. Greengauge estimates potential savings of 1 MtCO_2 by 2055 for a high-speed network assuming that grams of CO_2 per passenger kilometre fall from 29.5 to 1.3 within this timeframe (Greengauge, 2009). Interestingly, Akerman (2011) concludes that a long distance high-speed link between Stockholm and Copenhagen would achieve carbon saving, but this result is highly dependent on freight traffic switching from road. HS2 would be a passenger only service, with the majority of users forecast to switch from conventional rail services. Limited mode switch is expected.

Any emissions reduction from operation will be critically dependent on the decarbonisation of the electricity sector, the level of diversion from domestic aviation and load factors. Network Rail (2009) estimates that in the context of rapid decarbonisation of the electricity sector, emissions from the construction of new rail infrastructure could account for as much as 70% of total high-speed rail emissions. Much of this is related to concrete and steel so reducing emissions in these industries will be essential.

Support for further electrification is clearly indicated on carbon saving grounds and appears to be cost-effective. HSR is not expected to deliver significant carbon savings. It may yield carbon benefits in the more distant future – to minimise emissions attention must be paid to decarbonising construction.

Life Cycle Emissions of Infrastructure and Vehicles

As Preston (2009) notes with respect to HSR the share of construction costs in life cycle carbon emissions may add 18–35% to the operational CO_2 emissions. Moreover, as the electricity sector is progressively decarbonised this share will increase, as mentioned above, Network Rail estimates the share of construction at 70%. Chester and Horvath (2010) suggest that reducing the carbon intensity of rail construction (especially reducing the use of concrete in station construction) would be the most effective measure in reducing rail emissions in the US case. In the UK, Milford and Allwood (2010) examined rail track replacement using a life cycle analysis concluding that emissions from the annual replacement process could be reduced by up to 40% utilising an innovative double-headed rail. Even if this were not acceptable, other measures could yield smaller reductions.

Huang, Bird, and Bell (2009) suggest that for roads the share of construction costs in life cycle emissions is less than 10%. However, again this share would be expected to increase as emissions from vehicles fall. The Highways Agency (HA) Business Plan for 2009/10 (HA, 2009a) sets specific carbon reduction targets of 5% for their administrative operations and a 600 tonnes CO_2 reduction for network lighting. The midnight switch off of lights on various parts of the motorway network contributes to this (HA, 2009b). Interestingly, the RCEP (2009) in their report on artificial light in the environment argue for reduced outdoor lighting on the grounds of loss of amenity with respect to the night sky and stars and possible impacts on animals and plants. They also suggest that benefits from lighting in terms of road safety and crime reduction may be small in many locations. Other measures were indicated in the last business plan, such as the trial of an advanced telematics system in contractors' vehicles and the reduction in emissions from the office estates and business travel. However, the latest business plan does not indicate a precise reduction percentage objective (HA, 2011).

The choice of materials and technologies is important not only in reducing carbon emissions in construction but also in vehicle production. The SMMT (2010) reports significant improvements in production over the period 1999–2009 with per vehicle produced reductions in energy use of 24%, water use of 45% and CO_2 emissions of 41%. However, the reduced number of vehicles produced in the UK meant that the energy and water usage and CO_2 emissions per vehicle increased by about 20% between 2008 and 2009.

As production increasingly moves to more efficient internal combustion engine vehicles, hybrids, plug in hybrids and electric vehicles emissions in use will fall, this will – barring further productive efficiency gains – increase the share of production in life cycle emissions (Samaras & Meisterling, 2008). Life cycle assessment of the 2009 model Toyota Prius suggests that the level of emissions from production and end-of-life disposal is similar to a conventional vehicle. However, given the lower emissions from operation, the share of production and end-of-life emissions is higher at around one-third of the total (Toyota, 2009). The decarbonisation of the electricity sector should offset this increase to some degree by providing lower carbon intensity energy to the production process.

Although emissions from the construction of transport infrastructure and vehicles are formally allocated to sectors other than transport, clearly the scope for emission reduction needs to be pursued. This need is recognised by government – with the launch of a low carbon review of the construction industry (Department of Business, Innovation and Skills (BIS), 2009) and by the profession with the Institution of Civil Engineers (ICE) state of the nation report on low carbon infrastructure (ICE, 2009).

Smarter Choices

It is worth restating the insights offered by Cairns et al. (2004) as to the relative effectiveness of measures in the UK context (with added material from more recent sources where available):

- *Interventions at the destination to reduce car use*: through workplace and school travel plans resulting in 2–30% reductions in car journeys.
- *Intervention in the nature of access to cars*: car clubs and car sharing schemes. A car club is an organisation that owns a fleet of vehicles which are parked in designated bays and bookable for use by car club members. Each car club vehicle in the UK is estimated to substitute for 20 private cars (Harmer & Cairns, 2011). Car clubs are explored in more detail in a later section on ownership models.
- *Intervention to increase individual awareness of alternatives to the car*: public transport information and marketing, travel awareness campaigns and personalised marketing. 1.5–5% increase in bus use from public transport information and marketing, and 2–15% reductions in general car use from personalised travel plans.
- *Intervention to reduce the need or desire to travel*: teleworking, teleconferencing, home shopping. 10–30% reduction in business travel from teleconferencing, a reduction of 2–6 car journeys per teleworker from teleworking initiatives.

These estimates are from a necessarily limited number of case studies and there will always be a concern about the degree transferability of such evidence:

- The impacts may be overstated as first mover organisations may be more committed than later adopters. Moreover, they do not include public transport emissions for trips transferred from car, rebound effects or

induced traffic; such effects could be significant without additional complementary measures.
- The impacts may be understated as the case studies are relatively early examples of some measures and as such later implementations might be expected to have learnt from previous experience and be more effective. In the longer run, changes in culture and driving behaviour may lead to further reductions in a 'virtuous' circle.

The sustainable travel demonstration towns provide further evidence on the performance of smarter choices. This initiative focused on providing alternatives to the car and encouraging their use through individualised travel marketing. These towns range in size from Darlington with 85,000 people and Worcester at 94,000 to 137,000 in Peterborough. Initial surveys in Darlington suggested that around half of current car journeys made could be made by more sustainable modes (Socialdata & Sustrans, 2005). Preliminary results in 2007 suggested that reductions in car use in excess of 10% were being achieved with increases in the use of public transport, walking and cycling (Merron, 2007).

The final evaluations (Socialdata & Sustrans, 2009a, 2009b, 2009c) contain analysis based on surveys of around 4,000 individuals in each town in 2004 and 2008 using self-reported travel diary data for a nominated day for all household members. The reports do not attempt to analyse the impacts of other changes in the social and economic environment over the four-year period. However, the results are used to provide estimates of town level changes in travel behaviour and an estimate of reductions in CO_2 as shown in Table 2. A note of caution is advised as there are no confidence intervals available for the CO_2 saving estimates here. However, a later

Table 2. Sustainable Travel Demonstration Towns: Car Mileage Reduction and CO_2 Saving Per Annum.

Town	Million Car kms 2004	Million Car kms 2008	Difference in Car kms 2004 to 2008	CO_2 Savings Tonnes
Darlington	355.4	321.1	−34.3	7,000
Peterborough	610.7	579.8	−30.9	6,400
Worcester	427.1	407.8	−19.3	3,900
Total				17,300

Source: Socialdata and Sustrans 2009a, 2009b and 2009c. Average vehicle emissions assumed to be 207.5 g/CO_2 per km.

review for the DfT (Sloman et al., 2010) repeats the carbon saving figure (giving a slightly higher total of 17,510 tonnes).

In all three towns, car ownership increased over the period. Nevertheless, all were estimated to achieve net reductions in car kilometres ranging from 4.5% in Peterborough and 5.1% in Worcester to 9.6% in Darlington. This may be viewed as success especially when compared with an overall increase of 0.9% in car kilometres in Great Britain over the same time period (DfT, 2009e). The estimated reductions in car kilometres exceed the DfT estimate of a 3.7% reduction by 2020 from smarter choices implemented at a national level.

It is not possible to provide a cost-effectiveness estimate here that would be comparable with those in the DfT analysis. However, it is worth noting that in the DfT assessment of smarter choices extension the CO_2 savings form only 7.5% of the present value of benefits. The most important benefits arise from congestion reduction, fuel resource cost savings and noise, infrastructure and accidents with additional small savings in air pollutants. If a similar pattern is assumed here, the initiatives would be expected to be cost-effective. The cost of individualised travel marketing is often higher than other smarter choices measures. These experiments perhaps illustrate what may be achieved solely with 'pull' measures. If such measures were to be supplemented by 'push' measures, whether locally or nationally, one might expect a greater impact.

A parallel experiment was undertaken in the London Borough of Sutton, population 184,400. In this case, the changes were compared to control areas. One achievement was a 50% increase in cycling against static figures for Outer London and a fall in the control area. Although traffic flow fell by 2%, this is less than the falls in Outer London overall of 2.1% and in the control area 2.6% (Mayor of London, Sutton Borough and Transport for London, 2009). The fall in traffic levels will not necessarily fully reflect changes in residents' behaviour as it will include through traffic; however, larger falls in other areas suggest that the scheme is not as yet delivering on the objective of a 5% reduction in residents car use (Mayor of London, Sutton Borough and Transport for London, 2008). Other movements of interest that seek to facilitate behavioural change towards more sustainable practices are the Transition Towns (Hopkins, 2008) and the Slow Cities network (Cittaslow, 2011), both of which work with smaller communities where engagement is feasible.

Overall changes in excess of 20% saving in carbon or reduction in car use in response to single or combined measures or a concerted effort to consider change are unusual. Dietz, Gardner, Gilligan, Stern, and Vandenbergh (2009) suggest that in the US behavioural change in household energy and transport consumption could produce a 20% reduction in carbon emissions with no real impact on comfort – with a comparable reduction in Europe of about 10% given the greater levels of efficiency already achieved. It seems that individuals then experience real or perceived barriers to further change. Studies by Tight, Vicat, Bristow, Pridmore, and May (2007) and Lee-Gosselin (1989) suggest that in transport savings of around 20% are feasible but after that constraints prevent further voluntary change.

Governments are reluctant to base carbon reduction forecasts on behavioural changes because of the uncertainty in delivery and an evidence base that is not yet convincing with respect to achieving large savings.

Nevertheless, smarter choices were assessed by the DfT to be highly cost-effective delivering annual savings of 0.94 $MtCO_2$ at a negative net cost of £74 per tonne (reduced congestion and local environment improvements provide the greatest benefits). The DfT assessment recognises the need for complementary demand side measures such as congestion charging and parking controls and assumes that they are in place (DfT, 2009c). The CCC (2010) estimates a 2.7 $MtCO_2$ saving from smarter choices, which would involve a 5% fall in kilometres travelled by 2020. The effect of smarter choices should improve in line with higher investments in public transport infrastructure. This is an area where action is required alongside complementary measures. These measures may need to be implemented at a national level in the form of fiscal or trading instruments as a mass take up of congestion charging or parking schemes at a local level seems highly unlikely. Local Authorities have had powers to implement road user charging and workplace parking levies since the Transport Act 2000, yet the only application outside London has been a small-scale road user charging scheme running since 2002 in Durham (Durham County Council, 2011), and the workplace parking levy due to be introduced in Nottingham in 2012 (Nottingham City Council, 2011).

Traffic Management and Speed Restrictions

The DfT published a report on the costs and benefits of advanced signalling and traffic management systems in the motorway network in 2008 (DfT, 2008). Results from the M42 active traffic management pilot which included variable mandatory speed limits and hard shoulder running during congestion or incidents concluded that CO_2 emissions were 4% lower with the scheme than they would otherwise have been (HA, 2009c). The DfT estimates cost–benefit ratios for hard shoulder running schemes of around 7:1 whereas for motorway widening they are closer to 2:1 (DfT, 2008). CO_2 emissions are estimated to increase with both motorway widening and hard shoulder running. However, the increase is smaller with hard shoulder running as the increase in speeds is lower and mandatory variable speed limits are used (DfT, 2008).

The CCC (2010) estimates a 1.3 $MtCO_2$ saving from speed limit enforcement at 70 mph increasing to a 2.8 $MtCO_2$ savings if speed limits were reduced to 60 mph. The DfT impact analysis estimates an annual saving of 1.4 $MtCO_2$ from a 60 mph limit at a cost of £306 per tonne. Public acceptability is a critical issue in policy development. The speed limit reduction to 60 mph explored by the CCC received support of only 3 to 8% in national polls (DfT, 2009f; GfK Consumer Services, 2008). The Transport Secretary Philip Hammond announced his intention to consult on increasing the speed limit on motorways and other high standard dual carriageways on 3rd October 2011 (DfT, 2011c). Whilst forecasts of the impact of such an increase are diverse there is no doubt that carbon emissions would increase. The CCC (2011) estimates that such a limit could add 2.2 $MtCO_2$ to emissions, compared with the current situation. The initial analysis by the DfT, as reported in the Guardian (Carrington, 2011), suggests an increase of only 0.67 $MtCO_2$ and an additional 25 fatalities.

Research on Intelligent Speed Adaptation (ISA) based on extensive real world trials concluded that a voluntary scheme would deliver CO_2 reductions on 70 mph roads of about 3.4% and a mandatory scheme in reductions of 5.8%. On lower speed limit roads no savings were identified, indeed in some cases, 60 mph roads, for example small increases in emissions were recorded (Carsten et al., 2008). However, the key benefit attributed to the ISA is accident reduction (never less than 96% of the total benefits across a range of scenarios), and the authors report benefit–cost ratios over a 60 year assessment for a voluntary scheme of between 5:1 and 8.4:1, and for a mandatory scheme of between 10.3:1 and 16.7:1. Clearly any decisions to adopt ISA is reliant on the safety benefits, and any carbon savings would be marginal.

Pricing

Generally, pricing intervention can take the form of tax or permits and their potential effectiveness depends on a number of different factors. The theoretical effectiveness of both trading and environmental tax systems has been first demonstrated by Baumol and Oates (1971), whilst their differences were first formalised by Weitzman (1974). Carbon taxes generally take the form of excise taxes on the carbon content of fossil fuels and have been proposed and implemented over the years amongst the strategies to reduce emissions (Brannlund & Nordstrom, 2004; Jorgenson, Slensnick, & Wilcoxen, 1992). The concept of tradable permits finds its theoretical roots in the theory of property rights developed by Coase (1960) and subsequently by Dales (1968). These systems set a precise limit to emissions and allow marginal abatement costs to vary across sources and have been proposed and implemented over the years in order to allow exchange of permits amongst economic organisations, countries, energy and fuel producers and private citizens (Benz & Truck, 2009; Golombeck & Hoel, 2008; Pezzey, 2003).

A carbon tax on the entire economy has clear potential (Europe Economics, 2008) and was advocated in the RCEP report 'Energy the changing climate' back in 2000 (RCEP, 2000). In the absence of a carbon tax fuel duty may be utilised to lock in savings from technology and encourage further savings.

The CCC estimates that (CCC, 2010) a national road pricing scenario, which is additional to fuel duty, would yield a saving of 5.6 $MtCO_2$ per year by 2020. Although no cost estimates are provided, previous work as part of the Eddington review illustrated the net overall benefits of a scenario of marginal cost pricing, which also resulted in a small reduction in emissions (DfT, 2006).

Fuel Duty increases (over and above inflation) announced in the 2009 Budget were a 2 pence increase in September 2009 followed by annual increases of 1 penny from 2010 to 2013. These increases were estimated to save 2 $MtCO_2$ per year by 2020 and to bring in revenue of £1750 million per year by 2013 (HM Treasury, 2009). The 2011 Budget actually cut duty by 1 penny and introduced the 'fair fuel stabiliser' restricting increases in duty to inflation only when oil prices are above $75 a barrel, and Retail Price Index (RPI)+1p should prices fall below this threshold (HM Treasury, 2011a). No indication of cost-effectiveness of fuel duty increases is given but previous analysis of the Fuel Duty Escalator assessed this measure to be highly cost-effective at −£916.66 per tonne CO_2 (converted from the original −£250 per tonne carbon for consistency) (CfIT, 2007). In the analysis that follows a nominal cost estimate of −£1 per tonne CO_2 saved is applied as there is no current DfT estimate available. This is clearly an underestimate of the net benefits.

Fuel Duty increases are highly effective and will help to lock in other savings. Higher increases would help to avoid the estimated losses to the rebound effect of improvements in-vehicle technologies and reinforce the incentive to behavioural change. Although the effectiveness will fall over time as incomes rise, vehicles become more efficient and alternative fuelled vehicles penetrate the fleet, this is a very effective short to medium term measure. The recent Mirrlees review (Mirrlees et al., 2011, p. 276) suggests that 'in the absence of a major shift to congestion charging there is a case for increasing fuel duties in the UK'. However, they view the need to move to congestion based charging as urgent given the falling effectiveness of fuel duty with respect to the externalities of car use.

A problem in estimating the effect of fuel duty increases is the variation in the underlying price of oil. However, analysis undertaken for the CfIT, 2007 report (Anable & Bristow, 2007) indicated that an annual 2% increase in the price of fuel could secure CO_2 savings of 4.66 $MtCO_2$ by 2020 applying a conservative estimate of price elasticity of demand of −0.15 for the whole period.

Fuel duty is one of the few instruments capable of delivering large carbon savings. The analysis here assumes that the duty increases will lead to an increase in the real price of fuel.

Ownership Models

As efforts increase to encourage the development and adoption of electric vehicles in the UK different ownership models are starting to emerge. This is because the battery stacks are very expensive making the initial purchase cost significantly higher than that of a conventionally fuelled vehicle moreover, the range of an electric vehicle is also worse than that of a conventionally fuelled alternative (van Essen & Kampman, 2011). Business models then need to provide an attractive offer by reducing the upfront cost and/or providing a mobility guarantee, available models include:

- Separation of the ownership of vehicle and batteries, the consumer buys the vehicle at a lower price and rents the batteries or pays a fee for a range of services including batteries and charging. This includes the Better Place battery swap model which is nearing implementation (Better Place, 2011).
- Car club models, driven by manufacturers, for example Daimler's car2go which introduced its first battery only fleet in San Diego on 18th November 2011 (Quigley, 2011).

Car club membership in Britain has increased rapidly from 32,000 in December 2007 to 161,172 at the beginning of 2011. Harmer and Cairns (2011) estimate that:

- For every car club member 20 cars are either sold or not purchased.
- Car club vehicles emit 26% less carbon per kilometre than the average UK car.
- Taking into consideration car driver trips per person, car club members generate about 25% of the average CO_2 emissions of the average license holder. This figure reaches 50% if London's users are considered only.

Car clubs are clearly an area where further investment (TfL is providing £1 million in support) could yield significant and quick savings. In 2008 there were 34.7 million driving licence holders and 27 million private cars (DfT, 2010a) with the average vehicle covering 9,241 miles. Assuming a change from an average vehicle, halving emissions would save around 0.8 tonnes CO_2 per vehicle. This is probably too generous an assumption but even if this is halved to a saving of 0.4 tonnes CO_2 per person with membership of 200,000 that would indicate an annual saving of 0.08 $MtCO_2$. It doesn't seem unreasonable to imagine a membership of 1 million (less than 3% of licence holders) which could then yield an annual saving of 0.4 $MtCO_2$. This is clearly an area where more research is required to identify changes in behaviour and actual carbon savings; how this might change with greater take up; other benefits and costs of car clubs and the potential for the future.

Given that car clubs operate in the private sector and consumers choose to use them, there must be an element of consumer and producer surplus. Thus car clubs in their present form must be highly cost-effective in delivering reductions in carbon emissions.

Localism

Devolution of power and responsibility to local communities may assist in increasing both community resilience and carbon mitigation. Recent initiatives include the Low Carbon Communities Challenge (LCCC) (DECC, 2011) which aimed to test 'community-scale delivery of low carbon technologies, measures and approaches' (DECC, 2011, p. 3). The LCCC has provided funding of £10 million to 22 community initiatives in England, Wales and Northern Ireland for capital projects largely aimed at energy use in buildings. However, a small number of the 22 winning projects include small-scale transport investments in: a bio-diesel tank and 3 diesel pool cars (Hook Norton); a community electric vehicle (Ashton Hayes); a biodiesel pump (Awel Aman Tawe) and cycle parking (Haringey) (DECC, 2011). To date the only analysis of carbon emissions is for the baseline (GfK NOP Social Research, 2011). Clearly if this scheme succeeds there would be scope for increasing the transport content in any resulting policy initiatives.

The current government is committed to maintaining the scale and scope of the Post Office network of about 11,500 branches with stringent access criteria (BIS, 2010). There is recognition of the unique reach of the Post Office: high street banks have a total of about 10000 branches and the four largest supermarket chains about 4,000 branches. Moreover, 93% of the population rising to 99% in towns and cities are within 1 mile of a Post Office (BIS, 2010). Recent initiatives include establishing local branches in established small businesses which can offer longer opening hours (BIS, 2010). Thus the network is ideally placed to offer a wide range of local services that might reduce carbon emissions. These might include collection/delivery points for home deliveries and/or vehicle reception points where vehicles are unloaded and final deliveries made on foot, such measures can assist in reducing emissions from the expanding home delivery market (Zanni & Bristow, 2010).

Aviation

Emissions from domestic and international aviation were estimated to be 34.7 $MtCO_2$ in 2009 which is 5.8% of total UK emissions. Under central forecasts emissions are expected to increase to 47.6 $MtCO_2$ by 2030 and 49.0 $MtCO_2$ by 2050 (DfT, 2011d).

The inclusion of aviation into the EU Emissions Trading Scheme (ETS) is intended to avoid further growth in emissions in the EU and to achieve a limited reduction. The cap is set at 97% of average 2004–2006 emissions in 2012 and 95% in 2013. However, as aviation is part of the ETS the aviation sector will not necessarily reduce its own emissions – as the aim of the trading scheme is secure emissions reductions at least cost. A review of studies on the impact of the inclusion of aviation in the EU ETS by Anger and Köhler (2010, p. 42) concludes that 'despite including the aviation sector in the EU ETS, carbon emissions from the sector are estimated to increase and will roughly double from 2005 to 2050'. The authors also note that there appear to be cost-effective measures for carbon reduction within aviation that are not being adopted.

The DfT estimated positive NPVs from new runways at both Heathrow and Stansted – however, this was based on the old DEFRA shadow price of carbon rather than the new DECC carbon value which is considerably higher. A reanalysis by the Liberal Democrats suggested that the increased carbon costs alone are sufficient to move projects to a negative NPV in some scenarios (Hughes, 2009). There is not space here to rehearse arguments with respect to investment in airport expansion at any length. Boon, Davidson, Faber, Nelissen, and van de Vreede (2008) provide a detailed critique of the economic case for the expansion of Heathrow. It has to be noted, however, that Whitelegg, Haq, Cambridge, and Vallack (2010) estimate that not adding extra runway capacity to any London airport could bring about CO_2 emission saving of about 10% by 2050. The current government does not support a third runway at Heathrow.

Morris et al. (2009a) estimate a marginal abatement cost curve for the UK domestic aviation sector. They estimate that around 14% of CO_2 emissions could be reduced at negative cost to the airlines by 2012. Major contributions would be made by optimising landing and take-off practice (3.47%) (though this is dependent on air traffic control); reduced fuel tankering (3.45%) increased use of turboprops (2.45%) and better use of aircraft capacity (2.03%). The negative cost measures are almost wholly to do with operational practices, the sole exception being lightweighting of existing aircraft (using lighter weight materials in upgrades or retrofitting). The pattern for 2020 is similar but with a significant contribution from improved air traffic management practices. By 2020 they estimate that 24% of emissions could be eradicated at negative cost. Morris, Rowbotham, Angus, Mann, and Poll (2009b) identify a similar pattern for international aviation in Europe.

There would appear to be cost-effective measures in the aviation sector that have yet to be adopted. This would suggest that the incentives to do so are insufficient or that barriers exist. A remaining question is whether aviation should be treated differently from other modes of transport with respect to reducing carbon emissions. It is already treated more generously than surface transportation as it does not have to pay any tax on fuel. The current Air Passenger Duty (APD) is charged per passenger and increases with crude distance thresholds and does not therefore provide an accurate reflection of external costs or a clear incentive to airlines to increase efficiency. Moreover, APD does not apply to private aviation (business jets etc.) or dedicated freighters. The only planes that are subject to fuel tax are leisure flights under the Energy Products Directive (as indeed is leisure boating). The UK government has explored a per-plane duty and concluded that this could not be introduced without international agreement due to concerns over legality and feasibility (HM Treasury, 2011a). Government has recently consulted on simplification of APD and extension to business jets, neither of which is estimated to have anything more than a negligible effect on emissions (HM Treasury, 2011b).

Fig. 1 is taken from the SMMT, 2009 CO_2 report (SMMT, 2009). Whilst it effectively assumes that all these sources of revenue are intended to price carbon, it nevertheless provides a very interesting comparison of tax levels across the transport sector. The increased price of carbon of £51 (or £60 by 2020) for the non-traded sector would clearly bring the ETS phase 2 and red diesel within the cost-effective range. The weakness of the carbon price signal in the aviation sector is clear.

The inclusion of aviation with the ETS will start to provide a more adequate price signal and will encourage take up of cost-effective measures within the industry. However, reform of APD to reflect emissions more accurately appears unlikely in the short to medium term.

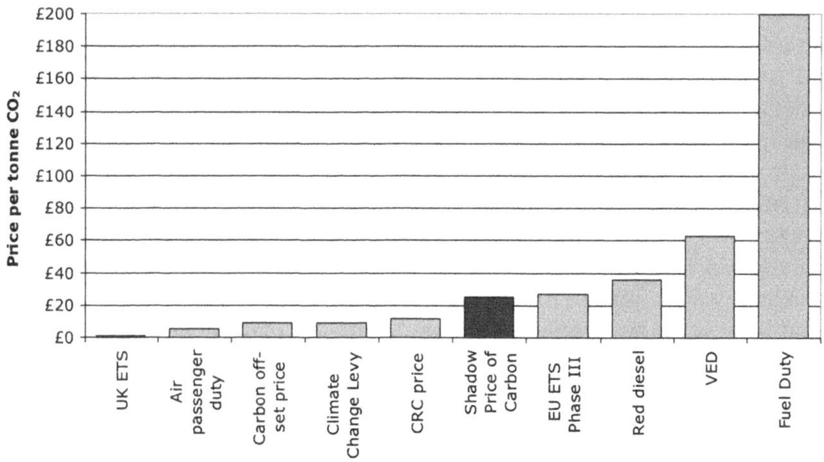

ETS = emissions trading scheme, CRC = carbon reduction commitment, UK ETS = current price, APD - short haul flight, 2tCO₂, carbon offset from JPMorganClimateCare, EU ETS phase III - 2012+ - Deutsche Bank, VED based on average 2007 car's CO₂ and fuel duty based on 60/40 petrol diesel split at 2008 average prices.

Fig. 1. Effective Price per Tonne CO_2 in Different Markets/Policies. *Source*: SMMT 2009.

Shipping

Whilst there is growing concern with respect to emissions from shipping the Environmental Audit Committee takes the view that little progress has been made towards international or EU action (EAC, 2009a). Carbon reduction from the shipping sector may be achieved through operational improvements (e.g. shifting to larger ships and operating at lower speed), technological improvements (improved propellers and aero and hydrodynamic design), cleaner fuels and shore-side measures (like utilising auxiliary engines located in the port to reduce energy consumption on board) (Whitelegg et al., 2010).

Current allocation methods based on bunker fuels suggest that around 6.8 $MtCO_2$ or 1.2% of total UK emissions were from shipping in 2006. In effect if allocated to the transport sector it would add 5% to the current total of 131 $MtCO_2$ emissions (which excludes international shipping and aviation). There will clearly be continuing discussion as to a fair allocation of shipping emissions between countries. Allocation based on bunker fuel probably understates the UK's responsibility here and government agrees this is the case (EAC, 2009b).

Government at present seeks to address the issue through the International Maritime Organisation (IMO) and the UNFCCC. Three strands are envisaged: to reduce emission through technical and operational measures; to establish a carbon price for shipping and a carbon reduction target for the global shipping industry (EAC, 2009b).

'Given the international nature of the shipping industry, and the relatively high costs of abatement, the government believes that the most effective way to tackle CO_2 emissions from ships would be through the implementation of a suitably designed emissions trading system in which shipping participates' (EAC, 2009b, pp. 7–8).

McKinsey (2009) identify cost-effective savings in the world shipping fleet assuming some baseline improvements in efficiency in hydrodynamics and machinery and load factors. Shippers would save money by implementing measures relating to technology and behaviour. Measures involving technology include: improved hydrodynamics through improved hull shape, tailor made propeller design, coating systems and stern flaps; improvements to machinery including engine optimisation and upgrades, waste heat recovery and multiple engines; and alternative fuels. Operational efficiencies include increased vehicle size, speed reductions and further increases in load factors. The net cost per tonne CO_2 saved is estimated to be −€5 falling to −€7 over time as fuel prices rise.

Crist (2009) examined a range of options and concludes that 'Combined technical and operational measures have been estimated to potentially reduce CO_2 emissions by up to 43% per tonne kilometre by 2020 and by up to 63% per tonne kilometre by 2050'. Some of these measures are found to have short commercial payback periods. A cost per tonne of CO_2 is only provided for speed reduction and estimated to cost \$20 to \$50. Corbett, Weng, and Winebrake (2009) estimate a cost per tonne saved of \$30 to \$200 for a mandatory speed reduction of 20%. A tax would probably reduce this cost by allowing shippers to respond in the most efficient manner.

Eide, Longva, Hoffmann, Endresen, and Daløøren (2011) concluded that 19% of current emissions could be reduced at negative or zero cost, for 2020 this increases to 24%. Many of the highly cost-effective measures are to do with operational processes and maintenance.

The Commission of the European Communities (CEC, 2009a) suggested that if no global agreement were reached by 2010 on shipping the EU should take action alone. In a further development they also proposed that revenues from market based measures applied to carbon emissions from international aviation and shipping could be used to support mitigation and adaptation in developing countries (CEC, 2009b). Action has been slow to say the least. However, the IMO announced in the summer of 2011 the adoption of regulations on energy efficiency in new ships from 2013 (IMO, 2011). In the first phase (2015–19) this will require a 10% improvement which will be tightened every five years, so by 2025–30 the improvement should be around 30% (CEC, 2011).

This is an area where unilateral action by individual governments is likely to achieve little. In the presence of open registers (or flags of convenience) any unilateral action is likely to encourage greater flagging out. It appears that the government is encouraging the development of the best policy set internationally. Whilst it appears that there are considerable savings to be made at relatively low cost it is difficult to see how these could be effectively incentivised in the short run.

Freight

Freight transport is estimated[4] to contribute about 20% of total transport emissions in the UK (DfT, 2010). A number of policy initiatives are at the disposal of national and local governments to reduce emissions generated by freight and logistics operations.

The DfT estimates that eco-driving for 90% of HGV drivers could deliver 3 $MtCO_2$ savings in the third carbon budget period from 2018 to 2022 (DfT, 2010) This is clearly a measure that could be implemented earlier and could reach the 90% target by 2020.

There are other cost-effective measures within the freight sector as illustrated by initiatives within the Freight Best Practice Programme which provides advice, case studies and publications to reduce environmental impacts via improved operational efficiency. The Freight Best Practice Impact Assessment estimated the cost per tonne CO_2 saved to be £8 (Lawson, Michealis & Waldron, 2007). The £8 is simply the cost to the public purse of securing the carbon reduction – nothing else is included. The £2 million cost of the scheme during the period they assessed was outweighed by cost saving attributed to the scheme of £83 Million amongst the sample of firms surveyed. This suggests a stunning cost–benefit ratio of 41.5:1 based on the cost of public funds. The assessment did not include any costs incurred by companies in terms of training or vehicle add-ons, which would reduce the net benefit. Nevertheless, it may be assumed that there is a commercial return and that the cost–benefit ratio including these costs is probably still in the very high range.

Major expansion of this scheme is indicated. This would permit greater one to one provision of advice. The idea put forward by McKinnon (2009) of some kind of accreditation scheme for hauliers reaching specific sustainability standards becomes very attractive here.

There are other modifications to vehicles that might be expected to be as (or nearly as) cost-effective as the LRRTs for HGV included in the LCTS, particularly those affecting the aerodynamics or rolling resistance (Baker, Cornwell, Koehler, & Patterson, 2009).

As Heavy Duty Vehicles account for about 24% of road transport emissions (Baker et al., 2009) or 22% of total transport emissions or about 29.282 $MtCO_2$ – what is the scope for reduction at low cost?

It seems possible to achieve around a 10% saving at low, zero or negative cost largely through extension of the Freight Best Practice Scheme. The addition of incentive schemes to encourage the purchase of lower carbon vehicles and add on kit would help as would an accreditation scheme. There are a range of other potential measures identified by McKinnon (2009) that could contribute further, for example consolidation centres, night time deliveries and infrastructure improvements.

Any measures to improve carbon performance in the UK will be undermined if they do not also apply to foreign lorries as they already have a fuel cost advantage. A charging system for foreign lorries that offered a reduced rate for accredited companies might be a very effective approach. This would also generate some revenue and assist in providing a more level playing field for UK hauliers.

Clearly savings are available at low or negative cost. Here a relatively conservative assumption is made of savings of 2 to 3 $MtCO_2$ per annum by 2020 largely from intensive application of Freight Best Practice.

TOWARDS A MORE COST-EFFECTIVE PROPOSAL

Table 3 orders policies in terms of cost-effectiveness to establish the least cost route to additional savings. At this stage only policies are included where DfT has produced estimates of the cost-effectiveness of measures.

The highly cost-effective measures in the revised baseline which achieve savings of 2.36 $MtCO_2$ are taken as given. Here, a more cost-effective approach to saving the 17.7 $MtCO_2$ in the LCTS is looked for in the first instance. If all measures with a price less than the 2020 cost of carbon are included the total saving is 11.93, leaving a gap of 5.77 $MtCO_2$ to be filled. With the measures in Table 3 this may only be achieved by including the 10% renewables saving a total of 19.23 $MtCO_2$. Table 4 shows Policy Package 1 which over achieves the target. The average cost per tonne of CO_2 saved is estimated for year 2020 using a simple weighted average.

Applying the same approach to the measures in the LCTS and using the DfT values per tonne CO_2 produced an average cost per tonne of £42.91[5] (including rail electrification which is a negative cost measure excluded from the DfT calculation). This result does not seem unreasonable as 11 of the 17.7 $MtCO_2$ come at a cost of £55 and above. The most effective single policy saving 7.3 $MtCO_2$ has the highest cost at £80.

The DfT approach of averaging costs over savings for the entire period would be preferred but cannot be replicated for all measures. However, the method used here is consistent between policy packages and suggests that the average cost of a tonne of CO_2 saved may be reduced by around £10 (or nearly a quarter) and an additional 1.5 $MtCO_2$ savings secured simply by:

- Excluding two very high cost complementary measures (more efficient air conditioning and tyre pressure monitoring systems) which deliver very small savings.

Table 3. Cost-Effectiveness of Carbon Mitigation Measures (based on values from DfT (2009a) except where stated otherwise).

Measures	$MtCO_2$ Per Year in 2020	Cost Per Tonne CO_2 (£)	Cumulative Savings
Negative or zero cost measures			
Fuel duty increases (2p 2009 and 1p 2010 to 2013)	2	−916.66[a]	2
Gear shift indicators	0.18	−114	2.18
SAFED for bus drivers	0.20	−82	2.38
Low rolling resistance tyres for HGVs	0.20	−77	2.58
Extend smarter choices	0.90	−74	3.48
Eco-driving for car drivers	0.20	−45	3.68
Electrification of 750 single track km of rail line	0.20	−40	3.88
Low rolling resistance tyres	0.15	−36	4.03
New car emissions 130 g CO_2/km by 2012–2015	3.30	−9	7.33
Low carbon emission buses	0.20	−7	7.53
Low viscosity lubricants	0.20	−5	7.73
Rail efficiency	0.36	−1[b]	8.09
Eco-driving freight sector	0.60	−1[e]	8.69
Measures costing less than the current cost of carbon			
Possible new EU van emissions regulation	1.90	11	10.59
Measures costing less than the 2020 cost of carbon			
New car emissions 95 g CO_2/km by 2020	3.70	55	14.29
Measures exceeding the cost of carbon			
10% transport fuel from renewable sources by 2020	7.30	80	21.59
More efficient air conditioning	0.18	173	21.77
Tyre pressure monitoring systems	0.10	225	21.87
Speed reduction and enforcement at 60 mph	1.4	307	23.27

[a]Based on the estimate of cost-effectiveness of the Fuel Duty Escalator.
[b]Own estimate – conservative.
[c]DfT estimate – does not sum due to rounding.
[d]Excluding rail cost-effectiveness.
[e]As the estimates for both car drivers and bus drivers are negative, eco-driving for freight sector drivers would also be expected to be negative – here the −£1 is a conservative assumption.

Table 4. Policy Package 1.

Measures	$MtCO_2$ Per Year in 2020	Cost Per Tonne CO_2 (£)	Cumulative Savings
Gear shift indicators	0.18	−114	0.18
SAFED for bus drivers	0.20	−82	0.38
Low rolling resistance tyres for HGVs	0.20	−77	0.58
Extend smarter choices	0.90	−74	1.48
Eco-driving for car drivers	0.20	−45	1.68
Electrification of 750 single track km of rail line	0.20	−40	1.88
Low rolling resistance tyres	0.15	−36	2.03
New car emissions 130 g CO_2/km by 2012–2015	3.30	−9	5.33
Low carbon emission buses	0.20	−7	5.53
Low viscosity lubricants	0.20	−5	5.73
Eco-driving freight sector	0.60	−1	6.33
Possible new EU van emissions regulation	1.90	11	8.23
New car emissions 95 g CO_2/km by 2020	3.70	55	11.93
10% transport fuel from renewable sources by 2020	7.30	80	19.23
Average/Total		32.99	19.23

- The addition of negative or zero cost behavioural measures (extending smarter choices, eco-driving for car drivers and HGV drivers).

The next question is whether it is possible to identify measures that would be more cost-effective than the 10% RTFO and deliver the 5.77 $MtCO_2$ required to fill the gap from 11.93 to 17.7. A number of measures have been discussed and here the focus is on low cost per tonne CO_2 options that could be implemented in the short run. Policy Package 1 is taken as the base but the most expensive measure the 10% RTFO commitment is removed. Policy Package 2 in Table 5 fills the gap from 11.93 to 17.7 as follows:

- Extensions to freight best practice 2.5 $MtCO_2$
- Extensions to smarter choices 0.9 $MtCO_2$
- Petrol price increases of 1% per annum to 2020 2.2 $MtCO_2$
- Further rail electrification 0.2 $MtCO_2$

Table 5. Policy Package 2.

Measures	$MtCO_2$ Per Year in 2020	Cost Per Tonne CO_2 (£)	Cumulative Savings
Gear shift indicators	0.18	−114	0.18
SAFED for bus drivers	0.20	−82	0.38
Low rolling resistance tyres for HGVs	0.20	−77	0.58
Extend smarter choices	0.90	−74	1.48
Eco-driving for car drivers	0.20	−45	1.68
Electrification of 750 single track km of rail line	0.20	−40	1.88
Low rolling resistance tyres	0.15	−36	2.03
New car emissions 130 g CO_2/km by 2012–2015	3.30	−9	5.33
Low carbon emission buses	0.20	−7	5.53
Low viscosity lubricants	0.20	−5	5.73
Eco-driving freight sector	0.6	−1	6.33
Possible new EU van emissions regulation	1.9	11	8.23
New car emissions 95 g CO_2/km by 2020	3.7	55	11.93
Extension to freight best practice	2.5	−1	14.43
Further work on smarter choices	0.9	−1	15.33
Additional increase in petrol price of 1% per annum	2.2	−1	17.53
Further rail electrification	0.2	−1	17.73
Average/Total		2.32	17.73

These measures should all involve a negative cost per tonne of carbon – but in the absence of cost-effectiveness estimates on a comparable basis to the DfT figures, they are assumed to have costs of –£1 per tonne CO_2 saved. Policy package 2, shown in Table 5, delivers 17.73 $MtCO_2$ savings at an overall net cost of £2.32 per tonne.

Policy Package 3 simply restores the 10% RTFO and results in average costs per tonne of £25.11 and a saving of 25.03 $MtCO_2$. The policy packages are compared in Table 6.

The ambitious package 3 seems most attractive in a carbon and resource constrained world. It does appear that measures are available to deliver the LCTS target of 17.7 $MtCO_2$ more cost-effectively. The majority of these highly cost-effective measures would require funding from government.

Table 6. Policy Packages and Performance.

Package	$MtCO_2$ Savings in 2020	Average Cost Per Tonne CO_2 Saved (£)	% of LCTS Costs Per Tonne CO_2
LCTS	17.7	42.91	
Policy Package 1 = LCTS minus TPMS and air con + extend smarter choices and eco-driving for car and HGV drivers	19.23	32.99	76.88
Policy Package 2 = PP1 minus RTFO 10% + extensions to freight measures; extensions to smarter choices, further rail electrification and fuel duty increases	17.73	2.32	5.41
Policy Package 3 = PP2 + RTFO 10%	25.03	25.11	58.52

Caveats: although the analysis here is simple and cannot provide accurate estimates, nevertheless it does allow comparison between packages.

However, the remaining measure – fuel duty increases – could be finessed to cover costs.

CONCLUSIONS

This chapter has examined a range of potentially cost-effective measures to reduce carbon emissions from transport. Measures that influence behaviour including smarter choices, eco-driving across modes, freight best practice and modest price increases are highly cost-effective. It appears that government targets could be delivered and indeed exceeded at lower cost than the current LCTS. However, policy development is influenced by a wide range of factors that help to explain why cost-effective measures are not exploited. These include

- The role of the European Union in determining policy. The RTFO and new car emissions measures are amongst the most expensive.
- Acceptability constraints and attempts to introduce congestion charging in UK cities have all failed with the exception of London and Durham. Similarly in recessionary times, there is strong resistance to increases in fuel duty.
- Uncertainties around delivery from behavioural measures.

Most significantly, highly cost-effective measures are identified across a number of areas that are simply not adopted.

There is a real need to explore in more depth the costs of comprehensive behavioural packages for personal, freight and service transport. In order to identify the most effective combinations of policies and facilities that can produce significant changes at least cost.

NOTES

1. The review has been criticised on a number of occasions, in particular for the magnitude of discount rates employed in the analysis and for the treatment of uncertainty, see, for example, Weitzman (2007) and Nordhaus (2007), and more recently, Aaheim (2010).
2. Although data is now available for 2009, we use 2007 for consistency with the LCTS.
3. For a more detailed discussion of the price of carbon, see Mandell (2011) and Tol (2005).
4. Providing accurate measures of freight emission is a complicated exercise. Please see McKinnon and Piecyk (2009) for a comparison of different methods and data sources.
5. This is higher than the DfT estimate which allows for phasing, however, it provides a reasonable basis for comparison.

ACKNOWLEDGEMENTS

This chapter draws on a think piece originally written for the Commission for Integrated Transport in 2009/10.The authors would like to acknowledge this support. The views and arguments expressed in this chapter (along with any errors) are solely the responsibility of the authors.

REFERENCES

Aaheim, A. (2010). The determination of optimal climate policy. *Ecological Economics*, *69*, 562–568.

Akerman, J. (2011). The role of high-speed rail in mitigating climate change – The Swedish case. *Transportation Research Part D*, *16*, 208–217.

Anable, J., & Bristow, A. L. (2007). *Transport and climate change: Supporting document to the CfIT report. September 2007*. Report prepared for the Climate Change Working Group of the Commission for Integrated Transport. Retrieved from http://www.cfit.gov.uk/docs/2007/climatechange/index.htm

Anger, A., & Köhler, J. (2010). Including aviation emissions in the EU ETS: Much ado about nothing? A review. *Transport Policy*, *17*, 38–46.

Anthoff, D., Hepburn, C., & Tol, R. S. J. (2009). Equity weighting and the marginal damage costs of climate change. *Ecological Economics*, *68*, 836–849.

Baker, H., Cornwell, R., Koehler, E., & Patterson, J. (2009). *Review of low carbon technologies for heavy goods vehicles*. June 2009. Report to the Department for Transport.

Baker, N. (2011). *Rail announcement*. Video, 24 March 2011. Department for Transport. Retrieved from http://www.dft.gov.uk/publications/rail-announcement

Baumol, W. J., & Oates, W. E. (1971). The use of standards and prices for the protection of the environment. *Swedish Journal of Economics*, *73*, 42–54.

Benz, E., & Truck, S. (2009). Modelling the price dynamics of CO_2 emission allowances. *Energy Economics*, *31*, 4–15.

Better Place. (2011) *Better Place introduces the first mass-market alternative to the conventional car, the Renault Fluence Z.E., together with membership packages that offer Israeli drivers up to 20% savings*. Press Release, May 15th 2011. Retrieved from http://www.betterplace.com/the-company-pressroom-pressreleases-detail/index/id/Better%20Place%20introduces%20the%20first%20mass%20market

Boon, B., Davidson, M., Faber, J., Nelissen, D., & van de Vreede, G. (2008). *The economics of Heathrow expansion*. CE Delft, February 2008.

Brannlund, R., & Nordstrom, J. (2004). Carbon tax simulation using a household demand model. *European Economic Review*, *48*, 211–233.

Bristow, A. L., Tight, M., Pridmore, A., & May, A. D. (2008). Developing pathways to low carbon land-based passenger transport in Great Britain by 2050. *Energy Policy*, *36*(9), 3427–3435.

Bristow, A. L., Wardman, M., Zanni, A. M., & Chintakayala, V. P. K. (2010). Public acceptability of personal carbon trading and carbon taxes. *Ecological Economics*, *69*, 1824–1837.

Cairns, S., Sloman, L., Newson, C., Anable, J., Kirkbride, A., & Goodwin, P. (2004). *Smarter choices – Changing the way we travel*. Final Report of the Research Project – The Influence of Soft Factor Interventions on Travel Demand. Report to the Department for Transport, London.

Cantarelli, C. C., Flyvbjerg, B., Molin, E. J. E., & van Wee, B. (2010). Cost overruns in large-scale transportation infrastructure projects: Explanations and their theoretical embeddedness. *European Journal of Transport and Infrastructure Research*, *10*(1), 5–18.

Carrington, D. (2011). *Speed limit rise would increase deaths and pollution, admits government, guardian.co.uk*, Friday 30th September, 2011. Retrieved from http://www.guardian.co.uk/environment/2011/sep/30/speed-limit-rise-deaths-pollution

Carsten, O., Lai, F., Chorlton, K., Goodman, P., Carslaw, D., & Hess, S. (2008). *Speed limit adherence and its effect on road safety and climate change*. Final Report. October 2008. Project funded by the Commission for Integrated Transport and the Motorists Forum, Leeds.

Chester, M., & Horvath, A. (2010). Life-cycle assessment of high–speed rail: the case of California. *Environmental Research Letters*, *5*, 1–8.

Cittaslow. (2011). *How Cittaslow works in the UK*. Retrieved from http://www.cittaslow.org.uk/page.php?Pid1=5&PLv=1

Coase, R. A. (1960). The problem of social cost. *Journal of Law and Economics*, *3*, 1–44.

Commission for Integrated Transport (CfIT). (2007). *Transport and climate change. Advice to government from the Commission for Integrated Transport.* London: HMSO, Commission for Integrated Transport.

Commission of the European Communities (CEC). (2009a). Towards a comprehensive climate change agreement in Copenhagen. COM(2009) 39 Final, Brussels.

Commission of the European Communities (CEC). (2009b). Stepping up international climate finance: A European blueprint for the Copenhagen deal. COM92009 475/3, Brussels.

Commission of the European Communities (CEC). (2011). *Commission welcomes the International Maritime Organization agreement to tackle CO_2 emissions.* Press release 18 July 2011. Commission of the European Communities. Retrieved from http://europa.eu/rapid/pressReleasesAction.do?reference=IP/11/895

Committee on Climate Change (CCC). (2008). *Building a low carbon economy – The UK's contribution to tackling climate change.* Report, Committee on Climate Change, London.

Committee on Climate Change (CCC). (2009). *Meeting Carbon Budgets – The need for a step change.* Report, Committee on Climate Change, London.

Committee on Climate Change (CCC). (2010). *The Fourth Carbon Budget. Reducing emissions through the 2020s.* Report, Committee on Climate Change, London.

Committee on Climate Change (CCC). (2011). *Meeting Carbon Budgets 3rd Progress Report to Parliament 30th June 2011.* Retrieved from http://www.theccc.org.uk/reports/3rd-progress-report

Corbett, J. J., Weng, H., & Winebrake, J. J. (2009). The effectiveness and costs of speed reductions on emissions from international shipping. *Transportation Research D, 14,* 593–598.

Crist, P. (2009). *Greenhouse gas emissions reduction potential from international shipping.* Discussion Paper No 2009 11 May 2009. Joint Transport Research Centre. OECD and the International Transport Forum, Paris.

Dales, J. (1968). Land, water and ownership. *Canadian Journal of Economics, 1,* 791–804.

Department for Business Innovation and Skills (BIS). (2009). *The low carbon industrial strategy,* Department for Business Innovation and Skills and Department of Energy and Climate Change, London. Available at http://www.bis.gov.uk/files/file52002.pdf

Department for Business Innovation and Skills (BIS). (2010). *Securing the post office network in the digital age,* November 2010. Retrieved from http://www.bis.gov.uk/assets/biscore/business-sectors/docs/s/10-1260-securing-the-post-office-network.pdf

Department for Environment, Food and Rural Affairs (DEFRA). (2006). *Climate Change the UK Programme 2006.* White Paper Cm 6764, London, TSO.

Department for Transport (DfT). (2006). *Transport demand to 2025 and the economic case for road pricing and investment.* Report, Department for Transport, London.

Department for Transport (DfT). (2008). *Advanced motorway signalling and traffic management feasibility study.* London: Department for Transport.

Department for Transport (DfT). (2009a). *Low carbon transport: A greener future. A carbon reduction strategy for transport.* Cm 7682, July 2009. Department for Transport, London.

Department for Transport (DfT). (2009b). *Transport Statistics GB, 2009 edition.* London: Department for Transport.

Department for Transport (DfT). (2009c). *Impact assessment of the carbon reduction strategy for transport, low carbon transport a greener future.* London: Department for Transport.

Department for Transport (DfT). (2009d). *Britain's transport infrastructure: Rail electrification. July 2009*. London: Department for Transport.

Department for Transport (DfT). (2009e) *Road traffic and congestion in Great Britain*, Q3 2009. 5th November 2009. Department for Transport, London.

Department for Transport (DfT). (2009f). *Public attitudes towards climate change and the impact of transport: 2006, 2007 and 2008*. February 2009 Report. Department for Transport, London.

Department for Transport (DfT). (2010). *Transport carbon reduction delivery plan*. Department for Transport, London.

Department for Transport (DfT). (2011a). *High speed rail: Investing in Britain's future*. Consultation document, February 2011. Department for Transport, London.

Department for Transport (DfT). (2011b). *Economic case for HS2: The Y network and London – West Midlands*. Report, Department for Transport, London.

Department for Transport (DfT). (2011c). *Government proposes 80MPH motorway speed limit*. Press release, Department for Transport, 3rd October 2011. Retrieved from http://www.dft.gov.uk/news/press-releases/dft-press-20111100

Department for Transport (DfT). (2011d). *UK aviation forecasts*. Report, Department for Transport, London.

Department of Energy and Climate Change (DECC). (2009). *Carbon appraisal in UK policy appraisal: A revised approach. July 2009*, Department of Energy and Climate Change, London.

Department of Energy and Climate Change (DECC). (2011). *Low carbon communities challenge*: Interim Report 2010/11, July 2011. Retrieved from http://www.decc.gov.uk/en/content/cms/tackling/saving_energy/community/lc_communities/news/news.aspx

Dietz, T., Gardner, G. T., Gilligan, J., Stern, P. C., & Vandenbergh, M. P. (2009). Household actions can provide a behavioural wedge to rapidly reduce US carbon emissions. *Proceedings of the National Academy of Science*, *106*, 18452–18456.

Durham County Council. (2011). *Durham road user charge zone* (congestion charge). Retrieved from http://www.durham.gov.uk/pages/Service.aspx?ServiceId=6370

Eide, M. S., Longva, T., Hoffmann, P., Endresen, Ø., & Daløøren, S. B. (2011). Future cost scenarios for reduction of ship CO_2 emissions. *Maritime Policy and Management*, *38*(1), 11–37.

Environmental Audit Committee (EAC). (2009a). *Reducing CO_2 and other emissions from shipping*. Fourth Report of Session 2008-09. HC 528. 1st June 2009. Environmental Audit Committee, House of Commons.

Environmental Audit Committee (EAC). (2009b). *Reducing CO_2 and other emissions from shipping: Government Response to the Committee's Fourth Report of Session 2008–09*. HC 1015. 23rd October 2009. Environmental Audit Committee, House of Commons.

Europe Economics. (2008). *A comparison of the costs of alternative policies for reducing UK carbon emissions*. Report for Open Europe. Europe Economics Chancery House, London.

Flyvbjerg, B., Skamris Holm, M. K., & Buhl, S. L. (2003). How common and how large are cost overruns in transport infrastructure projects? *Transport Reviews*, *23*(1), 71–88.

Gallagher, E. (2008). *The Gallagher review of the indirect effects of biofuels production*. Renewable Fuels Agency, St Leonards on Sea, East Sussex.

GfK Consumer Services. (2008). *DfT citizens' panel (Wave 4): Motorway driving*. Report to the Department for Transport, London.

GfK NOP Social Research. (2011). *REAP petite analysis of LCCC*. Retrieved from http://www.decc.gov.uk/assets/decc/11/tackling-climate-change/saving-energy-co2/2405-reap-petite-analysis-of-lccc.pdf

Golombeck, R., & Hoel, M. (2008). Endogenous technology and tradable emission quotas. *Resource and Energy Economics*, *30*, 197–208.

Greengauge. (2009). Fast forward – A high speed rail strategy for Britain. Kingston-upon-Thames. Available at: http://www.greengauge21.net/wp-content/uploads/fast-forward1.pdf

Hammond, P. (2011). *Intercity express and rail electrification*. Oral Statement. 1st March 2011, Department for Transport. Retrieved from http://www.dft.gov.uk/news/statements/hammond-20110301

Harmer, C. & Cairns, S. (2011). *Carplus annual survey of car clubs 2010/2011*. Report, Transport Research Laboratory, Workingham.

Hickman, R., & Banister, D. (2007). Looking over the horizon: Transport and reduced CO_2 emissions in the UK by 2030. *Transport Policy*, *14*(5), 377–387.

Highways Agency. (2009a). Business plan 2009/10. London.

Highways Agency. (2009b). *Looking forward*. Retrieved from http://www.highways.gov.uk/aboutus/24768.aspx

Highways Agency. (2009c). *M42 ATM monitoring and evaluation project summary report*. London.

Highways Agency. (2011). *Business plan 2011–2012*. Retrieved from http://www.highways.gov.uk/aboutus/documents/S100277_Business_Plan_2011-12_Web_4.pdf

HM Treasury. (2009). *Budget 2009 Building Britain's Future*. HC407 The Stationery office, London.

HM Treasury. (2011a). *Budget 2011*, HC 836 The Stationery office, London. Retrieved from http://cdn.hm-treasury.gov.uk/2011budget_complete.pdf

HM Treasury. (2011b). *Reform of air passenger duty: A consultation*, March 2011. Retrieved from http://cdn.hm-treasury.gov.uk/2011budget_airpassenger.pdf

Hopkins, R. (2008). *The transition handbook: From oil dependency to local resilience*. Totnes, Devon: Green Books Limited.

Huang, Y., Bird, R., & Bell, M. (2009). A comparative study of the emissions by road maintenance works and disrupted traffic using life cycle assessment and micro-simulation. *Transportation Research D*, *14*, 197–204.

Hughes, S. (2009). *Government must come clean on Heathrow climate costs – Hughes*. Press release. 24 September 2009. Hammersmith and Fulham Liberal Democrats. Retrieved from http://hflibdems.org.uk/news/000163/government_must_come_clean_on_heathrow_climate_costs__hughes.html

IMO. (2011). *Mandatory energy efficiency measures for international shipping adopted at IMO environment meeting*. Press briefing 42, 15July 2011. Retrieved from http://www.imo.org/MediaCentre/PressBriefings/Pages/42-mepc-ghg.aspx

Institution of Civil Engineers (ICE). (2009). *State of the nation: Low carbon infrastructure*. Report, Institution of Civil Engineers, London.

Jorgenson, D. W., Slensnick, D. T., & Wilcoxen, P. J. (1992). Carbon taxes and economic welfare. *Brookings Papers on Economic Activity: Microeconomics*, *1992*, 393–454.

Lawson, K., Michealis, C., & Waldron, D. (2007). *Freight best practice impact assessment final report*. London: Databuild for the Department for Transport.

Lee-Gosselin, M. (1989). *Voluntary and mandatory restraint on car use in Canada: What would people give up?* Proceedings of Seminar C, PTRC Summer Annual Meeting, pp. 181–192. University of Sussex, Brighton.

Malueg, D. A., & Yates, A. J. (2006). Citizen participation in pollution permit markets. *Journal of Environmental Economics and Management*, *51*, 205–217.

Mandell, S. (2011). Carbon emission values in cost benefit analyses. *Transport Policy*, *18*, 888–892.

Mayor of London, Sutton Borough & Transport for London. (2008). *Smarter Travel Sutton*, First Annual Report 2008, London.

Mayor of London, Sutton Borough & Transport for London. (2009). *Smarter Travel Sutton*, Second Annual Report 2009, London.

McKinnon, A. (2009). *Transport challenges and opportunities*: Briefing Paper on the Freight Transport Sector. November 2009, London.

McKinnon, A. C., & Piecyk, M. I. (2009). Measuring of CO_2 emissions from road freight transport. *Energy Policy*, *37*, 3733–3742.

McKinsey & Company. (2009). *Pathways to a low carbon economy*. Version 2 of the Global Greenhouse Gas Marginal Abatement Cost Curve.

Merron, G. (2007). *Gillian Merron's letter to chief executives on the success of the Sustainable Travel Towns*, 23 May 2007. Retrieved from http://www.dft.gov.uk/pgr/sustainable/demonstrationtowns/

Milford, R. L., & Allwood, J. M. (2010). Assessing the CO_2 impact of current and future rail track in the UK. *Transportation Research Part D*, *15*, 61–72.

Mirrlees, J., Adam, S., Besley, T., Blundell, R., Bond, S., Chote, R., ... Poterba, J. (2011, September). *Tax by design: The Mirrlees Review*. Oxford: Oxford University Press.

Morris, J., Rowbotham, A., Morrell, P., Foster, A., Poll, I., Owen, B., et al. (2009a). *UK aviation: Carbon reduction futures*. Final Report to the Department for Transport. July 2009. Manchester Metropolitan University and Cranfield University, London.

Morris, J., Rowbotham, A., Angus, A., Mann, M., & Poll, I. (2009b). *A framework for estimating the marginal costs of environmental abatement for the aviation sector*. Final Report January 2009. OMEGA, Cranfield University, Bedford.

Network Rail. (2009). *Comparing environmental impact of conventional and high speed rail*. New lines programme, Network Rail, London.

Network Rail, ATOC & Rail Freight Operator's Association. (2009). *Planning ahead*. Control Period 5 and beyond, Britain's railway from 2014. London. Available at http://www.networkrail.co.uk/browse%20documents/planning%20for%20cp5/planning%20ahead%20-%20control%20period%205%20and%20beyond.pdf

Nordhaus, W. D. (2007). A review of the Stern review on the economics of climate change. *Journal of Economic Literature*, *XLV*, 686–702.

Nottingham City Council. (2011). *Workplace parking levy: Information about Nottingham's Workplace Parking Levy (WPL)*. Retrieved from http://www.nottinghamcity.gov.uk/index.aspx?articleid=905

Pezzey, J. C. V. (2003). Emission taxes and tradeable permits. A comparison of views on long-run efficiency. *Environmental and Resource Economics*, *26*, 329–342.

Preston, J. (2009). *The case for high speed rail: A review of recent evidence*. Report number 09/128 October 2009. Royal Automobile Foundation (RAC), London.

Quigley, K. (2011). *Car2go launches all electric rental fleet, San Diego Business Journal*, Monday, 21 November 2011 Retrieved from http://www.sdbj.com/news/2011/nov/21/car2go-launches-all-electric-rental-fleet/

Royal Commission on Environmental Pollution (RCEP). (2000). *Energy the changing climate*. Royal Commission of Environmental Pollution, 22nd report Cm4749, London.

Royal Commission on Environmental Pollution (RCEP). (2009). *Artificial light in the environment*. Report of The Royal Commission on Environmental Pollution, London. Retrieved from http://eeac.hscglab.nl/files/RCEP_artificiallight%5B1%5D.pdf

Samaras, C., & Meisterling, K. (2008). Life cycle assessment of greenhouse gas emissions from plug in hybrid vehicles: implications for policy. *Environmental Science and Technology*, *42*, 3170–3176.

Sloman, L., Cairns, S., Newson, C., Anable, J., Pridmore, A., & Goodwin, P. (2010). *The effects of smarter choice programmes in the sustainable travel towns: Summary Report*. Report to the Department for Transport, Bristol, February.

Socialdata & Sustrans. (2005). *Darlington sustainable travel demonstration town: Travel behaviour research baseline survey 2004*. Report to Darlington Borough Council, Bristol.

Socialdata & Sustrans. (2009a). *Darlington – Sustainable travel demonstration town. travel behaviour research*. Final evaluation report for Darlington Borough Council, Darlington, March.

Socialdata & Sustrans. (2009b). *Peterborough – Sustainable travel demonstration town*. Travel Behaviour Research. Final evaluation report for Peterborough City Council, Peterborough, March.

Socialdata & Sustrans. (2009c). *Worcester – Sustainable travel demonstration town*. Travel Behaviour Research. Final evaluation report for Worcestershire County Council, Worcester, March 2009.

Society of Motor Manufacturers and Traders (SMMT). (2009). *New car CO_2 report 2009: Driving down emissions*. Society of Motor Manufacturers and Traders, London.

Society of Motor Manufacturers and Traders (SMMT). (2010). *11th Annual Sustainability Report*. Society of Motor Manufacturers and Traders, London.

Society of Motor Manufacturers and Traders (SMMT). (2011). *New car CO_2 report 2011: The 10th*. Society of Motor Manufacturers and Traders, London.

Stanley, J. K., Hensher, D. A., & Loader, C. (2011). Road transport and climate change: Stepping off the greenhouse gas. *Transportation Research A*, *45*, 1020–1030.

Stern, N. (2006). *The Stern review: The economics of climate change*. Report to HM Treasury, HM Treasury, London.

Tight, M. R., Vicat, A., Bristow, A. L., Pridmore, A., & May, A. D. (2007). An exploration of household response to personal travel carbon reduction targets. *International Journal of Sustainable Transportation*, *1*, 143–159.

Tol, R. S. J. (2005). The marginal damage costs of carbon dioxide emissions: An assessment of the uncertainties. *Energy Policy*, *33*, 2064–2074.

Toyota. (2009). *Prius and the environment*. Retrieved from http://www.toyotagb-press.co.uk/protected/vehicles/current/press_packs/prius/PriusEnvironmentalDeclaration.pdf

van den Bergh, J. C. J. M. (2011). Energy conservation more effective with rebound policy. *Environmental and Resource Economics*, *48*, 43–58.

Van Essen, H., & Kampman, B. (2011). *Impacts of electric vehicles* – Summary report. CE Delft, Delft, The Netherlands, April.

Weitzman, M. L. (1974). Price vs. quantities. *Review of Economic Studies*, *41*, 477–491.

Weitzman, M. L. (2007). A Review of the Stern review on the economics of climate change. *Journal of Economic Literature*, *XLV*, 703–724.

Whitelegg, J., Haq, G., Cambridge, H., & Vallack, H. (2010). *Towards a zero carbon vision for UK transport*. Report, Stockholm Environment Institute, York.

Zanni, A. M., & Bristow, A. L. (2010). Emissions of CO_2 from road freight transport in London: Trends and policies for long run reduction. *Energy Policy*, *38*, 1774–1786.

CHAPTER 10

A BEHAVIOURAL PERSPECTIVE ON THE RELATIONSHIP BETWEEN TRANSPORT AND CLIMATE CHANGE

Candice Howarth and Tim Ryley

ABSTRACT

Purpose – *To provide a behavioural perspective on the relationship between transport and climate change.*

Methodology/approach – *The factors influencing travel behaviour and the elements critical to behaviour formation are reviewed. The importance of behaviour change measures to reduce the impact of transport on climate change, and the application of behaviour change measures to increase the sustainability of transport, are examined.*

Findings – *There have been a range of travel behaviour measures implemented, such as individualised marketing programmes and travel plans, which have demonstrated some behavioural change impacts, in turn affecting climate change emissions, although they tend to be localised and small-scale.*

Social implications – *There is a real challenge to encourage individuals within society to exhibit more sustainable travel behaviour.*

Transport and Climate Change
Transport and Sustainability, Volume 2, 261–286

ISSN: 2044-9941/doi:10.1108/S2044-9941(2012)0000002013

Originality/value – *A range of behavioural issues still need to be resolved in terms of the relationship between transport and climate change, including a need to influence attitudes, to bridge the gaps between attitudes and both behaviour and intention, to make an impact at points of transition for individuals, to use cognitive dissonance as a way of harnessing social norms, and to understand more fully social pressure and group influence.*

Keywords: Sustainable travel; behaviour change; attitude-behaviour gap; sustainability; climate change

INTRODUCTION: A TRAVEL BEHAVIOUR PERSPECTIVE

Research in travel behaviour has increased over the last twenty to thirty years, incorporating more of the social sciences such as geography, psychology, economics and sociology into transport applications. Methodological examples include modelling the choices individuals make, in this case for transport, using both actual (revealed) and hypothetical (stated) preference data from behavioural economics, and sociology-based approaches such as social network analysis. The behavioural perspective focus of this chapter is through the use of psychological-based approaches such as the Theory of Planned Behaviour, a growing area within transport analysis over the last ten or so years. This approach, to understand more fully the travel behaviour of individuals, is set to continue given the growing concerns around the environmental impacts of transport such as climate change and the associated imperative for behavioural change towards more sustainable forms of travel.

Amongst the range of behavioural perspectives, evidence within this chapter primarily comes from psychology-based approaches, using predominantly UK examples, to understand travel behaviour and particularly land-based travel. Implications for other behavioural approaches are considered in the Conclusions section. Most travel behaviour research focuses on reducing unsustainable car use, including the examples presented in this chapter, although there are similarities in the approach to reduce unsustainable air travel (e.g. Davison & Ryley, 2010).

Factors Influencing Travel Behaviour

Drivers use cars as a means of mobility and as a way of maximising their utility by providing them with the most possible benefits. A fundamental human need for mobility ultimately drives the demand for travel. However, travel also has an intrinsic positive utility and is valued both for its necessity and recreational qualities, not just for being a means to reach a destination (Mokhtarian, Salomon, & Redmond, 2001). Adding to the positive utility of this intrinsic activity are activities undertaken while travelling such as reading, writing (as a passenger), talking and listening to music.

Car trips are undertaken for a variety of reasons including grocery shopping, the school run and long-distance travel; and the car has become a travel mode perceived as very hard to live without (Lucas & Jones, 2009). This perception is easily attributed to the benefits associated with car use such as convenience, independence, freedom, comfort, flexibility, security and status. However, it is also associated with disbenefits such as congestion, air quality, impacts on climate change, road accidents, maintenance costs and the emergence of feeling too dependent on this transport mode. The reasons behind the choice of trip for which the car is the principal mode are varied and complex and often down to personal preferences (some studies have grouped individuals into segments by transport preferences e.g. Anable, 2005; Ryley, 2006). This further increases complexities in changing these behaviour patterns. Policies aimed at reducing traffic congestion, for example, are often not fully effective due to the lack of incorporation of factors which influence traveller's preferences, subjective perceptions, desires and preferences in travelling (Cao & Mokhatarian, 2003).

Gärling (2005) characterises the core determinants of travel behaviour as being: activity choice, spatial organisation of the environment, travel demand and travel choice and that these are ordered according to the following: activity choice, destination choice, mode choice and departure time choice. A variety of factors influence these core determinants of travel behaviour and are often translated into an individual's need or desire to travel. These range from trip purpose (e.g. to get to work), social norms (e.g. abiding by 'rules' set by social structures), information (e.g. mass media), to habitual behaviour and beliefs, that may enhance or inhibit behaviours taking place. The factors will also vary according to the life stage of an individual. Three key life stages have been determined that affect travel needs and choices: (i) having children, (ii) gaining employment and (iii) retiring (Ryley, 2006).

Elements Critical to the Formation of Behaviour

There are several elements that exclusively, and among each other, interact producing a complex system in which behaviours are formed. This section demonstrates why an understanding of all factors that influence the formation of behaviour is required. It is not only necessary to know the reasons why an individual has carried out a specific action but also why they rejected others.

Awareness and Beliefs

Beliefs are a function of two elements: firstly direct observation (i.e. that a car is blue) leading to descriptive beliefs, and secondly the interaction with others leading to the formation of beliefs based on what is not directly observable (i.e. that a person is honest), also termed as inferential beliefs (Fishbein & Ajzen, 1975). The formation of one belief can lead to the development of another (i.e. someone believing buses enable the transport of many people and therefore believes public transport is a sustainable mode of transport) and the influence of awareness and information enhances the value of beliefs and consequently demonstrates the role of information in influencing beliefs about behaviour. Normative beliefs specifically encompass a set of regulations and rules formed within a social group and accepted among its members. These beliefs are particularly influential in justifying an individual's action following conformation that their beliefs are similar to those of his/her close social group (this may be composed of friends, family, teachers, doctors, etc.), in this case in relation to transport and climate change. A lack of awareness or misinterpretation of information can lead to the formation of beliefs: the role of adequate awareness, for example on climate change to ensure accurate understanding of its link with the impacts of personal travel, is a critical first step to creating sustainable behaviour.

A good understanding and awareness is important to ensure attitudes are well informed and lead to corresponding behaviour; however, solely educating people will not change their attitudes or behaviour as an individual's personal characteristics (such as age, gender and profession) also influence their awareness (Kollmuss & Ageyman, 2002).

Norms and Perceived Behavioural Control

Subjective norms refer to the influence of one's social environment on behaviour and intentions. These norms encapsulate how people close to them think they should, or should not, carry out a particular behaviour. These norms refer specifically to the influence of other people on one's behavioural intentions and hold a critical role in the decision-making process as, should an individual be prepared to behave in a certain way, this can still be affected and changed by the beliefs and attitudes of those around them. The more a person perceives that important peers believe or approve that they should perform the behaviour in question, the stronger the likelihood that this behaviour will take place, and vice versa (Ajzen & Fishbein, 1980). Norms are a strong determinant of behaviour and can be harnessed to reduce unsustainable practices and increase sustainable ones. An example of this is the household energy consumption study by Schultz, Nolan, Cialdini, Goldstein, and Griskevicius (2007).

A *perceived* behavioural control is the perceived feasibility of behaving a certain way and relates to the resources and opportunities available to an individual, increasing the likelihood of carrying out that behaviour. On the other hand, *actual* behavioural control relates to a particular behaviour being carried out (Dijst, Farag, & Schwanen, 2008). The latter is observable (i.e. inadequate public transport) and manageable, yet perceived behaviour control is personal and therefore not necessarily obvious or predictable. Perceived controls are produced by a merging of awareness, beliefs and norms which, depending on context, time and so forth, can lead to the reluctance of an individual to perform a particular behaviour (Lorenzoni, Nicholson-Cole, & Whitmarsh, 2007). Stradling, Anable, Anderson, and Cronberg (2008) demonstrate that car drivers show more concern for climate change than non-drivers and acknowledge that changes in behaviour would be beneficial, implying a positive response from consumers towards the reduction in car use, yet people continue to travel further and for longer distances (Department for Transport [DfT], 2009) highlighting the presence of barriers to behaviour change.

Attitudes and Intentions

Attitudes can be defined as an individual's evaluation of a range of aspects of the social world (Baron, Byrne, & Branscombe, 2006). Attitudes are influenced by awareness and knowledge and contribute to behaviour formation (Fazio, 2000). Individuals are usually ambivalent making them less flexible to change which, together with their complexities and sensitivity to human and natural influences, can often make attitudes hard to alter or influence. According to Fishbein and Ajzen (1975), attitudes are predispositions that guide or influence behaviour and the key element of an attitude is the evaluation of the response considered either favourable or unfavourable, which will determine whether such behaviour will be repeated. Attitudes are also influenced by a person's behavioural beliefs acquired by personal experiences: the stronger the belief, the higher the likelihood of its influence on the final behaviour (Kaplan & Fishbein, 1969).

A person's attitude towards a particular behaviour is determined by two things: salient beliefs that performing that behaviour leads to certain outcomes, and secondly his/her evaluation of these outcomes based on his/her motivations to comply with the beliefs of important others (Fishbein & Ajzen, 1975). An individual with strong environmental beliefs, for example, may want to travel more sustainably by using public transport on a regular basis. However, 'important' people around them (i.e. family members, teachers and peers) may consider using the bus as a statement of low status and choose to avoid this mode resulting in the initial sustainable behaviour intention not materialising. This can also occur in favour of sustainable behaviours where an individual reluctant to car share may be encouraged to do so by members of their community who actively collaborate in the school run or lifts to and from work, thereby highlighting and demonstrating the benefits of this initiative.

The best predictor of behaviour, according to Ajzen and Fishbein (1980) is a person's intention to perform that behaviour. This is through an evaluative process, influenced by the attitude towards that behaviour as well as subjective norms. These components are not equally weighted and the importance of each will vary according to the behaviour being carried out, the individual carrying it out, and the context in which this behaviour is to occur.

Intentions are presumed to be an immediate determinant of behaviour and the latter can be somewhat predicted if intentions are properly measured (Ajzen & Fishbein, 1980), but the stability of the intention itself is difficult to predict and control due to its dependence on subjective norms and attitudes. Intentions to act in a certain sustainable manner (i.e. purchase sustainable products) can be overruled by social norms exerted by the society in which an individual resides. Due to the complex interactions between all elements critical to the formation of behaviour such as behavioural control (whether perceived or actual), norms, values and beliefs, the predictability of intentions is, at times, ambiguous.

Cognitive dissonance is an individual's realisation that their behaviour is not consistent with their attitude (Baron et al., 2006) and refers to the existence of an inconsistency between at least two of the following three central variables: beliefs, attitudes and behaviour. This leads to the question that if one's cognitive dissonance is so strong, could this eventually lead to a behaviour change? Research by Aronson (1968) and Festinger (1957) has shown that people attempt ways in order to directly reduce this dissonance by:

- Trying to change their attitude or behaviour to reduce the gap between them;
- Finding information which supports their attitudes; and
- Deciding that the inconsistency is trivial and therefore does not matter.

To reduce feelings of dissonance, people are more likely to change attitudes to increase their consistency with behaviour (Aronson, 1968), and dissonance will be strongest when individuals have weak reasons to explain their inconsistent behaviour and vice versa (Leippe & Eisenstadt, 1994). Individual actions are therefore subject to influence by the norms expressed and followed by their peers and when these create inconsistencies with behaviour, this presents opportunities for influence.

Behaviour

According to Azjen and Fishbein's Theory of Planned Behaviour, combining perceived behavioural control with behavioural intention can be used directly as a tool to predict behaviour (Ajzen, 1991). Culture, society, language, past behaviour and mind all contribute to the formation of one's attitude and behavioural characteristics (Brown, 2004; Dunbar, 1998).

These separate variables rely on human emotional, motivational, perceptual and cognitive processes which in themselves are extremely complex. Even though they are unique, they are accessories in creating an individual attitude which, although may be shared by several people, may lead to distinctly different actions (Krech & Crutchfield, 1948). This transferability provides an opportunity to target behaviour changes without necessarily seeking to alter attitudes.

The formation of behaviour is not based solely on the influence of beliefs on attitudes and subjective norms. Countless external factors, as outlined above, interact within this process affecting the final behaviour. This is shown in Fig. 1 (note that dashed lines indicate influences from external variables).

Although attitudes are closely related to intentions, and the fact that a person's intention to perform a particular behaviour is determined by their attitude and subjective norms towards that behaviour, there is not necessarily a direct link between an individual's attitude and a given intention (Fishbein & Ajzen, 1975). However, an intention may be influenced indirectly by external variables and their impact on attitudes and subjective norms as long as they carry sufficient weighting on either attitude or norms (Fishbein & Ajzen, 1975).

An intention involves four different elements: the *behaviour* in question, the *target* object at which the behaviour is directed, the *situation*, and the *time* at which this behaviour takes place. For example, a person may choose to drive (behaviour) to get to work on time (target), has a car available (situation) and leaves home at 8 am (time). Research by Choocharukul, Tan Van, and Fujii (2007) highlighted how preference towards car use highly influenced residential location and that individuals preferring a lifestyle

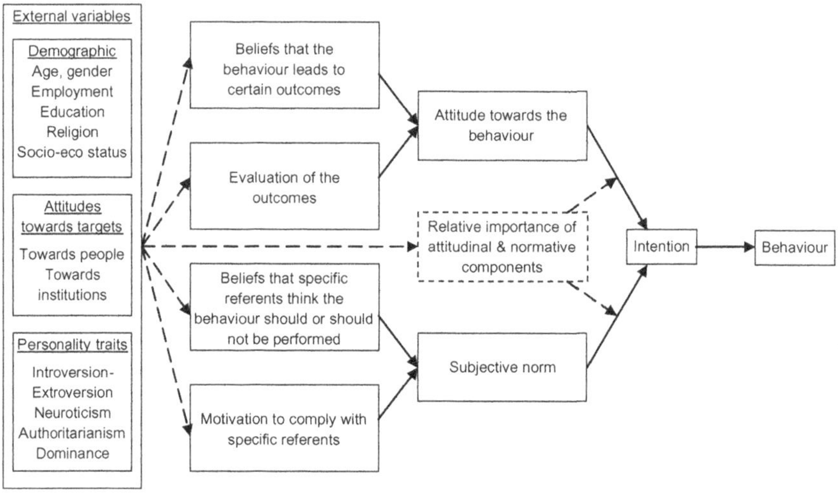

Fig. 1. The Stages Linking Attitudes to Behaviour. *Source*: Ajzen and Fishbein (1980, p. 84).

characterised by heavy car use would be less likely to reside somewhere in the future with good public transport and conversely, those who refrained from heavy car use expressed a lower intention to engage in heavy car use in the future. Existing situational context thereby also strongly influences behavioural intentions.

The issue of the predictability of the role of intention on behaviour has highlighted the existence of a gap in which intention to behave is not always realised (Sheeran, 2002). The Theory of Planned Behaviour expands on this and incorporates perceived behavioural control as an additional element explaining why this discrepancy between intentions and behaviour exists. However, travel behaviour is also habitual, implying that not all behaviours are a direct result of the formation of an intention (Gärling, Gillholm, & Gärling, 1998). In economic terms, as the number of car trips undertaken by an individual increases, the value of them starts to decline and they become habitual, contributing to a diminished marginal utility (Pearce & Turner, 1990). Consequently, habits can help explain the gap between intentions and actual behaviour change. Initiatives would therefore be more effective if they focused on impeding ability to perform existing behaviours and instead facilitate the formation of new behaviours and their transition into habits.

Behaviour can also be an adaptation process both in the short term and long term. Short-term changes refer to issues that may affect the individual within their lifetime in a way that would require a behaviour change, whereas long-term changes are those that would affect the basic survival properties of a population. Climate change and sustainable travel fit these definitions well with impacts and causes observable both in the short term (unsustainable travel behaviour and existing evidence of climate change) and in the long term (necessities to engage in sustainable travel behaviour and adopt new modes of travel with clear impacts of climate change). Increasing behavioural adaptation at the individual level will enable mitigation and adaptation to climate change to be achieved in the short and long term, subsequently helping to address political targets.

In order to influence and change behaviour, the multiple variables described above must be considered. In terms of making travel behaviour more sustainable, the main reason behind this requires an understanding of psychological and social interactions on the individual, interpersonal and community levels, as all of these are part of the factors influencing travel decisions.

THE IMPORTANCE OF BEHAVIOUR CHANGE MEASURES TO REDUCE THE IMPACT OF TRANSPORT ON CLIMATE CHANGE

Measures to Reduce Unsustainable Travel

Addressing emissions from the transport sector is extremely complex as it has numerous actors, consists of copious choices and functions according to diverse policies (Fig. 2). The UK's Department of Energy and Climate Change (DECC) and the Department for Environment, Food and Rural Affairs' (DEFRA) report (DEFRA/DECC, 2009) on 'Making the Right Choices for our Future' specifies that no single solution can reduce the causes and impacts of climate change but that '*using a mix of instruments to tackle climate change has several advantages, including the potential to deliver emissions reductions more efficiently and cost effectively than any single instrument alone*' (DEFRA/DECC, 2009, p. iii).

Strategies aimed at reducing unsustainable travel predominantly aim to target the act of travelling and focus on the use of hard measures related to political and technological advances. Relying solely on one measure to tackle transport's contribution to climate change is unrealistic as this will

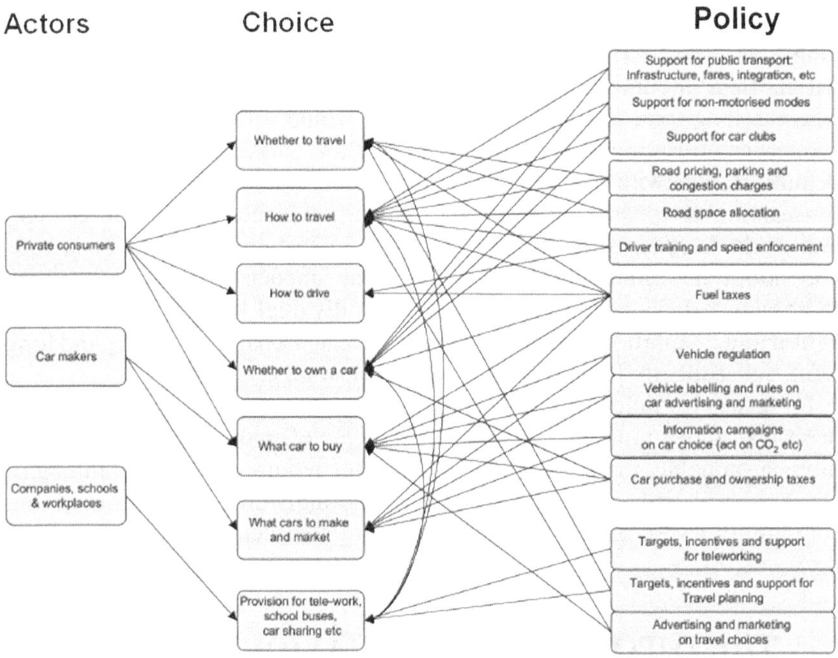

Fig. 2. The Range of Actors, Choices and Policies Transport. *Source*: Gross et al. (2009, p. vi).

not target all issues and will not achieve pre-determined targets. Combining policies, regulations, taxes and charges, technological innovation and demand restraint is the way forward and will benefit both transport and wider economies (Gross, Heptonstall, Anable, Greenacre, & E4Tech, 2009). The Intergovernmental Panel for Climate Change (IPCC) highlights the importance of '*eco-driving styles, increased load factors, improved maintenance, in-vehicle technological aids, more efficient replacement tires, reduced idling and better traffic management and route choice*' (IPCC, 2007, p. 326) as alternative behavioural methods to successfully 'de-carbonise' road transport.

Behavioural Approaches

Political and technological steps have the potential to address unsustainable transport and its link to climate change in the long term. However, it has been argued that the short and immediate terms can be addressed by behavioural changes that will have lasting effects (Anable, Lane, & Kelay, 2006). Using approaches targeted at altering specific behaviours, whilst simultaneously using existing political, technological and economic measures can increase the likelihood of sustained and long-lasting emissions reductions being achieved (European Environment Agency, 2010).

Social marketing programs, awareness campaigns and personalised marketing strategies can have a measurable impact on behaviour change and reduction in CO_2 emissions. In some circumstances, they have been found to achieve a 6–30% drop in car usage (Gross et al., 2009) and a reduction in CO_2 emissions of up to 35% (Fujii & Taniguchi, 2003). The benefits to car use and emissions reduction have been observed to last up to between two and four years, highlighting the benefits of these behaviour-targeting, awareness-raising schemes (Brög, Erl, & Mense, 2002).

The emission reductions required can be achieved by increasing the use of 'soft' measures with a focus away from infrastructure or engineering changes, road space control and so forth (Anable et al., 2006). Soft measures, known in UK policy terms as 'Smarter Choices,' relate to the psychological processes that lead to travel choices rather than the implementation of hard measures which provide an alternative to support unsustainable travel behaviour trends (DEFRA/DECC, 2009). They consist of (among others): workplace and school travel plans, personalised travel planning, travel awareness campaigns, public transport marketing, car clubs and car sharing schemes, teleworking and teleconferencing and home shopping.

These strategies have highlighted that within a ten-year time period, national traffic levels can be reduced by up to 11% and in urban areas at peak periods by up to as much as 21% (Cairns et al., 2008). Soft measures targeting behavioural choices aimed at reducing car use or encouraging more sustainable car use, for example, have the potential to significantly contribute to increasing the sustainability of transport. They, therefore, should be explored to determine the most effective method of implementation. Combining this range of behavioural measures with hard, political, regulatory and technological strategies could significantly contribute to a reduction of greenhouse gas emissions from the transport sector.

Targeting Unsustainable Travel Behaviour

Individual travel patterns vary considerably according to geography, demographics, culture, economic and social aspects. Such patterns also vary according to environmental awareness and concern, lifestyle choices, and availability and supply of energy sources (Fox, 1995; Nilsson & Küller,

2000). The UK Climate Change Programme shows that energy consumption is continuing to rise in the United Kingdom and specifically from the transport sector (DEFRA, 2006). In this light there are numerous options available to make transport at the individual level more sustainable (see Table 1). Many of these options are behavioural in nature.

Table 1. Suggestions for Increasing the Sustainability of Road Travel.

Training and awareness

- Direct feedback: measurements of emissions from car shown
- Driver training: compulsory environmental awareness training during driving education
- Promotion of on-board technologies: diagnostic equipment to identify and report problems; information systems to assist in maximising fuel economy; automated systems to improve fuel economy by maintaining certain vehicle functions

Demand reduction

- Congestion pricing and toll rings
- Workplace parking charge
- Carpooling
- Teleworking
- Fuel price increase
- Video conferencing
- Workplace travel plans

Individualised marketing

- Schemes providing tailored information

Attitude and behavioural shifts

- Flexible working: teleworking and teleconferencing
- Travel planning services
- Car sharing schemes
- Information provision
- Co-modality
- Integrating different modes of transport
- Travel blending

Standards, taxation

- Fuel efficiency standards
- Sliding scale tax on vehicles (Vehicle Excise Duty) to encourage purchase of environmentally friendly cars based on CO_2 emissions
- Carbon fuel tax to reduce congestion and CO_2 emissions
- Vehicle labelling providing more information on environmentally friendly options as well as CO_2 and fuel efficiency
- Parking charges
- Vehicle ownership taxes
- Reduce impacts (visual, audio, health) of transport on environment

Encouraging sustainable travel behaviour requires a tactical multi-faceted approach as achieving one particular change in travel behaviour (i.e. more public transport use) could lead to an increase in other behaviour (i.e. increase in travel overall). Table 1 provides a list of possible strategies to achieve changes in travel behaviour which could be utilised simultaneously not only as a mitigation measure but also as a form of adaptation. Increases in population, reduction of available resources and excessive pollution will cause restrictions on car use. Therefore, behaviour changes achieved now will facilitate future developments. A range of existing soft measures, aimed at reducing unsustainable travel behaviour, could provide potential to inspire long-term behaviour change, especially when combined with regulatory and taxation methods.

Training and awareness, not only of the environmental impacts of travelling but also of 'smart' driving techniques, would allow flexibility on the trips perceived as unchangeable or unavoidable and ensuring these are as sustainable as possible. Providing feedback on the savings (financial, gas emissions) achieved would also enable changes to occur in the long term.

Demand reduction seeks to address car use at the source by utilising a system in which unsustainable and unnecessary journeys (by car, for example) are discouraged. These are achieved by a combined method of hard and soft measures aiming to reduce unnecessary car trips (e.g. congestion charges) whilst increasing awareness of how the revenue derived from these methods is used to fund alternative means of transport or to provide supporting infrastructure. Demand reduction can target work-triggered travel by reducing car trips to and from the work place through the use of teleworking, video conferencing, car pooling and effective travel plans enabling a flexibility car trips and ensuring these are made as sustainable as possible.

Individualised marketing techniques enable the implementation of tailored and personalised travel information to increase awareness of alternative modes available, maximising community participation and ultimately highlighting how small behavioural changes can make a difference.

Making travel behaviour sustainable leads to attitude and behavioural shifts through flexible working schemes, increasing awareness of alternative methods of travel through the provision of relevant information. These methods target travel behaviour directly while addressing underlying issues which influence it (i.e. attitudes); this multifaceted approach allows for a flexible and attractive option targeting land-based travel behaviour which may be perceived as unchangeable. The term 'travel blending' specifically is used in two ways, originally coined by Rose and Ampt (2001) and referring to tailored household strategies consisting of packages aimed at encouraging sustainable use of transport. A more general definition however refers to the blending of trips and/or activities thereby reducing their overall environmental impact (Ampt & Rooney, 1998), commonly referred to in transport terms as trip chaining (Schmöcker, Fengming, & Noland, 2009).

'Kick starting' initial travel behaviour changes often require standards and taxation; soft measures are then used in conjunction to ease behaviour change and ensure long-term patterns occur. Fiscal measures targeting all aspects of car use, such as parking charges, fuel taxes and vehicle ownership taxes governed by vehicle CO_2 emissions must be related directly to emission reduction targets and then supported by measures aimed at facilitating behaviour changes in line with these standards and taxation strategies.

Enforcing a variety of technological and fiscal measures may divert attention away from the underlying issue, the continuously increasing vehicle fleet on roads. Even if CO_2 emissions per vehicle are reduced, and technological advances ensure a maximum sustainable system, this could be offset by car users driving more. It highlights the failure of existing measures in addressing the increase in car use. Research has shown that perceptions and attitudes of car drivers and travellers are crucial in influencing behaviour (Van den Bergh, Van Leeuwen, Oosterhuis, Rietveld, & Verhoef, 2006), and therefore must be explored to ensure sustainable travel grows. Attitudes towards travel behaviour, the impact of travel on the environment and the benefits of reducing unsustainable travel behaviour must be measured, evaluated and fully understood to ensure that a shift in travel behaviour is effective in both the long term and the short term.

Individuals respond in different ways for any policy measure relating to a reduction in private vehicle transport trips. Ryley (2010) states, using a road user charge example, that there are three travel behavioural options an individual faces in response to a transport policy measure: to continue to use private vehicle transport for a given trip with the same occupants; change at least one of the following aspects to the trip (mode, route, time of journey, destination, occupancy and/or frequency of travel), or decide not to make the trip.

Overcoming Perceived Barriers to Travel Behaviour Change

Travel behaviour in itself is a complex process where, as well as the interactions of multiple elements taking place, as described above, there is the intervention of the perceived behavioural control aspects. This results in the complex and, at times, inconsistent nature of travel behaviour. Issues such as travel mode, cost of alternative modes, travel time, travel purpose (leisure or work), number of passengers, comfort, convenience and so forth all play a key role in determining travel behaviour. All these elements, in the context of travel behaviour, have been examined to help determine reasons for specific behaviours and to influence change by investigating the role of changing attitudes (Anable et al., 2006). The general role of information influencing travel behaviour has been found to have a small impact (King et al., 2009), yet the extent to which *specific* information on climate change tailored to attitudinal and behavioural intentions, whilst simultaneously investigating how perceived behavioural control can be incorporated into sustainable travel policy, has not been fully analysed.

Perceived barriers can be responsible for a lack of behavioural actions, even if these are directly linked to an increase in well-being (Lorenzoni et al., 2007) and even if individuals demonstrate clear understanding of the benefits of changing behaviour (Stradling et al., 2008). Perceived social barriers can include a perceived lack of political, business and industry action, a lack of suitable alternatives (or information on alternatives) and a mistrust of information sources. Perceived individual barriers include a lack of knowledge, levels of scepticism, an externalisation of responsibility and blame, and a reluctance to change lifestyles or a feeling of helplessness (DEFRA, 2008; King et al., 2009; Lorenzoni et al., 2007; Lyons et al., 2008). The diverse nature of these perceived barriers presents a challenging task to instigate sustainable and long-lasting behaviour change.

Application of Behaviour Change Measures to Increase the Sustainability of Transport

Attitudes towards Climate Change and Sustainable Travel

Overall awareness of climate change in the UK is high (Norton & Leaman, 2004), yet knowledge and understanding of specific details such as the causes of climate change, the contribution of each individual, and more complex scientific processes, is more limited.

Identifying socio-economic groups whose knowledge, attitudes and understanding of climate change leads them to perceive it as a less pressing issue is key to initiate sustainable travel behaviour changes. The 2009 Euro Barometer on attitudes shows that climate change is considered the third most important issue facing the EU today after poverty food issues, with 67% of Europeans stating it is a very serious problem. There is a high level of optimism across the EU on the ability to combat climate change, but whether this translates into actual behavioural change is still questionable (Euro Barometer, 2009).

The inability for an individual to perceive the impacts of climate change, as well as envision the link between their personal activities (i.e. energy consumption) as a source of climate change strongly reinforce the existence of barriers to behaviour change (Howarth, Waterson, & McDonald, 2010). Salience of climate change, or being able to visualise its scope, is a significant indicator of perceptions of climate change and may help reduce self-denial to a level that can produce behaviour change (Lorenzoni et al., 2007). Confusion of the basic processes on the topic of climate change (Sterman and Booth Sweeney, 2002) can be attributed to a lack of personal connection to the issue in question: climate change is often perceived as being too abstract, whose related implications are too distant in time and space, with reduced credibility due to its uncertain and confusing nature, and lack of political commitment (Nicholson-Cole, 2005). Climate change is an area where the evidence is '*voluminous, the risks considerable, and the effects widely felt*' (Lewis, 2004), which can lead to a distancing from the issue and a sense of diminished responsibility leading to a perceived lack of urgency (Whitmarsh, 2004).

Attitudes to the use of different travel modes vary but show favouritism associated with the use of the private car. This mode of travel is perceived as a mode of travel reflecting independence and freedom and extreme practicality, while public transport services are viewed as inadequate (Van Vugt, Meertens, & Van Lange, 1995). This sense of freedom, convenience and flexibility is felt especially among women due to their complex life-styles revolving around work and children, reflected by a noticeable recent increase in car use by women. These factors inevitably lead to car dependence which can consequently inhibit personal independence and enhance habitual behaviour. It has been demonstrated that car ownership increases in line with income, yet a drop in income is not mirrored by a subsequent decrease in car ownership. This is due to the perceived necessity of car use and consequently, the costs of owning and running a car (fuel, tax, insurance, maintenance) are incorporated into household budgets (Dargay, 2001). In comparison with car use, public transport use is often determined by value for money, perceived levels of service, congestion and car parking issues which make the likelihood of using buses more likely (Tertoolen, van Kreveld, & Verstraten, 1997). The bus is also at times considered a sign of social status and an increase in patronage would only occur if the efficiency and cost effectiveness outweighs the benefits of car use (Fujii & Tan Van, 2009). It is therefore important to increase the appeal of alternative transport modes to encourage their use and simultaneously highlight the negative features of car use which current strategies fail to incorporate.

Individuals have a fairly good understanding of the impacts of each transport mode, and Lyons et al. (2008) demonstrate how a high level of awareness and concern for climate change is mirrored by acknowledgement that human activities contribute to this and specifically the transport sector. There also exists a willingness to change travel behaviour but only if these changes will be directly beneficial to the user (Hobson, 2001). The underlying message is that concern for the environment and the impact of transport on climate change exists. However this is equally met with a concern regarding necessary changes to existing lifestyles to accommodate changes. The Government could, therefore, take the leading role in making these changes as simple as possible through implementation of behavioural measures to supplement technological and political initiatives.

The majority of individuals believe that the transport sector contributes significantly to climate change with up to 70% of individuals sampled by DEFRA (2008) agreeing that road transport is a significant contributor and 52% of individuals believing that using a car less would have a major impact on reducing the UK contribution to climate change (DEFRA, 2007).

Over 75% of UK households own at least one car and 70% of individuals own a drivers licence implying that car use holds a fundamental place in UK individuals' lifestyles (RAC Foundation, 2010), with 80–90% individuals stating they would find it hard to adjust their lifestyles without the use of a car (Lucas & Jones, 2009). The car is thereby perceived as a very desirable travel mode due to its association with convenience, coverage, flexibility, wide access coverage, security and status building. Lucas and Jones (2009) highlight how it is not specifically the car itself but what it offers that is most attractive to individuals. Therefore, if alternative options were to offer at least the same benefits (that were made clearly visible), this could increase the likelihood of its usage. The combination of habitual travel with the strong perception associated with the car helps explain why intention is not always consistent with behaviour.

Evidence from by Sloman et al. (2010) highlights that the inflexibility and habit-driven nature of car use, as discussed previously, can in fact be altered. Results from the Sustainable Travel Towns initiatives in the United Kingdom showed that the use of carefully constructed social marketing techniques achieved an overall reduction in car trips by 9% and a reduction in distance travelled by car by 5–7% per person in the three towns (Darlington, Peterborough and Worcester). The largest reduction in car trips was achieved for medium and long-distance trips with a 45%

reduction. Large reductions were also achieved according to trip purpose through education-related trips (25%), work-related business trips (23%), shopping trips (14%) and leisure trips (12%). In addition to this, the most relevant changes were undertaken by individuals at a 'point of transition' in their life (i.e. attending college, seeking employment or recently retired), where personalised travel planning was aided by the provision of travel information tools about the surrounding areas and available transport options. Although personal associations with the car make travel behaviour change seem difficult to change, the use of properly tailored tools demonstrate that these barriers associated with the car and its purpose for fulfilling specific car trips can be overcome in certain circumstances.

Achieving Sustainable Travel Behaviour

Bridging the Attitude–Behaviour and Attitude–Intention Gaps

Studies have shown (Anable et al., 2006; De Young, 2000) that although pro-environmental attitudes can be conducive to sustainable behaviour, this only occurs under specific conditions where any *barriers* such as habit and peer pressure to behaviour or attitudinal changes are removed or reduced. However, information often considered to be adequate does not necessarily lead to pro-environmental behaviour (Howarth, Waterson, & McDonald, 2009).

It is vital to gain an understanding of people's perceptions and knowledge of climate change if sustainable behavioural changes are to be achieved (Bord, Fisher, & O'Connor, 1998). Identifying the characteristics of each typology of attitude will facilitate the establishment of a new scope for personalised targeting to increase the environmental awareness of each attitudinal group (Howarth, Waterson, & McDonald, 2011). DEFRA (2008) and Anable (2005) identify seven and six population segments respectively according to their environmental concern, behavioural changes, feelings of moral obligation, motivators and barriers to behaviour change, ecological world views, lifestyle choices and awareness. The DEFRA segmentation groups centred on environmental attitudes, varied in accordance with their willingness and their perceived ability to act, enabling suggestions of tailored behaviour changes as well as tailored strategies to implement these including for those falling into the group characterised by a low ability and willingness to act. The segmentation in Anable (2005) is established in relation to travel and the ability and willingness to change travel behaviour for environmental reasons. As identical behaviours can occur for different reasons (just as similar attitudes can lead to different behaviours), in order for policy developments to be successful an approach based on individually tailored incentives is required.

Anable et al. (2006) recognise that the complexities imbedded within the field of sustainable travel behaviour are so profuse that a strategic yet flexible and transparent approach is required to address all complexities and ensure an appropriate behavioural change. Where provision of information has been considered to be elementary and crucial in the attitude–behaviour influencing process, Anable et al. (2006) found two conflicting views on the matter. The first is that if individuals are sufficiently informed on climate change, then they will use this information accordingly to act in a more sustainable manner. The second view acknowledges that information is required but is only part of the solution and that a lot of attention is required on understanding fully all the dimensions (psychological, economical, social, etc.) behind behavioural change and to ensure they are applied appropriately across all levels of society. High levels of awareness, however, have not resulted in changes in travel patterns, suggesting that awareness is high, yet poorly linked to behavioural intentions and resulting in a lack of sustainable behaviour.

Combining environmental awareness with an increasing sense of moral obligation plays a role in counteracting the failures of increasing awareness on the unsustainable nature of car use. Population segmentation provided by DEFRA (2008) and Anable (2005) suggests that policy developments would be most successful if applied to groups who are open to using alternative modes of transport, and those currently using alternatives, as well as taking into account motivators and barriers to changes in each group. The impact of small incremental changes on groups of people who are already performing a desired behaviour is more likely to succeed than targeting groups of individuals who show no willingness or intention to change behaviour. However, it could be stated that each target group is not exclusively defined by its title and all individuals have elements required to increase behaviour changes, and should therefore not be excluded from behaviour-based initiatives.

The Importance of Points of Transition

Habit is important as it determines existing travel behaviour. It can mostly be altered at points of transition in life when key events occur. However, research has shown that chronic habit can occur in spite of interventions at these key life stages (Verplanken, Aarts, & Van Knippenberg, 1997). Although repeated behaviour often means past behaviour is a significant indicator of future behaviour, over and above attitudes and intentions, especially for travel behaviour, the influence of habit in the formation of travel behaviour can be weakened if the situational context is altered (Verplanken et al., 1997).

Prillwitz, Harms, and Lazendorf (2006) assume travel behaviour is habitual and therefore that changing travel behaviour can only occur when there are random and rare windows of opportunity when the context changes. This draws from the mobility biographies approach which deals with the development of travel behaviour over the lifespan of an individual. It states that although travel demand at an individual level is fairly stable in the medium term, changes can occur as a result of significant events and alter the context in which a behaviour is formed (Scheiner, 2007). According to Kloeckner (2004) and Van der Waerden, Timmermans, and Borgers (2003), key life events that are likely to have an impact on travel behaviour by unfreezing habits are:

- Change in number of individuals in a household
- Change of work situation
- Getting a car
- Getting a job
- Getting a licence
- Getting married
- Going to university
- Having children
- Moving home

These events are likely to have a significant impact on travel behaviour as they alter existing psychological constructs, break routine and require a re-orientation of travel-making decisions (Kloeckner, 2004). Triandis (1977) argues that intentions influence actual behaviour when these are not routine and the more a behaviour is repeated the more it is influenced by habitual behaviour than intention. In order, therefore, to maximise the effectiveness of existing travel behaviour change strategies they must be innovative in order to increase the influence of intentions.

Influencing Voluntary Travel Behaviour Change

Intention, although a significant determinant of behaviour, does not always translate into actual behaviour, implying that external variables influence this relationship. Research has shown that intentions are often influenced by habit where past behaviour, and its frequent repetition, means that behaviour occurs out of habit rather than following conscious deliberation.

Implementation intention, where a plan encompassing the when, where and how the behaviour change can occur is carried out by an individual (Gollwitzer & Brandstätter, 1997), is more likely to lead to actual behaviour. If this process of implementation is harnessed and utilised at points of transition (as described above), where habit has the least influence on behaviour formation, intention is most likely to translate into actual behaviour.

In order to get people to change behaviour, an individual's motivation to travel must be well understood and behaviour change must be targeted to ensure that change is achieved. Crucially, it is necessary to encourage travel behaviour changes that will actually appeal to people (Ampt, 2003). When a decision to change behaviour is made, it is more likely to be translated into actual voluntary travel behaviour change if there are sufficient options available to support this choice, as well as the existence of reinforcement in the form of personal benefits (i.e. accrued over time) or the recognition of infrastructure support. To encourage behaviour change, recognition must be made on the fact that all individuals are different and that each type of personality has a different travel behaviour preference. It must be noted, however, that similar attitudes can lead to different behaviours (Anable, 2005).

CONCLUSIONS

There is a strong desire amongst individuals for unsustainable travel (both motor car based, as discussed in this chapter, and air passenger travel). This chapter has highlighted a need for the deeper understanding of the travel behaviour complexities, including the myriad of determining factors, the varying impacts of underlying attitudes and social norms, concepts such as cognitive dissonance, and travel behaviour features such as habitual car use. This is particularly as, although awareness and concern for climate change is high amongst the public, individuals tend not to acknowledge their environmental impact or be willing to change their travel behaviour. There is, therefore, a real challenge to encourage individuals within society to exhibit more sustainable travel behaviour.

Using a psychology based behavioural approach, relevant elements critical to the formation of travel behaviour have been discussed, relating to deeper attitudes, beliefs and intentions of an individual, together with outside elements such as norms and perceived behavioural control at interpersonal and community levels. The Theory of Planned Behaviour, together with associated psychological constructs and approaches, has been presented within the chapter. Further developments of such frameworks are required to incorporate transport-related complexities such as habitual behaviour and the cyclical nature of behaviour change (e.g. relapse – a transport-related example of this is Thøgersen, 2005) as well as environmental aspects such as individual travel behaviour in response to climate change concerns.

The importance of travel behaviour change measures has been manifested in UK policy terms through programmes such as Smarter Choices and Sustainable Travel Towns. There have been a range of travel behaviour measures implemented, such as individualised marketing programmes and travel plans, which have demonstrated some behavioural change impacts, although they tend to be localised and small-scale. Despite current budgetary cuts, this looks set to continue with an increased role within UK government of influencing behaviour through 'nudging' (see Thaler & Sunstein, 2009), whereby individuals are subtly encouraged to adopt certain behaviours.

As demonstrated by this chapter, a range of behavioural issues still need to be resolved in terms of the relationship between transport and climate change, both in terms of mitigation measures, the primary focus of this chapter, and in behavioural adaptation. There is a need to influence attitudes towards climate change and sustainable travel; to bridge the gaps between attitudes and both behaviour and intention; to make an impact at points of transition for individuals; to use cognitive dissonance as a way of harnessing social norms; and to understand more fully social pressure and group influence. It is envisaged that more urgent travel behaviour policy actions, consisting of both hard, and the more behavioural, soft measures, will be required over the coming years as the climate change implications become more apparent.

REFERENCES

Ajzen, I. (1991). The theory of planned behaviour. *Organizational Behaviour and Human Decision Processes*, *50*, 179–211.

Ajzen, I., & Fishbein, M. (1980). *Understanding attitudes and predicting social behaviour*. Upper Saddle River, NJ: Pearson Education, Inc.

Ampt, E. (2003). *Voluntary household travel behaviour change – Theory and practice*. Paper presented at the 10th International Conference on Travel Behaviour Research, August 10–15, Lucerne.

Ampt, E., & Rooney, A. (1998, September). *Reducing the impact of the car – A sustainable approach: TravelSmart, Adelaide*. Paper prepared for the 22nd Australasian Transport Research Forum, Sydney, Australia.

Anable, J. (2005). 'Complacent car addicts' or 'aspiring environmentalists'? Identifying travel behaviour segments using attitude theory. *Transport Policy*, *12*, 65–78.

Anable, J., Lane, B., & Kelay, T. (2006). *An evidence base review of public attitudes to climate change and transport behaviour*. Department for Transport, London, 228pp.

Aronson, E. (1968). Dissonance theory: Progress and problems. In R. Abelson, E. Aronson, W. McGuire, T. Newcomb, M. Rosenberg & P. Tannenbaum (Eds.), *The cognitive consistency theories: A source book*. Chicago, IL: Rand McNally.

Baron, R. A., Byrne, D., & Branscombe, N. R. (2006). *Social psychology* (11th ed., 654pp.) Boston, MA: Pearson Education, Inc.

Bord, R. J., Fisher, A., & O'Connor, R. E. (1998). Public perceptions of global warming: United States and International Perspectives. *Climate Research*, *11*, 75–84.

Brög, W., Erl, E., & Mense, N. (2002, December). *Individualised marketing: Changing travel behaviour for a better environment*. Paper presented at the OECD Workshop: Environmentally Sustainable Transport, Berlin.

Brown, D. E. (2004). Human universals, human nature and human culture. *Daedalus*, *133*(4), 47–54.

Cairns, S., Sloman, L., Newson, C., Anable, J., Kirkbride, A., & Goodwin, P. (2008). Smarter choices: Assessing the potential to achieve traffic reduction using 'Soft Measures'. *Transport Reviews*, *28*(5), 593–618.

Cao, X., & Mokhatarian, P. (2003). The future demand for alternative fuel passenger vehicles: A preliminary literature review. UC Davis – Caltrans Air Quality Project. Retrieved from http://aqp.engr.ucdavis.edu/Documents/caoMokhAFVinterimreport72003.pdf

Choocharukul, K., Tan Van, H., & Fujii, S. (2007). Psychological effects of travel behavior on preference of residential location choice. *Transportation Research Part A*, *42*, 116–124.

Dargay, J. M. (2001). The effect of income on car ownership: Evidence of asymmetry. *Transportation Research Part A: Policy and Practice*, *35*(9), 807–821.

Davison, L. J., & Ryley, T. J. (2010). Tourism destination preferences of low-cost airline users in the East Midlands. *Journal of Transport Geography*, *18*(3), 458–465.

De Young, R. (2000). New ways to promote pro-environmental behaviour: Expanding and evaluating motives for environmentally responsible behaviour. *Journal of Social Issues*, 56, 509–526.

DEFRA. (2006). *Climate Change: The UK Programme 2006*. London: The Stationery Office.

DEFRA. (2007). *Guidelines to DEFRA's GHG conversion factors for company reporting. Annexes updated June 2007*. Department for Environment, Food and Rural Affairs. Crown Copyright: London.

DEFRA. (2008). *A framework for pro-environmental behaviours*. Department for Environment, Food and Rural Affairs. Crown Copyright: London.

DEFRA/DECC. (2009). *Making the right choices for our future: An economic framework for designing policies to reduce carbon emissions*. Crown Copyright: London.

Department for Transport (DfT). (2009). *Transport trends: 2008 edition*. London: The Stationery Office.

Dijst, M., Farag, S., & Schwanen, T. (2008). A comparative study of attitude theory and other theoretical models for understanding travel behaviour. *Environment & Planning A*, *40*, 831–847.

Dunbar, R. I. M. (1998). Behavioural adaptation. In G. A. Harrison & H. Morphy (Eds.), *Human adaptation* (pp. 73–98). Oxford: Oxford University Press.

Euro Barometer. (2009). *European's attitudes towards climate change*. Special Euro Barometer 313, Directorate General of the European Commission, Brussels, Belgium.

European Environment Agency. (2010). *Towards a resource-efficient transport system. TERM 2009: Indicators tracking transport and environment in the European Union* (Report No. 2/2010). EEA, Copenhagen, Denmark.

Fazio, R. H. (2000). Accessible attitudes as tools for object appraisal: The costs and benefits. In G. Maio & J. Olson (Eds.), *Why we evaluate: Functions of attitudes* (pp. 1–36). Mahwah, NJ: Erlbaum.

Festinger, L. (1957). *A theory of cognitive dissonance*. Evanston, IL: Row, Peterson.

Fishbein, M., & Ajzen, I. (1975). *Belief, attitude, intention and behavior: An introduction to theory and research*. Addison-Wesley Publishing Company, Reading, MA, United States.

Fox, M. (1995). Transport planning and the human activity approach. *Journal of Transport Geography*, *3*(2), 105–116.

Fujii, A., & Tan Van, H. (2009). Psychological determinants of the intention to use the bus in Ho Chi Minh City. *Journal of Public Transportation*, *12*, 97–110.

Fujii, S., & Taniguchi, A. (2003). Determinants of the effectiveness of travel feedback programs – A review of communicative mobility management measures for changing travel behaviour in Japan. *Transport Policy*, *13*, 339–348.

Gärling, T. (2005). Changes of private car use in response to travel demand management. In G. Underwood (Ed.), *Traffic and transport psychology. Theory and application*. Proceedings of the ICTTP 2004, Elsevier, Oxford.

Gärling, T., Gillholm, R., & Gärling, A. (1998). Reintroducing attitude theory in travel behavior research. *Transportation*, *25*, 129–146.

Gollwitzer, P. M., & Brandstätter, V. (1997). Implementation intentions and effective goal pursuit. *Journal of Personality and Social Psychology*, *73*, 186–199.

Gross, R., Heptonstall, P., Anable, J., & Greenacre, P., & E4Tech. (2009). *What policies are effective at reducing carbon emissions from surface passenger transport? A review of interventions to encourage behavioural and technological change*. London: UK Energy Research Centre.

Hobson, K. (2001). Sustainable lifestyles: Rethinking barriers and behaviour change. In M. J. Cohen & J. Murphy (Eds.), *Exploring sustainable consumption: Environmental policy and the social sciences* (pp. 191–209). Oxford, UK: Elsevier.

Howarth, C., Waterson, B., & McDonald, M. (2009). Public understanding of climate change and the gaps between knowledge, attitudes and travel behavior. Proceedings of the 88th Annual Meeting of the Transportation Research Board, Washington, DC.

Howarth, C., Waterson, B., & McDonald, M. (2010). Perceived behavioural control and the role of information on climate change in increasing sustainable travel. 41st Annual University Transport Study Group Conference, January 5–7, 2010, Plymouth, UK.

Howarth, C., Waterson, B., & McDonald, M. (2011). The importance of local road authorities in climate change awareness programmes. 42nd Annual University Transport Study Group Conference, Open University, January 5–7, 2011, London, UK.

Intergovernmental Panel on Climate Change. (2007). *Climate change 2007: Mitigation of climate change. Contribution of working group III to the fourth assessment report of the intergovernmental panel on climate change*. In B. Metz, O. R. Davidson, P. R. Bosch, R. Dave & L. A. Meyer (Eds.), Cambridge University Press, Cambridge, UK.

Kaplan, K. J., & Fishbein, M. (1969). The source of beliefs, their saliency, and prediction of attitude. *The Journal of Social Psychology*, *78*, 63–74.

King, S., Dyball, M., Webster, T., Sharpe, A., Worley, A., DeWitt, J., ... Jopson, A. (2009). *Exploring people's attitudes to climate change and travel choices: deliberative research.* Final report for the department for transport. People Science and Policy Ltd, London, UK.

Kloeckner, C. (2004). *How single events change travel mode choice – A life span perspective.* Paper presented at the 4th International Conference of Traffic and Transport Psychology and Behaviour, September 5–9, 2003, Nottingham, UK.

Kollmuss, A., & Ageyman, J. (2002). Mind the gap: Why do people act environmentally and what are the barriers to pro-environmental behaviour? *Environmental Education Research, 8*, 239–260.

Krech, D., & Crutchfield, R. S. (1948). *Theory and problems in social psychology.* New York, NY: McGraw-Hill.

Leippe, M. R., & Eisenstadt, D. (1994). Generalization of dissonance reduction: Decreasing prejudice through induced compliance. *Journal of Personality and Social Psychology, 67*, 395–413.

Lewis, J. (2004, March). *Science, relevance and recall.* Science and Public Affairs. Correspondence, p. 27.

Lorenzoni, I., Nicholson-Cole, S., & Whitmarsh, L. (2007). Barriers perceived to engaging with climate change among the UK public and their policy implications. *Global Environmental Change, 17*, 445–459.

Lucas, K., & Jones, P. (2009). *The car in British society.* London, UK: RAC Foundation.

Lyons, G., Goodwin, P., Hanly, M., Dudly, G., Chatterjee, K., Anable, J., Wiltshire, P., & Susilo, Y. (2008). Public attitudes to transport: Knowledge review of existing evidence. Final report to the Department for Transport. DfT, London.

Mokhtarian, P. L., Salomon, I., & Redmond, L. S. (2001). Understanding the demand for travel: It's not purely 'derived'. *Innovation, 14*, 355–380.

Nicholson-Cole, S. A. (2005). Representing climate change futures: A critique on the use of images for visual communication. *Computers, Environment and Urban Systems, 29*, 255–273.

Nilsson, M., & Küller, R. (2000). Travel behaviour and environmental concern. *Transportation Research Part D, 5*, 211–234.

Norton, A., & Leaman, J. (2004). *The day after tomorrow: Public opinion on climate change* (11pp.). London: MORI Social Research Institute.

Pearce, D. W., & Turner, R. K. (1990). *Economics of natural resources and the environment.* Harvester Wheatsheaf, London.

Prillwitz, J., Harms, S., & Lazendorf, M. (2006). Impact of life-course events on car ownership. *Transportation Research Record: Journal of the Transportation Research Board, No. 1985.* Transportation Research Board of the National Academies, Washington, DC, 71–77.

RAC Foundation. (2010). *The road ahead. A transport policy for difficult times.* London, UK: RAC Foundation.

Rose, G., & Ampt, E. (2001). Travel blending: An Australian travel awareness initiative. *Transportation Research Part D: Transport and Environment, 6*, 95–110.

Ryley, T. (2006). Use of non-motorised modes and life stage in Edinburgh. *Journal of Transport Geography, 14*, 367–375.

Ryley, T. (2010). A travel behaviour response to UK road user charging. *Proceedings of the Institution of Civil Engineers (ICE), 163*(2), 57–62. Transport Special Issue – Road User Charging.

Scheiner, J. (2007). Mobility biographies: Elements of a biographical theory of travel demand. *Erdkunde*, *61*, 161–173.

Schmöcker, J.-D., Fengming, S., & Noland, R. B. (2009). An analysis of trip chaining among older London residents. *Transportation*, *37*, 105–123.

Schultz, P. W., Nolan, J. M., Cialdini, R. B., Goldstein, N. J., & Griskevicius, V. (2007). The constructive, destructive, and reconstructive power of social norms. *Psychological Science*, *18*, 429–434.

Sheeran, P. (2002). Intention-behaviour relations: A conceptual and empirical review. *European Review of Social Psychology*, *12*, 1–36.

Sloman, L., Cairns, S., Newson, C., Anable, J., Pridmore, A., & Goodwin, P. (2010). *The effects of smarter choice programmes in sustainable travel towns: Summary report*. Report to the Department for Transport, London.

Sterman, J. D., & Booth Sweeney, L. (2002). Cloudy skies: Assessing public understanding of global warming. *System Dynamics Review*, *18*, 207–240.

Stradling, S., Anable, J., Anderson, T., & Cronberg, A. (2008). Car use and climate change: Do we practice what we preach? In A. Park, J. Curtice, K. Thomson, M. Phillips, M. Johnson & E. Clery (Eds.), *British Social Attitudes: The 24th Report*. London: Sage.

Tertoolen, G., van Kreveld, D., & Verstraten, B. (1997). Psychological resistance against attempts to reduce private car use. *Transportation Research Part A: Policy and Practice*, *32*(3), 171–181.

Thaler, R. H., & Sunstein, C. R. (2009). *Nudge: improving decisions about health, wealth and happiness*. Penguin.

Thøgersen, J. (2005). Understanding repetitive travel mode choices in a stable context: A panel study approach. *Transportation Research Part A: Policy and Practice*, *40*, 621–638.

Triandis, H. C. (1977). *Interpersonal behavior*. Monterrey, CA: Brooks/Cole.

Van den Bergh, J. C. J. M., Van Leeuwen, E. S., Oosterhuis, F. H., Rietveld, P., & Verhoef, E. T. (2006). Social learning by doing in sustainable transport innovations: Ex-post analysis of common factors behind successes and failures. *Research Policy*, *36*, 247–259.

Van der Waerden, P., Timmermans, H., & Borgers, A. (2003). The influence of key events and critical incidents on transport mod choice switching behaviour: A descriptive analysis. *10th International Conference on Travel Behaviour Research*, Lucerne, August 10–15, 2003.

Van Vugt, M., Meertens, R. M., & Van Lange, P. A. M. (1995). Car versus public transportation? The role of social value orientations in a real-life social dilemma. *Journal of Applied Psychology*, *25*, 258–278.

Verplanken, B., Aarts, H., & Van Knippenberg, A. (1997). Habit, information acquisition and the process of making travel mode choices. *European Journal of Social Psychology*, *27*, 539–560.

Whitmarsh, L. (2004, June). *Ignoring climate change*. Science and Public Affairs.

CHAPTER 11

TRANSPORT AND CLIMATE CHANGE POLICY IN THE UNITED KINGDOM: A SOCIAL JUSTICE PERSPECTIVE

Karen Lucas and Kate Pangbourne

ABSTRACT

Purpose – *The social dimensions of the relationship between transport and climate change are examined, in particular, the potential for unintended negative consequences to directly and/or indirectly arise from policies to reduce the climate change impact of the transport sector. It takes the example of current policies in the UK as its primary focus.*

Methodology/approach – *A combination of literature, policy review and the quantitative and qualitative evidence gathered through primary fieldwork research from a number of related studies.*

Findings – *It is identified that different social groups can experience very different outcomes in accessing transport and adapting to changes to the transport system, whether these are uniquely targeted towards certain individuals or more systemically applied across the whole population. For this reason, it is essential that policy makers fully understand the potential vulnerability and resilience of different social groups to policies*

Transport and Climate Change
Transport and Sustainability, Volume 2, 287–312

ISSN: 2044-9941/doi:10.1108/S2044-9941(2012)0000002014

that are intended to address transport-related climate change. The key component of social impacts should be systematically analyzed, by income, potential vulnerability and their spatial and temporal distribution, as well as according to resilience/adaptability to the proposed intervention. This continues to pose an important future challenge for research in this area of transport policy making.

Originality/value – *This chapter highlights the potential for unintended negative social consequences to directly and/or indirectly arise from policies to reduce the climate change impact of the transport sector.*

Keywords: Social equity/justice; transport; climate change; policy impacts

INTRODUCTION

This chapter is about the social dimensions of the relationship between transport and climate change policies, and the potential for unintended negative social consequences to arise from the implementation of these policies, if detailed social equity assessments are not undertaken during the design and implementation stages of the policy process. It has long been recognised that transport policies can have unintended negative effects within and outside of that sector, and this is also true for policies that address cross-cutting issues such as climate change. The aim of the chapter is to scope the likely social and distributional impacts (SDIs) of climate change policies within the transport sector with particular attention to their consequences for already disadvantaged population groups. The SDIs of climate change itself are only touched upon briefly, as there are a myriad of SDIs to consider in this context. Rather, focus is on detailed considerations of climate change policies for transport in the UK. Largely this is for reasons of opportunism, because the UK is unique in being the only country in the world that has introduced a legally binding framework for the reduction of CO_2 emissions via the 2008 UK Climate Change Act (Committee on Climate Change, 2009).

THE UK POLICY CONTEXT

Clearly, policy does not occur within a vacuum and so it is important to first understand the broader social and transport context in the UK into which climate change policies for transport are being introduced.

Current Social Trends

The basic currency of social trends in the UK is the persistence of inequality across a number of domains including

- Relative poverty and growing income disparities within and between different local areas and parts of the country;
- Changing geography of employment patterns and new flexible working arrangements and increased structural 'worklessness' in some regions;
- Health inequalities and reduced quality of life for low-income populations and certain vulnerable social groups, such as children, older people and some ethnic minorities;
- Diminishing access to key life-enhancing opportunities chances: employment, education, healthcare, work and leisure;
- Structural ageing of the population with increased financial insecurities into older age; and
- Inadequate housing provision and poor condition of rented housing stock leading to issues of fuel poverty.

These problems appear to continue despite considerable government efforts to reduce poverty and social exclusion over the 10-year period 1997 to 2007 (Hills & Stewart, 2007).

The Transport Dimension of Social Inequality

Social and physical mobility and the means to transport oneself or be transported are implicated in many current social issues. The relationship between poverty and travel behaviour is a major topic, which one chapter alone cannot address in all its detail, but this relationship has now been well established within the wider literature. For example, the Social Exclusion Unit (2003) report describes how people who are already experiencing social and economic disadvantage also tend to experience the worst environmental and social impacts of the transport system. Banister (2007) also notes that most transport infrastructure and service investment, wherever it takes place, benefits the rich more than the poor. Transport inequalities tend to occur in two dimensions: on the one hand, living in poverty results in less access to transport because of the cost and this can result in an intensification of social exclusion for certain areas, as shown in Fig. 1.

Ironically, encouraging more 'local lifestyles' is a key climate change policy option for tackling transport's greenhouse gas emissions. However,

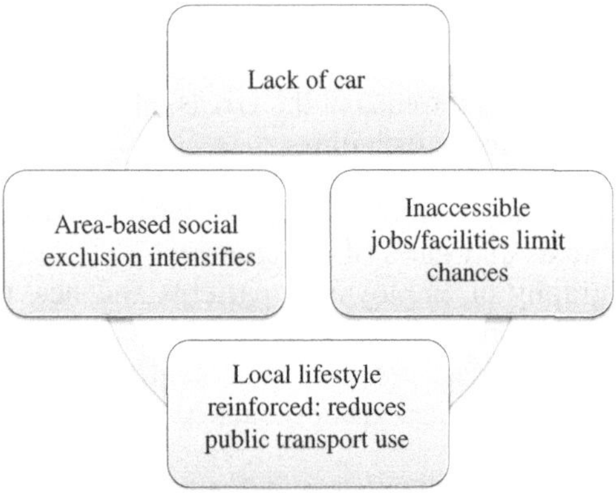

Fig. 1. Transport Exclusion Resulting in Intensification of Social Exclusion. *Source*: Pickup and Giuliano (2005; Fig. 4.1, p. 41).

this flies in the face of the considerable pressure on all households to improve access to life-chances and basic facilities through car ownership. Evidence shows that car ownership in the United Kingdom has increased dramatically amongst the *lowest* income households, from 34% in 1997 to 49% in 2008 (Department for Transport, 2008). This creates a related vicious cycle, illustrated in Fig. 2, which has further complex impacts on marginalized groups. For example, they could be effectively forced to live in cheaper housing in peripheral areas and experience higher transport costs as a result. Individuals who remain carless in these circumstances are subjected to an increasing 'mobility gap' (Pickup & Giuliano, 2005).

In rural areas, there is additional structural dependence on the car, as there are few or no practical transport alternatives, as well as reduced opportunities to undertake activities locally due to the absence of services and facilities. Therefore, rural populations may experience further deprivation as a result of having to pay a higher proportion of their income on maintaining an ability to travel by car (Dargay, Gately, & Sommer, 2007; Gray, Farrington, Shaw, Martin, & Roberts, 2001). It is for these reasons that policies to facilitate mobility have been featured in tackling social inequalities, for example through the 'wheels to work' schemes (Countryside Agency, 2002; Steer Davies Gleave, 2005; Wright, Nelson, & Cooper, 2009).

The second key dimension of transport inequalities derives from the ill-health effect of the transport system. Particular problems are air quality and noise-related pollution and exposure to pedestrian accidents in areas

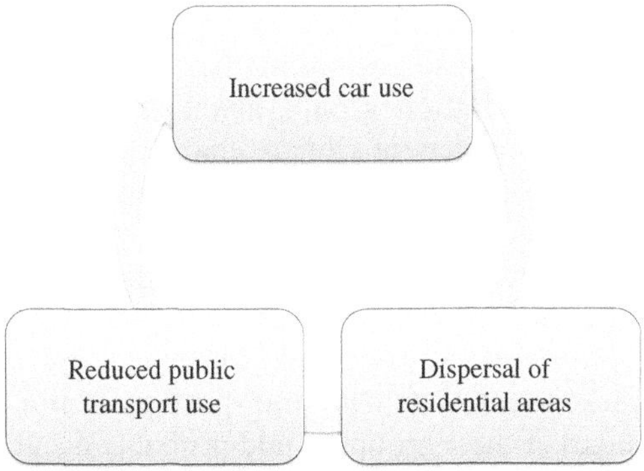

Fig. 2. Vicious Car Dependency Cycle of Increasing Car Use. *Source*: After Pickup and Giuliano (2005; Fig. 4.1, p. 41).

adjacent to busy traffic routes. In Britain, the poorest fifth of households are concentrated in areas that produce the least pollution from car exhausts, due to their low levels of car ownership, but inhale the most from them, due to their close proximity to and higher exposure to traffic on busy roads (Mitchell & Dorling, 2003). Paradoxically, the regulation of vehicle pollution can also have a regressive outcome in that it is lower income households that are more likely to own the older (and 'dirtier') vehicles and tend to be less able to maintain them effectively. Thus, otherwise laudable policies that aim to reduce vehicle emissions can also further increase exclusion from mobility by removing the older and hence cheaper vehicles from the road and market (Pickup & Giuliano, 2005).

Pre-Existing Difference Between an Individual's Transport Accessibility and Their Travel Experiences

As Dorling (2010, p. 13) highlights:

> Within each country who gets to consume and pollute the most is again very unevenly distributed. Men tend to pollute more. Men drive more and fly more often, children pollute the least and are most affected by pollution in the long run, not least because their lungs are smaller and they breathe in at car exhaust pipe level.

Persistent inequality is considered a 'social bad' within any self-proclaiming egalitarian society and there is a moral obligation to tackle social inequities through the policy process and to ensure that policy interventions do not have compound negative effects on already disadvantaged populations. It is therefore extremely important that policy makers fully understand exactly which population groups and local areas are likely to be *most* vulnerable to the impacts of climate change policies, and to develop better systematic assessment of the nature, extent and severity of those impacts.

As noted earlier, there are already socially and economically disadvantaged groups within the UK, whose adverse situation may be compounded by their inability to access key services, which may be partly or wholly attributable to the current system of transport provision. The transport-related disadvantages of these groups would probably occur with or without the introduction of policies to reduce greenhouse gas emissions, but such policies have the potential to further exacerbate or mitigate this condition.

Clearly, every person's circumstances are unique, but their associated travel behaviour is determined by a number of different key factors, which can combine in various ways according to life stage and employment status/type, as identified in Fig. 3.

These differences and their knock-on social consequences have been examined in detail elsewhere (see, e.g. Lucas, 2004, pp. 45–46). In this instance, it is sufficient to note that in some cases, differences in an individual's travel behaviour simply reflects their different personal circumstances, particularly life stage, and may not always be indicative of reduced well-being or inequality (e.g. in general people of working age travel more than those who are retired). However, when differences in travel outcomes *are* a reflection of inequalities of access, achievement, opportunity and/or ability, this should be of concern to policy makers. It can be suggested that the factors that contribute to individual travel behaviours could thus potentially be used as an initial framework for more systematic analysis of the SDIs of climate change policies.

Additionally, there can be a geographical or spatial determinant to an individual's experience of the transport system; the usual distinction made is between people living in urban and rural areas. It is an acknowledgement of the greater dispersion of population and services and the associated increased distances travelled by rural dwellers, and some of these impacts have been alluded to. However, the differences are often more subtle, as the key element is the quality of the transport services and/or infrastructure that serves a particular area. For example, low-income estates on the urban periphery are often poorly served by public transport and can be overlooked in behaviour change treatments for walking and cycling. Consideration of these area-based effects is likely to be more relevant for the evaluation of

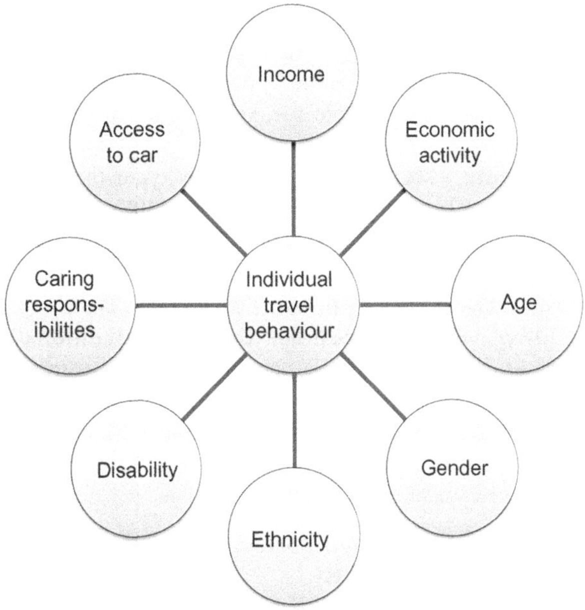

Fig. 3. Key Socio-Demographic Factors for Considering Differences in Travel Behaviours (Developed by the Authors).

individual transport projects rather than overarching transport policies and programmes.

Therefore, it can be contended that precisely because of these pre-existing transport inequalities, it is extremely important to ensure that any new policies do not at worst serve to exacerbate these inequalities and at best aim to redress the balance, in the interests of egalitarianism. However, the SDIs of climate change policies are currently poorly understood and conceptualised within the transport sector and even less often mitigated against.

IDENTIFYING THE SOCIAL AND DISTRIBUTIONAL IMPACTS OF CLIMATE CHANGE POLICIES FOR TRANSPORT

The transport literature does not often consider the social equity and/or fairness of new transport policies and interventions, but this has been even less evident in the case of transport policies to encourage CO_2 emission reductions. Reducing carbon and other greenhouse gas emissions within the transport sector may at face value appear to be quite a simple task (i.e. simply reduce the amount of people's travel in order to reduce their carbon emissions).

However, 21st century society is predicated on hyper-mobility: the need to travel to participate in everyday activities (Urry, 2007), coupled with a high degree of physical and emotional attachment to private automobiles (Lucas, Blumenberg, & Weinberger, 2011). This is reflected in the rapid increase in car ownership and use, as already noted. This increase is particularly evident amongst the lower income households that have traditionally relied on public transport and walking as their primary modes of transport (Lucas & Jones, 2009).

Thus, it is clear that policies intended to reduce climate change in the transport sector are likely to have a number of significant and diverse equity effects on these households, including

- Changes to the quality of transportation available, affecting people's opportunities and quality of life;
- Transport planning decisions affecting the location and type of development that occur in an area, and the resulting accessibility, land values and developer profits;
- Indirect and external costs arising from transport facilities, activities and services, such as congestion delay and accident risk imposed on other road users, infrastructure costs not funded through user fees, pollution, and undesirable land use impacts;
- Price structures can significantly affect financial burdens as transport expenditures represent a major share of most household, business and government expenditures;
- The significant land take for transport facilities is generally exempt from rent and taxes, representing an additional but hidden subsidy of transport activity; and
- Transport planning decisions can stimulate employment and economic development which have distributional impacts (Litman, 2011, p. 2).

In order to better understand the extent to which these social impacts might be relevant in the UK context, the Department for Transport recently commissioned a knowledge review of the social and distribution impacts of its key transport policies for climate change (AEA Group, 2011). The report identifies four important findings in terms of the likely equity effect of its current policy strategy. Firstly, that on average, policies which aim to reduce car use through fiscal measures (e.g. congestion charging and parking charges) can broadly be considered as *progressive*, in that higher income groups own and use cars more than those on low incomes. However, there is one group of car users that is potentially most adversely affected: those low-income drivers who do not have alternatives to using their car (e.g. those whose jobs are in a charging zone, but who live outside of this area).

Secondly, current policies that aim to reduce CO_2 emissions through modal shift are more likely to improve the ability of those on low incomes (particularly lone parents, younger, older and disabled people) to access employment opportunities and services when they focus on improvements to bus infrastructure and services. Conversely, improving rail or cycling infrastructure is most likely to benefit economically active, higher income males and, in the case of cycling, to least benefit older and younger women, and renters with limited space for bicycle storage.

Thirdly, policies focusing on stimulating the purchase or use of energy-efficient private vehicles could marginally reduce the affordability of new car purchases for some low-income households. There could also be short-term, knock-on effects on the second-hand car market, as second-hand cars might also become more expensive, again adversely affecting lower income households.

Finally, policies such as the regulation of fares, concessionary fares, car clubs and car sharing have the greatest potential to improve travel affordability and would enable targeted audiences to gain better access to employment and other opportunities and services. However, they could have rebound effects by providing a perverse incentive of encouraging more overall travel by all groups, which may not be desirable if the overall aim of the policy is to reduce CO_2 emissions from the transport sector as a whole.

The Scottish government also recently commissioned research to establish the relevant policies for climate change mitigation in transport (mitigating transport's climate change impacts (MTCCI) by Atkins and University of Aberdeen, 2009). This exercise involved assessing carbon abatement potential, technical feasibility, cost of implementation, political and public acceptability, as well as accessibility and social impacts. An important additional element was to identify whether or not the main levers or drivers for the policy interventions lay with the Scottish government or whether they were reserved at the UK level or affected by EU level directives.

A total of 22 policy options were identified as being (a) devolved to Scotland and (b) of greatest benefit to mitigate the climate change impact of transport. These options were divided into seven categories, and are shown in the table below. The report itself aggregated SDIs, which obscures differential effects. Thus, here potential SDIs of each identified option are highlighted to indicate the potentially complex outcomes from the application of these packages. It is emphasised that the detailed *design* of instruments is crucial to safeguard socially vulnerable groups (Table 1).

The study found that the car demand management (smart measures) policy option was the category with the greatest potential to reduce emissions. Travel planning was particularly singled out for its potential, but it should be noted that this term covers a wide range of approaches, from workplace travel plans to individual travel marketing. The scale, that is rate of implementation and level of intensity with which it is applied, is also a significant factor in its overall potential to reduce CO_2. One important finding is that fiscal policy options for car demand management offer significant potential for emissions abatement, but the infrastructure options are significantly less effective. From the point of view of abatement, schemes involving large-scale public transport investment are towards the bottom of the list. This is a significant finding in relation to the potentially negative social impacts of climate change policies in the transport sector.

The report also reiterates the importance of protecting the most vulnerable and least resilient citizens from the 'worst shocks' from the system of transport provision, such as dramatic increases in oil prices, flooding, heat waves and other weather effects, as well as fundamental changes to policy.

Considering Resilience and Capacity for Adaptation in Relation to SDIs of Climate Change Policies

Work has been undertaken on theoretical responses to change, and work on behaviour and revealed preferences, in order to understand people's resilience and capacity for adaptation in relation to changes of policy. In relation to how people might respond to an involuntary 'cost of reduction' scenario, Doherty, Lee-Gosselin, Burns, and Andrey (2002, p. 179) note that even relatively straightforward response strategies can often lead to important primary and secondary effects on the activities of individuals. It can be noted that previous research has shown that initially some households simply accept the change and absorb the additional costs or other burdens created by the car use reduction policy.

However, over the medium term, differences in ability to adapt can accrue as a result of time constraints preventing rescheduling of activities, or

Table 1. The 22 Transport Policy Options Available to the Scottish Government for Climate Change Mitigation.

Package	Policy Options	Potential SDIs
Technology	1. Electric car technology and network development 2. Procurement of low-carbon vehicles	Unaffordability for low-income groups, but could have air quality benefits at point of use (but negative effects at point of generation)
Driving style	3. Active traffic management 4. National motoring package 5. Speed reduction in trunk roads	Less beneficial for low-income groups, apart from potential for improved fuel efficiency
Car demand management	Fiscal/Infrastructure 6. Bus/rapid/mass transit infrastructure investment (inc. bus priority) 7. Cycle infrastructure investment 8. High-speed rail links 9. National network of car clubs 10. National road user charging 11. Introduce/increase public parking charges 12. Rail investment 13. Introduce/raise residential/private parking charges 14. Bus/LRT fares reductions 15. Walking infrastructure investment 16. Workplace parking levy	(6, 12, 17 & 18) likely to be beneficial to low-income groups if the design appropriately (i.e. right areas, right ticketing and service types) (7) Evidence suggests these interventions favour fit professional males at present (8) Likely to be exclusive due to cost of use (9) Potential for benefiting lower income groups who cannot otherwise afford to use a car (10) Whether this is social progressive or regressive is dependent on instrument design (11, 13 & 16) regressive for low-income/car-dependent groups
	Smart measures 17. Bus quality contracts/statutory partnerships 18. Widespread implementation of travel plans 19. Provide community hubs	(14) Beneficial, but may offer perverse incentives to increase overall travel amongst all groups (15) Should be beneficial (19) Potentially boosts accessibility but location important
Freight	20. Freight best practice	Should be beneficial if leads to air quality improvements and optimises loading (reduces vehicle movements)
Land use planning	21. Urban density increases	Potential for exclusion if segregation and gated communities emerge
Aviation	22. Improve public transport surface access at airports	Potentially beneficial for the large numbers of low-paid employees at airports, but overall SDIs hard to identify

Source: Expanded from Atkins/University of Aberdeen (2009).

increases in cost leading to economic and social hardship, or a loss of independence. Over the longer term, more radical changes in activity patterns (such as responding by increasing teleworking or teleshopping), car ownership and lifestyle can be adopted. Gärling, Gärling, and Loukopoulos (2002, p. 99) speculate on a similar set of likely impacts of reduced car use and produced a list of potential coping strategies and second-order effects over time. While this listing of effects is plausible, there are two major complications. First, effects are likely to be policy specific, whether quota-based, financial [dis]incentives, personalised travel planning or otherwise. Further, the timescales of impacts are likely to depend on the degree of advance knowledge of the policy in general and its perceived permanence, blurring whether a response is short, medium or longer term.

Direct understanding of people's behavioural responses to congestion (i.e. revealed behaviour) is not so prominent in the literature. An early report by Mokhtarian, Raney, and Salomon (1997) on city workers in San Diego, California, had three key findings. Firstly, individuals are more likely adopt low-cost coping strategies before they adopt higher cost strategies. Secondly, individuals who face increasing congestion view the choice of alternative coping strategies in ways that are partly dependent on their socio-economic and demographic characteristics. Finally, measures designed to reduce travel may have a smaller impact than expected, as individuals try first to maintain current levels of travel while reducing the personal impacts of congestion.

With these considerations to the fore, and with the awareness that some of the adaptation responses of individuals are actually the *intended* outcome of transport and climate change policies, and that the evidence that they have negative impacts on well-being and life-chances cannot be definitively inferred, some case study examples are explored to illustrate the potential social consequences of policies aimed at reducing people's car-based travel. These contrasting studies, whilst conducted for different purposes, reveal how a range of transport-related SDIs can be uncovered.

EMPIRICAL EVIDENCE FROM CASE STUDIES

Case Study 1: Car Dependence Study (England)

In 2008, just after an increase of around 6 pence per litre in the cost of a tank of petrol (SpeedLimit, 2011), a small exploratory scoping study was undertaken, using a qualitative methodology, of the likely economic and social consequences of enforced car use reduction in the United Kingdom (Lucas & Jones, 2009).

Research Design

The research design consisted of five focus groups in two case study areas, two in Nottingham and three in Banbury, with randomly selected members of the public recruited according to a pre-determined set of criteria to reflect the issues to be pursued in connection with the implementation of potential policies for car restraint. The groups were as follows:

1. *Voluntary switchers* – these were people of a range of ages who had made a voluntary switch from their cars to public transport for some journeys as a result of reduced fares or free bus tickets that were recently introduced by Nottingham City Transport.
2. *Banned drivers* – these were people of a range of ages that had received an enforced driving ban of over six months in the last year living in the Nottingham area.
3. *Drivers aged over 75 years* – these were elderly drivers living in and around Banbury in Oxfordshire, who might be considering giving up driving in the near future for reasons of their age.
4. *New drivers* – these were young people of between 18 and 25 years living in and around Banbury, who had passed their driving tests within the previous year.
5. *Non-car owners* – these were people living in and around Banbury in non-car owning households and without regular access to a car.

The focus groups took place during the first two weeks of September 2008 and each discussion was approximately two hours in duration. Their key aim was to gain insight into the participants' perceptions of their current travel behaviours in terms of whether these largely took place out of choice or necessity and to understand how they might be able to respond to three different future 'car-use constraint' scenarios, namely:

1. Where the cost of motoring is significantly more (at least double) as a result of continuing increases in petrol prices due to global barrel price increases;
2. Where additional cost is placed on the use of cars through either road user or place-based congestion charging or parking charges;

3. Where their car use is restricted due to petrol rationing or the introduction of a personal carbon allowance.

There was particular interest in how participants thought their lifestyles would be affected and what they saw as the key benefits and dis-benefits for themselves, their families and dependents and wider society of a transition to a less car-reliant lifestyle. There was a desire to better understand the cost of this adjustment economically and socially. These specific groups and locations were identified because it was felt they would help to provide insights on the activity and lifestyle effects of policy measures to reduce car use as a result of their own recent personal experiences.

Personal Preferences and Travel Choices

Most people in the groups who could drive and had cars available to them used them for most trips. Those who did not currently have cars generally wished to drive and relied heavily on lifts from family and friends for most of their trips (it should be noted that this group was recruited from a semi-rural village). One or two people in each of the groups could be described as 'reluctant drivers' in that they felt unconfident driving or did not drive very much for reasons of cost or for fitness reasons or simply because they did not much like driving. Most people generally preferred to drive or be driven as a passenger over using public transport and often said that they preferred to walk than take a bus or train.

There was some discussion about the benefits of not driving, such as improved health and fitness, more social networking opportunities and cost savings; there was much greater discussion about the dis-benefits of not driving. The focus was mostly centred on the inadequate, inconvenient and/or over-priced alternatives.

> You can't rely on other people, if you want to shop at our age, you don't want to walk miles to it. In the country, you really do need a vehicle, a lot more than in the town. (Male participant, over 75 years group, Banbury)

Many participants cited loss of personal control and maintaining their independence as a big reason for not reducing their car use. Several of the participants in the 'non-car owners' group alluded to feeling isolated and having to deal with difficult *'personal issues'* (mostly depression) as a result of the isolation.

> I hate that terrible burden you know that you, you tend to think that they're going oh Christ, B's going to want a lift again but they're not but you always get that feeling that that's what they're thinking. (Male participant, non-car owners group, Banbury)

Participants across all the groups had a strong sense of what journeys were essential to their economic and social well-being and where they were using the car for reasons of 'laziness' or 'general convenience'. There was the general sentiment across all the groups that once you have a car you tend to use it, even when it's not necessary.

Some participants, but actually very few across all the groups, expressed a feeling that they were emotionally (and possibly irrationally) attached to their cars. Travel to work and servicing family needs such as shopping were the prime motivations for using the car across all the groups and also the journeys that people felt they could least do without in most cases. Time pressures were also considered a big factor in choosing to use the car and particularly the burden of children's escort trips were seen as a significant factor in parents' reliance on their cars in all of the groups:

> Without ferrying children all over the place, yes, I could get away without using the car at all. (Female participant, voluntary switchers group, Nottinghamshire)

Gender Differences

There were gender differences in attitudes to driving across all of the groups, but these were not necessarily along usually expressed stereotypical lines. For example, both women and men enjoyed the speed and skill of driving and it was a young woman in the 'new drivers' group that was particularly involved with doing up and customising her car. However, in the 'over 75 years' group, it was only the women who talked about their increasing nervousness about driving:

> To be honest, it's my age. I feel like I've become a little bit of a coward, of getting in it sometimes and taking it very long distances whereas I used to do much longer journeys. I'm a little bit frightened of driving too far. (Female participant, over 75 years group, Banbury)

Car Restraint Measures

There were mixed responses to the three increased car restraint measures scenarios in all of the groups. Generally, people were more resistant to place-based pricing charges and road user charging, although they said this would be most likely to stop them using their cars in certain places or at given times. Most people had already been affected by increases fuel costs and many had already made changes to their travel behaviour, such as changing to a more economical vehicle, car sharing and trip chaining.

> A lot of my friends don't drive and they don't appreciate how much money it costs you, even just to drive to the next village to pick them up and then go back to where you're

> going. I bet over the year I've been driving I've had to spend quite a few hundred pounds just going to pick friends up. (Male participant, new drivers group, Banbury)

Generally, people found it quite difficult to imagine the effect of such increases on their lifestyles and activities and were unable to respond to these questions in any more than a very superficial way. It was suggested that this was largely a result of the focus group method, which did not allow for detailed budgeting and trade-off exercises with the participants. This would need to be considered in any future pursuit of the 'cost of adjustment' issue.

Such exercises can offer only a snapshot of people's perceptions of the current situation and are in no way intended as a true representation of the wider picture. The group participants were purposely sampled to reflect a particular set of personal and situational circumstances and this clearly had an influence on their responses. Nonetheless, our focus groups provided us with some useful practical insights into the degree of car dependency that people feel they have currently, as well as what they might do if their cars were less available to them.

It was clear from the discussions that most people are able to differentiate between car reliance for certain trips and the pleasure and convenience of having a car. The main reasons participants gave for using their cars seemed to be habit, but this was underpinned by practical considerations of cost, convenience and a sense of pragmatism, given most people's busy lifestyles and time commitments. There was also a strong feeling from the majority of participants that most of the places where they lived simply did not have sufficient alternatives in place for them to reasonably undertake their daily activities without a car (apart from the Arnold area of Nottingham, where the voluntary switchers group were recruited and people reported that considerable targeted efforts had been made to provide better public transport services).

On the whole, people's responses to both the increased cost and petrol-rationing scenarios were quite unsophisticated. For example, it was difficult to get the participants to make proper budgetary trade-offs without offering some sort of base-case situation upon which to calculate an adjustment and going into a great amount of detail about personal finances, which would have been inappropriate in a group situation. As a result, the main impacts people talked about were largely travel based, although some interesting quality of life and wider societal impacts were noted.

> It's like a few years ago when we had that petrol strike it had a knock-on effect where I worked, people weren't coming in because they said well I'm not going on public

> transport, I'm not spending hours to get to work (a) I can't afford it and (b) I haven't got the time, so the sickness rate really went up through the roof. (Female participant, voluntary switchers group, Nottinghamshire)

Case Study 2: Intelligent Transport Systems: the Health and Quality-of-Life Implications for Older People (Scotland)

Information and communications technology (ICT) is increasingly being employed in intelligent transport systems (ITS) as one element in travel demand management interventions. ICT also has a profound effect on access to services and activities in ways that have unpredictable effects on the geography of our mobility and the temporal organization of everyday lives (Lyons, 2009). Felstead (2009), for example, describes how mobile and ICT makes 'anywhere' a place of work. For older people, at the transition from employment to retirement, their already documented lower levels of technical skills and practices are at risk of profound marginalization as the younger, technologically literate see only the financial benefits, and not the social costs, of switching service access to digital platforms. Nevertheless, many have hoped for ICT to bring beneficial virtual mobility to counter mobility-related social exclusion (e.g. Kenyon, Lyons, & Rafferty, 2002).

A four-fold typology of ICT effects on travel can be found in early work by Salomon (1986) and Mokhtarian (1990), although other effects have since been identified. These are as follows:

1. Substitution – ICT decreases travel.
2. Stimulation – ICT increases travel.
3. Efficiency – ICT improves transport operations.
4. Indirect – ICT impact on other areas of life changes mobility patterns.
5. Supplementation – a variation of substitution (Kenyon et al., 2002).
6. Redistribution – ICT alters mode, geography and timing of travel (Lyons & Haddad, 2008).
7. Enrichment – ICT facilitates mobile office or provides other improvements to the experience of travel (Lyons & Urry, 2005; Mokhtarian & Salomon, 2001).

In relation to our argument on understanding the SDIs of climate change mitigation and adaptation policies, many elderly people are already 'at risk' of social exclusion as a result of low levels of personal mobility and difficulties accessing public services, particularly healthcare at a time of greater need. A specific focus on access to healthcare in conjunction with transport builds on the understanding that older age groups have a greater need for healthcare services, and are also more dependent on public transport to meet all transport needs than the general population (Cloke, Milborne, & Thomas, 1997; Social Exclusion Unit, 2003). The elderly are also the biggest single group experiencing poverty in rural areas (Shucksmith, 2003). Scotland's rural areas are also experiencing a long-term trend in increasing health inequalities (Levin & Leyland, 2006), all of which was a justification for this case study research.

Research Design

The study, held in 2008, was also exploratory, qualitative and based on focus groups. The aim was to explore issues of *digital* inclusion and quality of life for primarily rural older people in relation to their experiences of and attitudes towards using ICT in meeting their transport and healthcare needs (Pangbourne, Aditjandra, & Nelson, 2010), that is focusing on the potential substitution, efficiency, indirect and redistribution effects of ICT. Four focus groups were held in total, two for each age group 55–64 and 65–74 years. Two were held in Aberdeen City and two in Aberdeenshire (one for each age group in each local authority area). Although not all volunteers were recruited to focus groups, all were surveyed with a questionnaire, resulting in 48 useable responses. The survey also captured views from the 75+ age group and covered a wider, largely rural, area. The focus groups were not filmed, and so it was not always possible to identify the gender of the speaker for the quotations below.

Heterogeneity in Travel and ICT Choices and Attitudes

Analysis of the focus group material and the questionnaires traced the complex relationships between individual attitudes and needs in accessing the services provided by the three relevant policy sectors of transport, healthcare and communications technologies. It could be observed that older age groups in both the focus groups and the surveys showed a slight tendency to be less comfortable with ICT and therefore to integrate it less into daily life, something which is in line with other studies (Kridel, Rappoport, & Taylor, 1999). Nevertheless, nearly three-quarters of the survey respondents have broadband access at home. The data suggests that the differences in enthusiasm for or use of ICT between individuals occur through interactions between knowledge, practices, and factors relating to an individual's physical and mental experience of ageing more than availability of opportunity; thus, measures of broadband availability are not a proxy for digital exclusion.

In Focus Group 1, held in Aberdeen for the 55–64 years age group, there was a mix of attendees from Aberdeen and the surrounding area (up to 30 miles away). This group's discussion touched on several aspects of healthcare, such as general accessibility of the service and using ICT for information, but also specific transport issues, particularly buses, community transport services and parking. ICT-specific issues included social benefits and dis-benefits, skills issues and finding information, particularly about transport.

Focus Group 2, held in Aberdeen for the 65–74 years age group, again with a mix of participants from Aberdeen and the surrounding area (up to 50 miles), touched on similar issues, but discussed mobile phones much more in relation to arranging transport and making and changing arrangements for accessing appointments. Nevertheless, several people in the group either did not have a mobile phone or had one but did not know how to use it. This age group had more of a perception that ICT is impersonal and that voice communication was important.

> Losing the personal touch, everything's become impersonal ... it was nice to be recognised. (Female participant, aged 65–74 years, Focus Group 2)

This highlights an importance of social contact for well-being and quality of life for older people, and how the increasing emphasis on ICT can create a sense of exclusion:

> Can't find what I want on-line, I'm scared of it, cut off ...

The individualised nature of this experience is emphasised by another participant, however:

> I get more contact on-line, I'm not cut-off, I'm part of it ...

This group contained more 'remote rural' participants, and they were generally more familiar with a sense of feeling excluded, as a result of rurality, in which ICT access is only one of many issues, including post office closures, library and community education services, and distance to hospital and medical facilities.

Gender and Older Age Differences

In Focus Group 1, a general consensus emerged that whilst there would always be some *older* people who would not engage with ICT, for a variety of reasons, there was considerable scope for reaching out to the interested but disconnected. A key point was that basic information on using ICT for accessing health and transport services has to be delivered by non-ICT means:

> Supermarkets are underutilised as an information point – go where people are, print information on the backs of receipts. Having information at the hospital and libraries only reaches people who go there. (Female participant, aged 55–64 years)

Subsequent to the focus groups, a participant contacted the researchers with further reflections relating to the age-related exclusionary effects of public transport for accessing healthcare, something which older age groups have to do more frequently.

> For an elderly person to get from this area [2.5 miles from Ellon] to Aberdeen to hospital would involve hiring a taxi to the park-and-ride, getting off in Aberdeen and walking some considerable distance to get on another bus (no integration of transport) to the ARI or whatever part of AH Trust they are to attend. That could easily take 2–3 hours, with the same performance on the return journey. (Female participant, aged 55–64 years)

ICT-Based Accessibility Measures

A concern for exploration in the focus groups was that a policy reliance on ITS (of the informational kind or supporting demand-responsive transport services) or virtual access to healthcare services (such as NHS Direct/NHS 24) may exacerbate the social exclusion threat of an ageing demographic and increased longevity if the normalisation of ICT in older people's lifestyles does not occur.

> I think many older people feel like second-class citizens when it comes to accessing services through 'new' technology as a goodly minority have not the skills or other means to do so, nor do some wish to learn. If too much emphasis is made on using technology, what happens to e.g. our over 85s who may not have a relative, friend or advocate to help with the technology? If we rely too much on technology and cut out human interaction is this going to further undermine the health of some people (mental and emotional as well as physical)? (Female participant, aged 55–64 years)

This suggests that the development of ITS needs to take account of the particular cognitive and physical needs of older people as well as their geographic context, in order to avoid unintentional exclusion and ideally to enhance quality of life. Whilst fast ICT access has become substantially more pervasive in UK society, those who are not 'connected', for whatever reason, may become a smaller social group overall but experience a more profound and widening exclusion from any policy area or service that switches to a digital platform.

The biggest ICT users in our study were in the 65–74 years age group, had the highest educational attainment, but were also the most critical of public sector efforts to push citizens too far and too fast into new ICT-reliant delivery modes for public services. A particular concern was a lack of sufficient attention to issues such as older citizen vulnerability in an on-line setting. Participants felt that on-line public services not only left older people more vulnerable to identity theft and fraud but increased social vulnerability by reducing their contact with 'real humans', thus isolating them.

Our recommendations focus on using ITS to facilitate access to healthcare, as that was the main concern of the participants, due to the psychological stress of travelling to healthcare appointments in circumstances which are inherently stressful:

> [It is] terrifying getting a bus when you are not feeling great. (Participant, aged 55–65 years, Focus Group 4: Aberdeenshire)

In this instance, participants did see ICT as a facilitator, but felt strongly that the 'patient' should not have to negotiate themselves, in that healthcare administration should be able to make realistic appointment times using automated information about patient location and transport availability. In general, the focus group participants were not making use of the abundance of transport information available through the Internet both at home and on mobile devices. This finding was confirmed by the subsequent questionnaire survey, to which the following response was common:

> Speaking for myself neither [mobile phone and Internet] are of any use to me, and, even if I had the use of them, I would not use them for the purposes stated. I prefer a personal 'hands-on' approach, which I trust more than the technology. (Male respondent, over 75 years, Aberdeen City)

Design of the user interface is also an issue, as cognitive and physical capacity is highly diverse in older age groups, making it difficult for ICT, healthcare and transport providers alike to achieve a good fit between personal circumstances and operational need. It will be difficult to ensure that older age groups, particularly those in rural areas and with a need to access healthcare, are not adversely affected by climate change mitigation transport policies utilising ICT.

Overarching Lessons from the Case Studies

The two empirical examples presented highlight the diversity of social contexts, policy areas and individual factors that are implicated in our social constructions and experiences of mobility.

A Preference for Car-Based Travel

In Case Study 1, the preference for car travel, either as driver or passenger, was strong, and those who were legally excluded from driving experienced a sense of isolation. In relation to climate change policies that might restrict car use, such isolation may not be felt in the same way, as a driving ban as a result of a driving offence is somewhat different in social nature to driving reduction stimulated by climate change policies.

In Case Study 2, there was also a preference for car-based travel, and a general reluctance to be forced to access services via digital media, due in part to lack of confidence, but also to a sense that it reduces social contact for older people.

Qualitative research of this nature can *alert* policy makers that there could potentially be certain social consequences arising from climate change policies in transport. The qualitative aspects of travel behaviour are not well captured by existing appraisal methods (Jones, 2009) and the heterogeneous mobility and accessibility needs of the growing older age group already challenge the transport profession. However, this approach does not *specifically* nor *systematically* identify the precise nature of the SDIs of a given policy or intervention or which groups will be affected, to what extent, and how it can be mitigated.

The Need for Context-Specific Understandings

Assessing the SDIs of transport policy in general, and in relation to climate change in particular, needs to become considerably more sophisticated, contextually specific and quantitatively robust if such vulnerable groups are to be properly protected. This requires leading expertise and research activity from across a range of disciplines to more holistically focus on the uneven production and experience of energy vulnerabilities within and outside the transport sector, and the likely consequences of emerging low-carbon energy systems for the changing nature and distribution of equity across time and space and in different social and geographical contexts. This is a basic requirement of achieving social justice for presently excluded and vulnerable groups and is work that is already on-going within other disciplines and policy sectors (e.g. energy and fuel poverty, environmental justice).

CONCLUSIONS

The SDIs of transport decision making, whether in the United Kingdom or elsewhere, are generally less well considered and understood in comparison to economic and environmental factors. Our chapter identifies that different social groups can experience very different outcomes in accessing transport and adapting to changes to the transport system, whether these are uniquely targeted towards individuals or more fundamentally and systemically applied. It can be considered that the nuances of vulnerability and resilience are of critical importance in tackling SDIs for all policies, but particularly for policies intended to address climate change, as to be effective, these policies will need to be comprehensive and hard hitting and applied to all activities and to all populations sectors.

It is also essential that policy makers realise that climate change policies within transport will occur concurrently and cumulatively with climate change policies in other sectors. Untangling this web of policy interactions requires more disaggregation of the different SDIs for *each* policy intervention, and a better appreciation of how the SDIs are likely to change over time as well, something that is not generally done at present. It is proposed that adopting a more systematic framework within which to assess the SDIs of climate change policies, in which the key social impacts (e.g. wealth, health, accessibility) are distributionally analysed by income quintile/decile, as well as by potentially vulnerable groups (e.g. grouped by the social factors highlighted in Fig. 3) and spatially and temporally, as well as according to resilience/adaptability to the proposed intervention (i.e. according to the alternative transport options which are available to them and their ability to access these options, including cost, information, suitability, scheduling). It is clear that robust and easy to apply metrics will need to be developed to undertake such analyses and this is out with our own current research efforts at this time, but is certainly a research challenge to be considered in the future.

A need for a step change in the effectiveness of transport governance can be suggested, if climate change policies for transport are to avoid negative social consequences (see Marsden & Rye, 2010; Pangbourne, 2010). For example, our case studies have shown that health, social inclusion, well-being and sustainable development objectives are all cross-cutting policy areas that are not well integrated with either transport or spatial planning and that are often not entirely 'goal compatible' with each other. It is proposed that identifying and appraising the SDIs of transport policies in a disaggregated, yet more systematic and integrated manner, would make abstract policy strategies such as climate change reduction more relevant to the people who may be most affected and yet least able to readily adapt. The greater transparency would also make policy makers better able to design better policy responses in this area.

It is suggested that there is an urgent need for the design and implementation of a systematic framework to analyse the SDIs of climate change policy for transport. This chapter has outlined some potential ways forward in this respect, but this remains an important challenge for future research and policy making not only in the United Kingdom but also internationally.

REFERENCES

AEA Group. (2011). *Knowledge review of the social and distributional impacts of department for transport climate change policy options*. Retrieved from http://assets.dft.gov.uk/publications/knowledge-review-of-sdis-of-climate-change-policy-options/knowledge-review-of-sdis-of-climate-change-policy-options-full-report.pdf

Atkins and University of Aberdeen. (2009). *Mitigating transport's climate change impact in Scotland: Assessment of policy options Scotland: The Scottish Government*. Retrieved from http://www.scotland.gov.uk/Publications/2009/08/26141950/0

Banister, D. (2007). Cities, mobility and climate change. *Journal of Industrial Ecology*, *11*(2), 7–10.

Cloke, P., Milborne, P., & Thomas, C. (1997). 'Living lives in different ways? Deprivation, marginalisation and changing lifestyles in rural England'. *Transactions of the Institute of British Geographers, New Series*, *22*, 210–230.

Committee on Climate Change. (2009). *Annual report and accounts*. London: The Stationery Office.

Countryside Agency. (2002). *Two wheels work: A good practice guide for developing and implementing wheels to work schemes*. Cheltenham: Steer Davis Gleave report for the Countryside Agency.

Dargay, J., Gately, D., & Sommer, M. (2007). Vehicle ownership and income growth, worldwide: 1960-2030. *Energy Journal*, *28*(4), 163–190.

Department for Transport. (2008). *Transport trends: Current edition 2009*. Retrieved from http://www.dft.gov.uk/pgr/statistics/datatablespublications/trends/current/section4vptas.pdf

Doherty, S., Lee-Gosselin, M., Burns, K., & Andrey, J. (2002). Household activity rescheduling in response to automobile reduction scenarios. *Transportation Research Record*, *1807*, 174–181.

Felstead, A. (2009). Detaching work from place: Charting the progress of change and its implications for learning. *Beyond Current Horizons Challenge Paper*, a programme organized by the FutureLab and funded by the Department for Children, Families and Schools, the Department for Innovation, Universities and Skills, and Foresight. Retrieved from http://www.beyondcurrenthorizons.org.uk/wp-content/uploads/final_felstead_detachingworkfromplace200901162.pdf

Gärling, T., Gärling, A., & Loukopoulos, P. (2002). Forecasting psychological impacts of car use reduction: A challenge to an environmental psychology of transportation. *Applied Psychology: An International Review*, *51*(1), 90–106.

Gray, D., Farrington, J., Shaw, J., Martin, S., & Roberts, D. (2001). Car dependence in rural Scotland: Transport policy, devolution and the impact of the fuel duty escalator. *Journal of Transport Geography*, *17*, 113–125.

Hills, J., & Stewart, K. (Eds.). (2007). *A more equal society? New labour, poverty, inequality and exclusion*. Bristol, UK: The Policy Press.

Jones, P. (2009). Changing perspectives: Challenges for research and practice. *Proceedings of the Scottish Transport Applications and Research Conference*. Glasgow, 1st May 2009.

Kenyon, S., Lyons, G., & Rafferty, J. (2002). Transport and social exclusion: Investigating the possibility of promoting inclusion through virtual mobility. *Journal of Transport Geography*, *10*, 207–219.

Kridel, D. J., Rappoport, P. N., & Taylor, K. D. (1999). An econometric study of the demand for access to the internet. In D. G. Loomis & L. D. Taylor (Eds.), *The future of the telecommunications industry: Forecasting and demand analysis*. The Netherlands: Kluwer Academic.

Levin, K. A., & Leyland, A. H. (2006). A comparison of health inequalities in urban and rural Scotland. *Social Science and Medicine*, *62*, 1457–1464.

Litman, T. (2011). *Evaluating transport equity: Guidance for incorporating distributional impacts in transportation planning Victoria Transport Planning Institute*. Retrieved from http://www.vtpi.org/equity.pdf

Lucas, K. (Ed.). (2004). *Running on empty: Transport, social exclusion and environmental justice*. Bristol, UK: The Policy Press.

Lucas, K., Blumenberg, E., & Weinberger, R. (2011). *Auto motives: Understanding car use behaviours*. Bingley, UK: Emerald Publishing Ltd.

Lucas, K., & Jones, P. (2009). *The car in British society*. London: RAC Foundation.

Lyons, G. (2009). The reshaping of activities and mobility through new technologies. *Journal of Transport Geography*, *17*, 81–82.

Lyons, G., & Haddad. (2008). Commute displacement or commute replacement: The rise of part-day homeworking. Paper presented at the 87th Annual Meeting of the Transportation Research Board, Washington, DC, January 2008.

Lyons, G., & Urry, J. (2005). Travel time use in the information age. *Transportation Research*, *39A*, 257–276.

Marsden, G., & Rye, T. (2010). The governance of transport and climate change. *Journal of Transport Geography*, *18*(6), 669–678.

Mitchell, G., & Dorling, D. (2003). An environmental justice analysis of British air quality. *Environment and Planning A*, *35*, 909–929.

Mokhtarian, P. L. (1990). A typology of relationships between telecommunications and transportation. *Transportation Research*, *24A*(3), 231–242.

Mokhtarian, P. L., & Salomon, I. (2001). How derived is the demand for travel? Some conceptual and measurement considerations. *Transportation Research*, *35A*(8), 695–719.

Mokhtarian, P. L., Raney, E., & Salomon, I. (1997). Behavioral responses to congestion: Identifying patterns and socio-economic differences in adoption. *Transport Policy*, *4*(3), 147–160.

Pangbourne, K. (2010). *The changing geography of Scottish transport governance*. Unpublished Ph.D. thesis, University of Aberdeen, Aberdeen, UK.

Pangbourne, K., Aditjandra, P. T., & Nelson, J. D. (2010). New technology and quality of life for older people: Exploring health and transport dimensions in the UK context. *Intelligent Transport Systems – IET*, *4*, 318–327.

Pickup, L., & Giuliano, G. (2005). Transport and social exclusion in Europe and the USA. In K. P. Donaghy, et al. (Eds.), *Social dimensions of sustainable transport: Transatlantic perspectives* (pp. 38–49). London: Ashgate.

Salomon, I. (1986). Telecommunications and travel relationships: A review. *Transportation Research, 20A*(3), 223–238.

Shucksmith, M. (2003). *Social exclusion in rural areas: A review of recent research London: Department for Environment Food and Rural Affairs*. Retrieved from http://alpsknowhow.cipra.org/main_topics/governance_capacity/pdfs/shuksmith.pdf

Social Exclusion Unit. (2003). *Making the connection; final report on transport and social exclusion*. London: Social Exclusion Unit.

SpeedLimit. (2011). *Petrol prices Website*. Retrieved from http://www.speedlimit.org.uk/petrolprices.html

Steer Davies Gleave. (2005). *Evaluation of wheels to work schemes*. Cheltenham: Final report by Steer Davis Gleave for Countryside Agency.

Urry, J. (2007). *Mobilities*. Cambridge: Polity Press.

Wright, S., Nelson, J. D., & Cooper, J. M. (2009). An evaluation of the transport to employment (T2E) scheme in Highland Scotland using social return on investment (SROI). *Journal of Transport Geography, 17*, 457–467.

CHAPTER 12

THE ROLE OF INFORMATION IN REDUCING THE IMPACTS OF CLIMATE CHANGE FOR TRANSPORT APPLICATIONS

Owen Waygood, Erel Avineri and Glenn Lyons

ABSTRACT

Purpose – *To investigate the potential future role of information delivery in terms of reducing transport-related emissions. The assumed response of travellers to information and the notions of content and context are provided, together with factors that can determine an individual's propensity to seek and engage with travel information.*

Methodology/approach – *A review of information use, the behavioural stages of change and context design is undertaken, incorporating a research case study in the use of framing.*

Findings – *Recent theoretical and empirical insights into behavioural change provide new insights into the role of information in reducing transport-related emissions. Traveller information offers many benefits to the user and could be influential in affecting travel behaviour change, although there are socio-psychological factors outside the control of information service designers that also need to be considered, such as the 'stage of change' people are at in relation to a change in behaviour.*

Transport and Climate Change
Transport and Sustainability, Volume 2, 313–340

ISSN: 2044-9941/doi:10.1108/S2044-9941(2012)0000002015

Originality/value – *The chapter examines how travel information could go beyond the provision of trip choice data and incorporate contextual elements to make it more effective. Concepts related to the Transtheoretical Model (stages of change) and the latest thinking in behavioural economics are discussed in relation to an increase in the likelihood of environmental travel alternatives being considered and chosen.*

Keywords: Information; content and context; stages of change; behavioural economics; rational and irrational response

INTRODUCTION

It has long been held that information has an important part to play in supporting and influencing the travel decisions of individuals. Further, it has been widely hoped, if not assumed, that improvements in information provision will yield changes in travel behaviour that are to the advantage of the transport system as a whole (whether they be reductions to externalities like congestion or carbon dioxide). Information serves a number of important roles (Lyons, 2006): making the individual aware of the travel options available to them; enabling 'better' judgement in selecting between the options; and assisting in being able to plan, undertake and successfully complete the travel option pursued. As the information age has taken hold, the nature and extent of availability of information has evolved considerably. Alongside this, research into understanding the role of information in influencing behaviour has advanced insights into how the content and context of information provision may exert influence. This chapter examines factors that may influence the efficacy of information provision, with particular reference to how the environmental dimension to people's travel choices can be addressed.

The provision of travel information about transport-related attributes, such as travel time, travel costs and, more recently, transport-related carbon emissions might be seen not only as a service provided to the public, but also as an instrument intended to change their travel behaviour. However, some expectations of big effects from new information systems have remained unrealised (Chorus, Molin, & van Wee, 2006). It can be argued that the design of travel information systems has often ignored the psychological, ergonomic and cognitive processes of choosing to seek and then retrieving and using information. While rational man theory suggests that individuals base choices on the attributes of the choice set ('*information content*'), the way information is being presented ('*information context*') can also have a strong effect on travellers' use of and reaction to information – a particular focus of this chapter. Following recent studies in behavioural economics and cognitive psychology, we illustrate how 'choice architecture', through the design and incorporation of small features in the environment of choice making ('*nudges*'), might help individuals to overcome cognitive biases, and to highlight the 'better' choices for them.

INFORMATION USE

If information provision is intended to influence travel behaviour change, then an appropriate starting point is to understand its use and the choice making processes related to travel information.

Rational and Irrational Determinants of Travel Choice

The behavioural assumptions on responses of individuals to travel information, commonly applied in the design of Advanced Traveller Information Systems (ATIS), can be traced back to economic theory and the paradigm of rational man. Ben-Akiva and Lerman (1985, p. 32) describe the theory of choice as a collection of procedures that define the following elements: (i) decision-maker; (ii) alternatives; (iii) attributes of alternatives; and (iv) decision rule. The attractiveness of an alternative in the mind of the traveller can be described as 'utility' related to the values of the travel attributes.

A choice to take a mode can relate to many things, and some of the reasons to take a mode could be: independence, privacy, social interaction, convenience, time (total, portion), cost (real/perceived), comfort, experience, environmental impact, social impact (e.g. safety, congestion), or health. As information concerning the poor performance of a mode can stimulate change (Chorus, Molin, & van Wee, 2006), providing information on things that people value, but that cars perform poorly at in comparison to other alternatives may help swing the pros and cons of the decisional balance. One example of this would be the Brighton and Hove travel planner, Journey On[1] (Fig. 1).

As travellers are expected to act as rational human beings, and specifically to exhibit consistency and transitivity in their choices (Ben-Akiva & Lerman, 1985, p. 38), then the *way* alternatives and attributes are presented

Mode	Distance	Duration	Calories	Cost	CO_2 impact
Walk	1.44 miles	29 min	137	£0 ☺	0 kg ☺
Cycle	1.40 miles	9 min	94	£0 ☺	0 kg ☺
Public Transport	1.30 miles	23 min	25	£1.50	0.21 kg
Car	1.53 miles	6 min	0 ☹	£1.07	0.49 kg ☹

Fig. 1. Example Trip Adopted from the Brighton and Hove Travel Planner Including Design Features Such As Smiley Faces. That Act As Injunctive Norms (Schultz et al., 2007). *Source*: www.journeyon.co.uk

to the traveller should not matter greatly. However, as it will later on be argued, it is not just the information *content* that may be influential; travellers may also be affected by the manner in which information is presented (its *context*).

A strategic review of travel information research commissioned by the UK Department for Transport (Lyons, Avineri, Farag, & Harman, 2007) examined evidence that travellers do not necessarily respond to the information in the way ATIS designers may believe they should. It has been argued within this literature that travellers' limited cognitive resources have a strong effect on travel-choice behaviour, and evidence on systematic deviations from rational models have emerged from studies of the processes of gathering travel knowledge, interpreting travel information, and processing it in real time (Avineri & Prashker, 2004; van de Kaa, 2010; Avineri & Waygood, forthcoming). Even when travel information is simple, reliable and accurate, travellers tend to interpret and value this information in ways that systematically violate the main assumptions of rational behaviour. Although informed travel choices made by individuals should not be seen as 'irrational', it appears more weight should be given to what are sometimes called 'irrational' motives and behaviours in the design of information systems, recognising the importance of context, emotions, habit, impulses, social influence and a range of socio-psychological factors that might substantially vary between travel information users.

Information Acquisition

It is asserted, with support from empirical findings, that most people, most of the time, do not consult information because their everyday behaviours are routinised and habitualised – they are sufficiently familiar and predictable for the individual to make a (conscious) decision based on prior experience and typically yield a satisfactory outcome from that decision. Correspondingly people have a greater propensity to seek information when they face unfamiliar and/or uncertain circumstances. Such circumstances apply when individuals make a trip for the first time (Chorus, Molin, Arentze, Timmermans, & van Wee, 2006). It can also concern people who are less established in a mode (i.e. they do not use it out of habit) (Kenyon & Lyons, 2003; Lyons et al., 2007) and who need more information for various modes before making a decision (Verplanken, Aarts, & van Knippenberg, 1997).

In light of such findings, the city of Malmö, Sweden, created a travel information programme through which new residents were given travel information for modes other than by car (SMILE, 2006). Rather than waiting for the individual to seek information (that would most likely relate to their usual/preferred mode), the programme actively provided information. The study found that 95% of people contacted wanted and accepted transport advice and 15% of regular car users reported that they had changed some common travel patterns. Similarly, Bamberg (2006) reports a study where new residents were given a free bus ticket and personalised travel information. In comparison to the control group, the study group had statistically higher public transit use. Those findings suggest that information provided when new travel decisions are being made can lead to more sustainable travel.

People who are satisfied with their current mode are less likely to check information, especially on other modes (Chorus, Arentz, & Timmermans, 2006; Chorus, Molin, et al., 2006; Farag & Lyons, 2008; Kenyon & Lyons, 2003; Lyons et al., 2007). In such circumstances it may be necessary to 'push' information to them, unsolicited. A promising medium in this regard in relation to motorists is the radio – it has an enduring popularity with motorists (Lyons et al., 2007) and the individual can be a passive recipient of information that is delivered as 'infotainment'. Similarly, roadside variable message signs can be used to push formation to motorists (and has been done so, e.g. in The Hague to convey local air quality information[2]). However, here it should be noted that in a UK study, only a third of motorists passing a sign displaying immediate warning information were found to have even seen the sign (Chatterjee, Hounsell, Firmin, & Bonsall, 2002). In other words, pushing information to travellers is no guarantee that they will connect with the information.

In a review of the role of information on car use, Chorus, Arentz, et al. (2006) concluded that, 'Information provision on the performance of the currently chosen alternative ... may help change car-drivers' choices in the long run'. Their review found that ex-post information was particularly valuable for changing subsequent trips. With consideration to climate change it may be necessary to incorporate such information into cars, as many people are unlikely to seek this out independently considering the findings related to habit and preferred choice. However, this information may still be ignored.

Even if information is available, there may be low awareness of the information source, or even if people are aware of it, they may not use it (Coulter, Clegg, Lyons, Chatterton, & Musselwhite, 2007; Farag & Lyons, 2008; Kenyon & Lyons, 2003; Lyons et al., 2007; Peirce & Lappin, 2005). The use of such resources relates to the discussions above. Farag and Lyons (2012) in an empirical study of online public transport information found that propensity to consider public transport had a stronger influence on consulting public transport information than the other way around. What follows is that raising awareness of information availability may not increase its use – attention must be given to the need to raise awareness of public transport as a viable option and it is this that is likely to generate greater use of information.

One further point about information acquisition relates to the source. Not everyone owns a smartphone, nor is everyone comfortable or able to access online information. It is important to think who the target is and where they obtain information. In the UK, older people were found to rely on more traditional sources of paper-based information, whereas younger people more commonly used Internet sources (Farag & Lyons, 2008). However, this can vary across countries, as in Japan mobile phones are available that are specifically tailored to older people.

An appreciation of how people (are assumed to) make decisions and when they are prompted to consult or engage with travel information leads us now to consider further the heterogeneity of the travelling public in terms of social-psychological factors that are pertinent to the prospects of, and the requirements for, information influencing travel decisions.

PSYCHOLOGICAL AND BEHAVIOURAL STAGES OF CHANGE

The framing of this chapter is the notion that people are exhibiting current travel behaviours that are more or less detrimental in environmental terms – notably some forms of travel yield more or less greenhouse gas emissions than others. At a system level it is desirable to see travellers moving collectively to more sustainable travel choices – that is those that reduce the carbon footprint of travel. At the level of the individual traveller, behaviour will be governed *in part* by the extent to which the person considers 'environmental impact' to be an attribute of travel alternatives that has relevance to her selection of a preferred option. '*In part*' is stressed because more sustainable travel choices may in some instances arise as a consequence of individuals considering other impacts such as monetary or health – that is I may coincidentally travel more sustainably because I have chosen to walk and cycle more and drive less for health reasons, irrespective of my views regarding the environmental impact of my travel.

In this section, the Transtheoretical Model (TTM) is considered in order to conceptualise how different individuals can be at different stages psychologically and in turn behavioural change leading from one behaviour to another. It is noted how such interpretation of context can help considerations of the appropriate offering of information content to assist behaviour change in the direction desired.

TTM was developed in the health field to help address addictive behaviour, specifically smoking (Prochaska & Norcross, 2001), but it has also been applied to a wide range of health behaviours (Prochaska, Redding, & Evers, 2008; Prochaska et al., 1994). The model 'uses stages of change to integrate processes and principles of change across major theories of intervention' (Prochaska et al., 2008, pg. 97). It was built to assist health practitioners identify what type of help was required to move an individual towards successfully changing their problem behaviour. It is considered here as a theory to guide the provision of transport information to help individuals move from less sustainable to more sustainable behaviour. It has been applied previously in this regard in relation to drivers' willingness to reduce car use (Beatty, Meadows, & White, 2002) and cycling to work (Gatersleben & Appleton, 2007). It has also been applied to projects on information campaigns (TAPESTRY[3]) and mobility management (MAX[4]).

The model is composed of six temporal *stages of change* and a number of *processes* that can help move the target from one stage to the next. By understanding what the particular stage of change is, information can be tailored to best assist the target person or population move to the next stage. This involves understanding how the target views the pros and cons of changing travel behaviour in the decision process (Prochaska et al., 1994) so that the right triggers, or motivations, are addressed. The sequential stages of change are summarised as follows:

Pre-contemplation – The target individual's behaviour is a problem, but the individual does not consider it to be (or may wish to change but exhibits no consideration of doing so).
Contemplation – The individual is aware of the problem and is considering a change (i.e. knowledge of the problem may not be enough to cause change).
Preparation – The individual is preparing for a change, possibly trialling actions, but not as a sustained level (e.g. relapse to problem behaviour occurs frequently).
Action – An action has been chosen and sustained for up to six months.
Maintenance – An action has consistently been carried out for at least six months – lower risk of relapse, but the individual may still be consciously making the choice to act in this way.
Termination – An action has become habit (i.e. an automatic, not consciously considered behaviour) and the threat of relapse is negligible.

The first four of these stages are discussed in more detail in relation to moving from 'less sustainable' to 'more sustainable' transport behaviours and consider the ways in which individuals might be 'informed' to encourage movement through the stages. Information provision here will be seen to be broader than merely the concept of a journey planning facility that provides information on attributes of the alternatives. This further underlines the breadth of the notion of 'information context'.

Pre-contemplation

In the pre-contemplation stage the target may not consider their behaviour to be a problem and may be acting in a habitual manner. Here the challenge is one of consciousness raising and an awareness of the causes, consequences and 'solutions' to the problem. Unlike smoking where the behaviour is directly a problem to the individual, it can be more difficult for an individual to subscribe to the view that their unsustainable travel behaviour is a problem for them or indeed that acting to achieve a 'solution' may benefit them. Due to numerous such psychological hurdles related to climate change (see Swim et al., 2009 for examples), it may be necessary to focus on more individual or local impacts of car use (Anable, Lane, & Kelay, 2006; Swim et al., 2009).

It has been suggested (see TAPESTRY, 2003) that after raising awareness of the general problem, it is necessary for the target individuals to accept responsibility. Due to the scale of climate change, individuals may feel that any change that they might effect would have little overall impact (Swim et al., 2009; Waygood & Avineri, 2010), thus they may feel reduced responsibility. Initiatives such as Transportation Town[5] or the Housing and Transportation Affordability Index[6] are attempting to address this problem of scale by contextualising the individual's impact in their provision of information on greenhouse gases (see section on context).

The process of *dramatic relief* suggests that emotional experiences should be triggered. People are more motivated by emotive stories of individuals than they are by statistics (Slovic, 2010), so developing campaigns that highlight success stories or focus on impacts to an individual may be effective motivators. However, the impact of climate change on other individuals that are geographically and culturally removed may not be as effective (Swim et al., 2009).

It will be appreciated from the above that to take individuals out of pre-contemplation may be a formidable challenge and may 'only' support contemplation of change.

Contemplation

At this stage an individual accepts that there is a problem with their personal behaviour and is contemplating some change. It is necessary to help the target understand the pros and cons of behaviour change and to shift the decisional balance away from the problem behaviour. It will also be important for the individual to believe that they *can* change.

The process of *environmental re-evaluation* encourages the individual to think about the impact his problem behaviour has on others. Information here may relate to personal benefits (e.g. health, monetary savings), community benefits (e.g. reduced traffic, NO_2), or global ones (e.g. climate change) such as the example of Brighton and Hove's travel planner in Fig. 1. However, consideration to the design issue of framing (discussed later) could improve the effectiveness of such information. The end result is to reduce greenhouse gases, but people do not necessarily need to be motivated by environmental concerns to reduce their transport impacts.

Gatersleben and Appleton (2007) found that information that supports cycling as a flexible transport option rather than explicitly in terms of the health or environmental benefits may be more effective at creating behaviour change. Such results highlight that although the end goal is to achieve sustainable transport behaviour, an individual's motivation(s) to do so does not need to be environmentally based.

Self-liberation is the belief that one can change. This is related to self-efficacy where confidence exists that despite problems, one will succeed in changing behaviour. With consideration to the framework that the MAX project created, Gatersleben and Appleton (2007) found that pre-contemplators had the highest personal barriers, which might be addressed through individualised travel programmes that focus on removing perceived barriers such as Individualised Marketing (Brog & Erl, 2008), Travel Blending (Rose & Ampt, 2001; Taylor & Ampt, 2003), or the Travel Feedback Program (Taniguchi, Hara, Takano, Kagaya, & Fujii, 2003). Such programmes often identify people who are already considering change.

Preparation

In the preparation stage, individuals are deciding how they might change their problem behaviour. The process of *counterconditioning* should here be applied by highlighting different relevant transport options. Practical information and tacit know-how (rather than conceptual information), are useful in facilitating behavioural change (De Young, 1996). People with less knowledge or experience may desire prescriptive information, while more experienced individuals may desire descriptive information that allows them to make a refined choice themselves. In a state of low knowledge, more prescriptive advice may also be more useful than a range of options (Chorus, Arentz, et al., 2006).

The process of *helping relationships* suggests that social connections will be an important source of support and information, alongside formal information sources – that is making plans in front of others has a pledge component which is seen as psychologically meaningful. Research on descriptive norms suggests that using information on how others behave can be an effective way to change behaviour. Despite people not believing that information about how others behave would affect their personal behaviour, research has found that it does (Nolan, Schultz, Cialdini, Goldstein, & Griskevicius, 2008; Schultz, 1999; Schultz, Nolan, Cialdini, Goldstein, & Griskevicius, 2007).

Action

At this stage an individual has chosen an action to address their problem. The cons of the problem behaviour outweigh the pros and thus such information is not the focus. For this stage, the process of *contingency management* is relevant. Here the appropriate information will be feedback on their chosen action related to their motivations to change (e.g. calories burnt if it is a health motivation, greenhouse gas emissions if it is environmental). Prochaska et al. (2008) suggest that positive feedback with rewards is relied upon more than punishment for self-changers. Appropriate feedback can help the individual feel that they are accomplishing something, or help them adjust their actions to make improvements.

There is no clear-cut theory on how feedback works. Darby (2006b) suggests that it is part of a learning process that contributes to behaviour. In some situations, feedback is intrinsic (such as touching a hot stove), and so behaviour is learnt in certain ways (such as not touching hot objects). However, in other situations such as CO_2 emissions, feedback is hidden or disconnected from the action. As a result, tools like carbon calculators attempt to complete the link between one's actions and the resultant CO_2 produced (Avineri & Waygood, 2010; Coulter et al., 2007).

In terms of feedback, continuous and directly available information is best (Abrahamse, Steg, Vlek, & Rothengatter, 2005; Darby, 2006a). General information has been found ineffective in creating behavioural change (but may be useful earlier in the consciousness raising process), but tailored and authoritative information does have an impact. One aspect of providing personal feedback is through *ongoing measuring and monitoring*. The act of measuring resource flows makes invisible consumption visible, and allows participants to expose it to conscious scrutiny (Avineri & Goodwin, 2010; De Young, 1996; Staats, Harland, & Wilke, 2004). Botrill (2007) highlighted ongoing measuring and monitoring as a distinctive feature of what was termed 'second generation' carbon calculators. Those tools allowed users to register and track use over time. However, such tools are typically accessed by people who are highly concerned about climate change already and their personal impacts (Coulter et al., 2007).

Eco-driving is an area where ongoing measuring and monitoring is being applied in encouraging drivers to become more energy efficient in their driving styles. In several hybrid vehicles more efficient driving is encouraged through feedback that has a game-like aspect. For example, in more recent versions of Ford's SmartGuage (Ford Media, 2010), the tool now incorporates a long-term (four to six weeks[7]) driving pattern feedback in the form of growing flowers. If the driver maintains more efficient driving over a longer period of time, then flowers will sprout. Such feedback may help establish more energy efficient driving patterns rather than just short term responses to the previous feedback of growing leaves in response to short-term driving efficiency.

THE IMPACT OF CONTEXT DESIGN

The preceding sections have highlighted some of the challenges and opportunities associated with influencing behaviour through information provision. They underline that while information content itself has an importance, the significance of *context* in terms both of the nature of the individual's travel (habit, familiarity, certainty, etc.) and social-psychological considerations (the stage of change they are at, according to the TTM) cannot be understated. In this section we examine further how individuals can be affected by a further contextual issue – information context design. By this, the meaning is not the content of information itself (i.e. 'the facts') but the way in which those facts are portrayed to the individual.

People are influenced by images, symbols and context, that is the manner in which information is being presented to them. For example, a textual message coloured in green would carry an additional connotation besides the text content. It is widely recognised that colours can have an impact on our emotions and feelings. Green, for example, has been generally found to have a relaxing effect; it has both positive impressions such as refreshment, quietness, naturalness, and negative impressions such as tiredness and guilt (Davey, 1998; Mahnke, 1996; Saito, 1996, cited in Kaya & Epps, 2004).

Although individuals seem to be largely unaware of colour effects (and other effects of visual, textual and contextual design of information on their perceptions and choices), they are influenced by their use in specific contexts; fashion designers, product designers and marketers have incorporated such insights in the design of products and goods to influence consumer choices. For example, warm colours (such as red and orange) have been found to be stronger, more exciting and more arousing than cool colours (such as blue and green) that are considered calming (Bellizi & Hite, 1992; Davidoff, 1991; Kreitler & Kreitler, 1972). This concept has been applied in a variety of situations in an attempt to influence individual behaviour, an example of arousing consumers by using warm colours is the use of the colour red in gambling environments to stimulate gambling (Griffiths & Swift, 1992; Spenwyn, Barrett, & Griffiths, 2010).

A critical feature of the mind's cognitive architecture is the 'modularity of the mind' (Fodor, 1983). The mind is composed of an array of interacting, specialised subsystems with somewhat limited flows of intercommunication. Many perceptual and cognitive processes work independently of each other and specialise in processing specific inputs from the environment. This concept can be illustrated by optical illusions – where previous 'objective' knowledge on a visual image might be in conflict with its perception. Brain scientists and cognitive psychologists have discovered that the brain functions as if it had two systems of decision-making; one is very fast and automatic, while the other one is reflective (Epstein, 1994). It appears that while the reflective system processes the information *content*, and applies rather systematic and rational 'algorithms', a parallel process takes place in our brain by the automatic system which processes the *context* of information – such as the visual environment of the main message, applies heuristic 'short-cuts', and attaches emotions and feelings.

This model of information processing contrasts sharply with the old idea that any piece of relevant information would be accessed by a rational cognitive process, while irrelevant information would be ignored by the reflective system. However, both systems work in parallel in the processing of information, and sometimes generate different outcomes. Optical illusions demonstrate the modularity of the mind: even when aware of the illusion, individuals cannot simply have it 'turned off'.

An informed design of the context of travel information might be another important dimension in promoting sustainable behaviour. In applying a holistic approach to the design of travel information systems, simultaneous use should be made of different methods that influence and stimulate both the reflective and the automatic system, and which are pulling in the same direction. Such an approach (one that combines content design and context design) would be seen as more effective than a traditional approach to the design of travel information systems that focuses on information content only.

Research in behavioural sciences, especially cognitive psychology, indicates that individuals' choices in a wide range of contexts deviate from the predictions of economic theory which largely assume a rational behaviour. Some of these deviations are systematic, consistent, robust and largely predictable. Evidence on systematic deviations from rational models have mainly emerged from studies on consumer behaviour, financial behaviour, health behaviour and more recently – travel behaviour

(Avineri & Prashker, 2003; Avineri & Prashker, 2004; Kahneman & Tversky, 1979; Tversky & Kahneman, 1974; Tversky & Kahneman, 1991). The 'predicted irrationality' (a term coined by Ariely (2008) of individuals could (and some argue that it should) play a role in the design of behavioural change interventions. Thaler and Sunstein (2008) advocate the use of 'choice architecture' to influence behavioural change: 'nudges', small features designed in the environment of choice making, could help individuals to overcome cognitive biases, and to highlight the better choices for them – without restricting their freedom of choice, and without making big changes to the physical environment, the set of choices, or the economic attributes of the choices.

Despite the considerable work done by researchers in behavioural sciences and practitioners in a range of applied fields such as marketing, product design, urban design, public health and education, it can be argued that the planning and design of travel information systems is often made without much reference to existing theories and evidence from behavioural studies on the influence of context. The main purpose of this section is to illustrate and discuss the potential application of some of the insights from behavioural sciences and their relevance to the design of travel information systems. There has been so far only limited literature on context effects in the design of traveller information systems, and choice architecture has not been applied and evaluated in this context. Therefore, its effectiveness remains an open question, which is not addressed in this chapter.

Three specific concepts of context design are explained in detail and illustrated in this section: design of default choices, framing of textual and numeric messages, and the contextualisation of travel-related CO_2 emissions in the design of choices.

Defaults

People are influenced by 'defaults' set for them by the designers of information systems. A default choice is what happens if no effort is made by the individual to change the outcome. For example, a default action may be to receive a plastic bag when shopping and the customer must make some effort to change that outcome.

It is often impossible for private and public institutions to avoid picking some option as the default. Picking 'smart' defaults is a way to nudge towards specific choices. Enrolling people automatically into savings plans or organ donation, while allowing them to opt out, is an example of a nudge based on default design. Well-chosen default rules are examples of helpful 'choice architecture'. For example, the 'Save More Tomorrow' plan (Thaler & Benartzi, 2004), which allows employees to commit themselves now to increasing their savings rates later, when they get raises, has been remarkably successful. Enrolling people automatically into savings plans, while allowing them to opt out, is an example of a successful default choice (Madrian & Shea, 2001). Organ donation policies that make use of opt-in defaults are more successful than others (Thaler & Sunstein, 2008).

In the context of ATIS design, in response to a user's interactive query on travel alternatives, the system could present by default (or prioritise through 'semantic framing') the more desired choices (i.e. sustainable modes such as walking, cycling or public transport will be presented by default; eco-driving choices presented by an in-car system). For example, in the UK, Transport Direct[8], a national-scale web-based journey planner, provides travellers pre-trip information about both public transport and car alternatives to get from a specified origin to a destination. Ticking off the public transport option, the user would receive only information related to car transport. However, it can be assumed that many 'passive' users of the system would not make the cognitive effort to tick off an option and will be provided with information on the (more sustainable) public transport, even if many of them have the original intention of using a car for the journey. Kenyon and Lyons (2003) termed such systems *Integrated Multimodal Traveller Information* and found that it could produce a modal shift amongst some users.

Highlighting the alternative ways to travel, and providing additional information about their co-benefits (such as the expected reduction in carbon emissions) might contribute to the level of awareness, change attitudes and intentions, and might even provide an opportunity to bring about behavioural change among car users. In some situations, some 'rational' utility maximisers among car users might find that sustainable choices, with their co-benefits, might provide a better option for them and be motivated to change their intended choice (or at least be encouraged further into stages of contemplation and preparation). Although relevant information about sustainable alternatives is provided, many users of travel information systems would not consult it as they might not find it relevant to them, or not be willing to pay the cognitive effort in pursuing such information.

The main argument in favour of incorporating defaults into travel information systems would be that even a user who does not have the original intention of pursuing information on sustainable choices, would be exposed to such information. Another argument would be the relatively marginal cost in designing such defaults options (they do not require 'new' content of information or much addition to the source code). As information on other choices is provided and accessible, and users choices are not restricted in any way (not even by 'friendly advice') this could be seen as a gentle push in the right direction, and not as choice enforcement – in line with the 'libertarian paternalism' approach advocated by Thaler and Sunstein (2008).

Framing

Framing is the semantic manipulation of rationally identical problems. In *semantic framing* only alternative phrasing is used, while in *valence framing*, information is put in either a positive or negative light (Hallahan, 1999). Semantic framing may reorganise a sentence so that a key point comes at the beginning rather than the end. Valence framing attempts to put something in a positive or negative light (e.g. the spin on a story) such as renaming 'tax cuts' as 'tax relief' because 'relief' has positive connotations.

Related to valence framing and semantic framing is a framing technique that highlights gains or losses associated with the choice outcome, called *gain/loss framing*. It has generally been observed that people tend to feel and behave differently when information is presented (or 'framed') in terms of gains or losses (Kahneman & Tversky, 1979; Thaler, Tversky, Kahneman, & Schwartz, 1997; Tversky & Kahneman, 1991). This observed behaviour, called gain/loss asymmetry, refers to the fact that people tend to be more sensitive to decreases in their wealth than to increases. Roughly speaking, losing £100 produces a negative feeling which is twice as intense as the positive feeling experienced when gaining £100.

In Prospect Theory (Kahneman & Tversky, 1979; Tversky & Kahneman, 1991) the choice outcomes evaluated by the decision-maker are perceived as 'gains' or 'losses' relative to some reference point. Such a reference point may be the current asset position, but may be influenced by the presentation of the outcomes or expectations of the decision-maker. Recent studies provide evidence that travellers exhibit aversion to loss and have a strong tendency to avoid choices associated with losses (Avineri, 2006; Avineri & Prashker, 2004; van de Kaa, 2010).

One of the considerations in the design of information is the semantic presentation, or framing, of information. In such a case, the information

Table 1. Framing Effects: Three Ways of Presenting the Same Travel Time Information (from Avineri, 2011).

A	Commuting by car will take 20 minutes Cycle-commuting will take 15 minutes
B (gain framing)	Commuting by car will take 20 minutes Choosing to commute by bike, your journey will be *5 minutes shorter*
C (loss framing)	Cycle-commuting will take 15 minutes Choosing to commute by car, your journey will be *5 minutes longer*

remains the same, but how it is presented differs. Information on the attributes of alternative choices could be re-organised so that positive or negative aspects are highlighted. In presenting information on the attributes of different travel choices (e.g. journey time), the information could be presented in absolute time units, or one of the alternatives as the default choice – therefore implying its outcome as the value of the reference point, and highlight to the individual the relative gain or loss associated with the other alternatives.

Table 1 illustrates the application of loss aversion in the framing of travel choices. It presents three ways of presenting the same information on the journey time of two options: car and cycle commuting.

Under the rational choice model, the format of the information should not matter. The information presented by all three formats in Table 1 has the same content; it is the difference in how it is presented (B with cycling as a gain; C with car use as a loss) that could influence the choice – in a way that cannot be explained or predicted by classical economics.

None of the information formats impose a restriction of the travel options for the traveller and therefore can be seen as a 'soft' measure. Through the loss-framing applied in framing alternative C, the traveller is encouraged, or 'nudged', to choose the option which is considered to be preferable.

Studies from the health field (Kühberger, 1998) and home energy use (Gonzales, Aronson, & Costanzo, 1988) have found that framing affects choice. In the health field a meta-analysis of 136 empirical papers found that framing is a reliable phenomenon (Kühberger, 1998). However, both those fields deal with losses to the individual and with relatively familiar topics such as financial loss or negative health impacts. Climate change can be considered a social dilemma, rather than an individual problem like home expenses or deteriorating health from some behaviour. The latter point suggests that people may not have as strong a tendency to 'avoid' increases in CO_2 emissions, thus lowering the effectiveness of loss framing. Further to that, most CO_2 emissions information is presented as mass (Avineri & Waygood, 2010), but research suggests that such information may not be useful as people may not be able to interpret it as being a 'good' or 'bad' amount (Coulter et al., 2007; Waygood & Avineri, 2010, 2011). If individuals have little experience dealing with CO_2 emissions information and the dominant format of mass is not facilitating understanding, then it may be difficult for individuals to notice a difference between amounts. However, considering the effectiveness of loss framing in other fields, it may be possible to use that technique to increase the likelihood of an individual perceiving a difference despite the different levels of understanding and relevance to an individual.

An Environmental Impact-Oriented Case Study of Framing

Avineri and Waygood (forthcoming) examined whether the design consideration of loss framing could increase the perception of differences about the mass of travel-related CO_2 emissions. The hypothesis on the relative effectiveness of loss framing, based on the empirical evidence in other fields, is that it will increase the likelihood of a loss being perceived. Different than most studies on loss aversion their research focuses on the effect of framing on perception rather than its direct effect on choice. Perceiving differences between the outcomes of choices is believed to be important in choices made in a social dilemma context. Therefore, incorporating loss-framing as a nudge to highlight the differences between alternative choices is a preliminary stage in behavioural change required before or in parallel to a change to one's motivation and intention of behaviour change towards sustainable choices.

To test the effectiveness of gain and loss framing, two comparison sets were semantically constructed by Avineri and Waygood (forthcoming) so that gains and losses were distinctly presented. The comparison sets were based on the per passenger amounts of CO_2 produced over a five mile trip by a bicycle, full car, and single occupancy 4×4 (the amounts were 132 g, 500 g and 3400 g, respectively). After semantically reorganising the comparison sets, four sets of statements about CO_2 emissions generated by the transport modes resulted. The statements were as follows:

Loss framing for the first comparison set:

Mode X produces 132g of CO_2 for a 5 mile trip.
The amount produced by mode Y is 368g higher (i.e. worse).

Gain framing for the first comparison set:
Mode X produces 500g of CO_2 for a 5 mile trip.
The amount produced by mode Y is 368g lower (i.e. better).

Loss framing for the second comparison set:
Mode X produces 500g of CO_2 for a 5 mile trip.
The amount produced by mode Y is 2900g higher (i.e. worse).

Gain framing for the second comparison set:
Mode X produces 3400g of CO_2 for a 5 mile trip.
The amount produced by mode Y is 2900g lower (i.e. better).

Each of the participants was asked to respond to one set with gain framing and one with loss framing. The participants were asked to choose whether the two CO_2 emission amounts were: about the same; slightly different; or much different. The experimental findings reported in Avineri and Waygood (forthcoming) suggest that there are impacts from both loss framing and the scale of the CO_2 being compared on how people perceived difference between alternative travel choices. Although behaviour change has not been evaluated in a direct way, it is assumed that as the perception of the difference between CO_2 amounts is larger for loss framing, there would be an increased likelihood of the lower CO_2 emissions choice being made by the individual who has been exposed to loss framing than the one exposed to gain framing. The findings suggest that loss framing might help to improve the design of information-based initiatives as it works as a 'nudge' to improve the effect of CO_2 information and is therefore likely to increase the choice of sustainable options. The potential application of loss framing would be for governments (and other travel information providers) to highlight and 'nudge' more sustainable travel choices, and being as transparent as possible. This approach could be applied in the 'choice architecture' of ATIS (such as web-based journey planners, navigation and way-finding devices, and in-vehicle information systems) or as part of other behaviour change measures (such as personal travel planning). Another question still to be addressed however is what happens if the individual is aware that they are subject to framing? A government-provided travel information service exposed by the media to be using framing *may* be discredited and results from focus groups related to this question highlight that when informed of the technique, people feel that they are being manipulated (Avineri & Waygood, forthcoming). Such results do suggest that although effective, further investigations and arguments around their use will likely arise.

Contextualising Travel-Related CO_2

Presented with travel-related CO_2 emissions information by mass, people may appreciate that less is better than more, but to what degree and what the desirable thresholds are may not be clear.

Due to the relative 'newness' of CO_2 information, people are likely acting in a state of uncertainty about how sustainable their motorised travel choices are. In a high state of uncertainty other alternatives could make certain choices appear more or less sustainable.

Despite the increasing presence of CO_2 information, there seems to be little research on whether people understand that information. Coulter et al. (2007) found that both users and non-users of carbon calculator websites reported not really understanding the results when presented in units of mass. The interpretation of information depends on the users' background knowledge and the provided context.

Through its contextualisation, information on CO_2 emissions can be better grasped, and its presentation would be more likely to be effective in motivating behaviour change. One way to contextualise greenhouse gas information would be to represent it as something more familiar to people. This might be showing the volume of gas as a balloon, the energy consumption as hours of a common household appliance, or something that interacts with the gases such as trees. Although graphical representation is beneficial in many regards, consideration must be given to what it conveys. A common image associated with CO_2 emissions is that of the carbon footprint. However, the image itself does not convey any injunctive information; that is, information that would allow the user to know if the amount was desirable from a societal perspective. Coulter et al.'s (2007) study highlighted that visual equivalents displaying impacts (negative and positive) were more desirable than simply CO_2 mass.

Other presentation styles, or formats, that have some degree of context could be considered as an alternative to presentation of transport-related CO_2 emissions in mass units. For example, the presentation style of tree equivalents (the number of trees required to sequester the CO_2 output) visualises the rather abstract CO_2 emissions. The presentation style of earth equivalents visualises the CO_2 information and attempts to show what is sustainable by showing how many earths would be required if all people behaved as the user. Carbon emissions presented as percentages of a carbon budget or an 'allowance' represented a cap and trade system. To increase the relevancy of CO_2 information to people, governments are contemplating cap and trade systems or tax systems related to CO_2 output.

Mass:	**"680 kg of CO_2/year"**
Trees equivalent:	
Earths equivalent:	
Carbon budget:	**"27% of yearly budget"**

Fig. 2. Information on Travel-Related CO_2 (Typical Annual Per Passenger Amount For A Full Mid-Sized Car), Presented in Four Different Formats.

Cap and trade systems could change behaviour by giving an acceptable limit and through market mechanisms. The acceptable limit set by the cap would help to create a reference point for people to judge whether a choice would be a gain or loss (related to the concept of a reference point and gains/losses associated with travel choices, see the previous discussion on framing).

Fig. 2 illustrates how the same information about travel-related CO_2 emissions, a typical yearly amount of emissions produced over a five mile trip by a car, can be presented in four formats: mass, trees equivalent, earths equivalent and a carbon budget. The yearly amount is based on the UK average mileage of 6923 miles.[9] The tree equivalents used are based on results from American Forests.[10] A carbon allowance of 3.6 tonnes was used based on results from the World Wildlife Fund's footprint calculator.[11]

Another consideration to equivalents might also be the dual process theory's experiential versus analytic systems (Epstein, 1994). Certain equivalents might work on an emotional level rather than in the slower analytic system. Equivalents such as the number of trees necessary to absorb the CO_2 might act in the experiential system as it would encode the information as an image and metaphor rather than purely a number.

Along with contextualisation, there may be further benefits to visualising CO_2 mass. In a review of transport information use, graphical information was found to be the format that is most easily understood (Chorus, Molin, & van Wee, 2006). Images assist with the encoding of information (Gick, 1989), are good at concretising abstract concepts (Stenning & Oberlander, 1995), and improving the comprehension of scientific material (Large, Beheshti, Breuleux, & Renaud, 1995). Images can also assist comprehension when the individual has low prior knowledge (Mandl & Levin, 1989; Waygood & Avineri, 2011). Through those various benefits, visualisation of the CO_2 information may improve the usability of visual equivalents and improve people's ability to rank travel alternatives on a sustainability scale.

The findings of Coulter et al. (2007) and Waygood and Avineri (2011) suggest that the dominant format of using CO_2 mass to inform decision-makers is not very effective. Waygood and Avineri (2011) found that the best format to present CO_2 was the carbon budget, based on a cap and trade system where a limit was set. In post survey focus groups, individuals commented that it was the easiest to interpret as percentages and budgets were something that they dealt with regularly (Waygood & Avineri, 2010). Coulter et al. (2007) also found that people desired a 'target allowance' that would work like calorie counting. That format likely worked best because it gave context to the information, and it gave the decision-maker a reference point to work with.

CONCLUSIONS

This chapter has sought to offer an insight into some of the challenges and opportunities associated with information provision in relation to encouraging a move towards more sustainable travel behaviour. What is evident is the considerable diversity of factors that are likely to contribute in practice to an individual's encounter with information provision and the behavioural influence, if any, that may result.

At present, it can be argued a number of elements converge: (i) the information age has made possible a much greater capacity and flexibility in the gathering, processing and dissemination of information; (ii) academic research in the field of traveller information is now giving much more attention to the 'homo psychologicus' as opposed to 'homo economicus'; and (iii) there are new societal concerns regarding the impacts of travel behaviour and thus the need to influence behaviour that go beyond congestion to now include health and climate change. Taken together, there is both an opportunity and an imperative to position the use of information to be more effective in its role of supporting and influencing behaviour.

However, as the chapter also reveals, while many insights to pursuing this goal are apparent, the picture is complex and far from being strongly understood. There is still rather limited evidence of putting theories into practice and establishing the efficacy of conceptual wisdom in reality.

Travel information services that are focused on the preparation and action stages are substantially developed and common place in many people's lives – available on a variety of platforms and offering both pre-trip and en-route information, particularly in relation to tangible factors such as time, distance and cost. While it is known that most people most of the time do not consult such services this reflects the local nature of much travel and (corresponding) familiarity and predictability. Meanwhile information services are able to demonstrate appreciable levels of use overall suggesting that they meet a demand for those instances where individuals need, in some capacity, to further inform their decision-making. An important question then becomes: how much further can information provision go in relation to influencing travel behaviour? The chapter has highlighted on the one hand that some social-psychological factors are outside the control of the information service designers – the 'stage of change' people are at in relation to any move from one behavioural focus to another. On the other hand, it has highlighted that in some respects information context design could make a difference though this remains to be tested. Whether it makes a difference may depend upon how marginal the actual differences are between travel options. Considerations such as the value of one mode over another to society must be made outside of this chapter's focus on the use of information to affect choices.

What then of the new environmental impact dimension to travel behaviour and consideration of changing behaviour? In terms of behavioural effect of information there are two notable situations: (i) choice attributes other than environmental impact change in terms of content or context such that they influence behaviour in a way that is desirable in terms of environmental impact; or (ii) environmental impact becomes a (stronger) choice attribute because of social and personal norms and values and thus more directly exerts influence on choice outcomes resulting in more sustainable travel behaviours. It is likely that both situations are simultaneously at work in practice and their relative importance may be changing. It is this aspect of change that is perhaps most important in terms of opportunity. If it becomes the case that more people are moving forwards along the pathway as depicted by the TTM (for economic, health or environmental reasons at a personal level) then greater prospect arises for demand for information increasing and in turn the need to ensure information design is best placed to support and further positively influence the movement through stages of change.

NOTES

1. http://www.journeyon.co.uk
2. http://www.denhaag.nl/lucht
3. http://www.transport-research.info/web/projects/project_details.cfm?id=13774 &page=results
4. http://www.max-success.eu
5. http://transportationtown.com
6. http://htaindex.cnt.org
7. http://green.autoblog.com/2010/06/04/2011-lincoln-mkz-hybrid-gets-updated-smartgauge-with-apple-bloss/. Accessed on 28 April, 2011.
8. http://www.transportdirect.info
9. http://www.dft.gov.uk/pgr/statistics/datatablespublications/trends/current/section2ptbm.pdf
10. http://www.americanforests.org
11. http://footrprint.wwf.org.uk

REFERENCES

Abrahamse, W., Steg, L., Vlek, C., & Rothengatter, T. (2005). A review of intervention studies aimed at household energy conservation. *Journal of Environmental Psychology*, *25*(3), 273–291. doi:10.1016/j.jenvp.2005.08.002

Anable, J., Lane, B., & Kelay, T. (2006). *Review of public attitudes to climate change and transport: Summary report* (Contract No: PPRO 004/006/006). London: The Department for Transport.

Ariely, D. (2008). *Predictably irrational*. New York, NY: HarperCollins.

Avineri, E. (2006). The effect of reference point on stochastic network equilibrium. *Transportation Science*, *40*(4), 409–420.

Avineri, E. (2011). Applying behaviour economics in the design of travel information systems. *The 43rd Annual UTSG (The Universities' Transport Study Group) Conference*, Milton Keynes, UK.

Avineri, E., & Goodwin, P. (2010). *Individual behaviour change: Evidence in transport and public health*. London: The Department for Transport. Retrieved from http://www.dft.gov.uk/pgr/scienceresearch/social/behaviour-changes/pdf/transport-and-health.pdf

Avineri, E., & Prashker, J. (2003). Sensitivity to uncertainty: The need for a paradigm shift. *Transportation Research Record*, *1854*, 90–98.

Avineri, E., & Prashker, J. (2004). Violations of expected utility theory in route-choice stated preferences: Certainty effect and inflation of small probabilities. *Transportation Research Record: Journal of the Transportation Research Board*, *1894*, 222–229.

Avineri, E. & Waygood, E. O. D. (2010). *Behavioural inception report. Carbon Aware Travel Choices (CATCH)*. Retrieved from http://www.carbonaware.eu/fileadmin/user_upload/

Deliverables/CATCH_DEL_DOC_D1.1_20Behavioural_20Inception_20Report_orginal_V1.pdf

Avineri, E., & Waygood, E. O. D. (forthcoming). Applying valence framing to enhance the effect of information on transport-related carbon dioxide emissions. *Transportation Research Part A*.

Bamberg, S. (2006). Is a residential relocation a good opportunity to change People's travel behavior? Results from a theory-driven intervention study. *Environment and Behavior*, *38*(6), 820–840. doi:10.1177/0013916505285091

Beatty, S., Meadows, M., & White, D. (2002). Stages of change in drivers willingness to reduce car use before and during the UK fuel crisis. In G. Lyons & K. Chatterjee (Eds.), *Transport lessons from the fuel tax protests*. Aldershot, UK: Ashgate Publishing.

Bellizi, J. A., & Hite, R. E. (1992). Environmental colour, consumer feelings and purchase likelihood. *Psychological Marketing*, *9*(5), 347–363.

Ben-Akiva, M. E., & Lerman, S. R. (1985). *Discrete choice analysis: Theory and application to travel demand*. Cambridge, MA: MIT Press.

Botrill, C. (2007). Internet-based carbon tools for behaviour change. *European Council for Energy Efficient Economies (ECEEE) Summer Study 2007*, Côte d'Azur.

Brog, W., & Erl, E. (2008). Global problems need global solutions – Behaviour change in three continents. Transportation Demand Management 2008, July 16–18, Vienna.

Chatterjee, K., Hounsell, N. B., Firmin, P. E., & Bonsall, P. W. (2002). Driver response to variable message sign information in London. *Transportation Research Part C: Emerging Technologies*, *10*(2), 149–169. doi:10.1016/S0968-090X(01)00008-0

Chorus, C. G., Arentz, T. A., & Timmermans, H. J. P. (2006). *Observing the making of travel choices under uncertainty and information: Validation of travel simulator*. Paper presented at the 85th annual meeting of the Transportation Research Board, Washington, DC.

Chorus, C. G., Molin, E. J. E., Arentze, T. A., Timmermans, H. J. P., & van Wee, G. P. (2006). Travelers' need for information: An empirical study into the role of knowledge. Paper presented at the 85th annual meeting of the Transportation Research Board, Washington, DC.

Chorus, C. G., Molin, E. J., & Wee, B. v. (2006). Travel information as an instrument to change car drivers' travel choices: A literature review. *European Journal of Transport and Infrastructure Research*, *6*(4), 335–364. Retrieved from http://www.ejtir.tbm.tudelft.nl/issues/2006_04/index.asp

Coulter, A., Clegg, S., Lyons, G., Chatterton, T., & Musselwhite, C. B. A. (2007). *Exploring public attitudes to personal carbon dioxide emission information*. London: Department for Transport.

Darby, S. (2006a). *The effectiveness of feedback on energy consumption: A review for DEFRA on the literature on metering, billing and direct displays*. Oxford: Environmental Change Institute, University of Oxford.

Darby, S. (2006b). Social learning and public policy: Lessons from an energy-conscious village. *Energy Policy*, *34*(17), 2929–2940. doi:10.1016/j.enpol.2005.04.013

Davey, P. (1998). True colors: The glorious polychromy of the past suggests a strong historical need for color, despite current reductive fashions. *The Architectural Review*, *204*, 34–36.

Davidoff, J. (1991). *Cognition through Color*. Cambridge, MA: The MIT Press.

De Young, R. (1996). Some psychological aspects of reduced consumption behavior. *Environment and Behavior*, *28*(3), 358–409. doi:10.1177/0013916596283005

Epstein, S. (1994). Integration of the cognitive and the psychodynamic unconscious. *American Psychologist*, *49*(8), 709–724. doi:10.1037/0003-066X.49.8.709

Farag, S., & Lyons, G. (2008). Explaining public transport information use when a car is available: Attitude theory empirically investigated. *Transportation Research Record*, *2069*, 85–92.

Farag, S., & Lyons, G. (2012). To use or not to use? An empirical study of pre-trip public transport information for business and leisure trips and comparison with car travel. *Transport Policy*, *10*, 82–92.

Fodor, J. A. (1983). *Modularity of mind: An essay on faculty psychology*. Cambridge, MA: Bradford Books, MIT Press.

Ford Media. (2010). *Lincoln SmartGauge with EcoGuide: Taking drivers to the next level of engagement*. Retrieved from http://media.ford.com/images/10031/SmartGauge_MKZHyb.pdf

Gatersleben, B., & Appleton, K. M. (2007). Contemplating cycling to work: Attitudes and perceptions in different stages of change. *Transportation Research Part A: Policy and Practice*, *41*(4), 302–312. doi:10.1016/j.tra.2006.09.002

Gick, M. L. (1989). Two functions of diagrams in problem solving by analogy. In H. Mandl & J. R. Levin (Eds.), *Knowledge acquisition from text and pictures* (Vol. 58, pp. 215–230). Advances in Psychology. New York, NY: Elsevier Science Publishers.

Gonzales, M. H., Aronson, E., & Costanzo, M. A. (1988). Using social cognition and persuasion to promote energy conservation: A quasi-experiment. *Journal of Applied Social Psychology*, *18*(12), 1049–1066. doi:10.1111/j.1559-1816.1988.tb01192.x

Griffiths, M. D., & Swift, G. (1992). The use of light and colour in gambling arcades: A pilot study. *Society for the Study of Gambling Newsletter*, *21*, 16–22.

Hallahan, K. (1999). Seven models of framing: Implications for public relations. *Journal of Public Relations Research*, *11*(3), 205.

Kahneman, D., & Tversky, A. (1979). Prospect theory: An analysis of decision under risk. *Econometrica*, *47*(2), 263–291.

Kaya, N., & Epps, H. H. (2004). Relationship between color and emotion: A study of college students. *College Student Journal*, *38*, 396–405.

Kenyon, S., & Lyons, G. (2003). The value of integrated multimodal traveller information and its potential contribution to modal change. *Transportation Research Part F: Traffic Psychology and Behaviour*, *6*(1), 1–21. doi:10.1016/S1369-8478(02)00035-9

Kreitler, H., & Kreitler, S. (1972). *Psychology of the arts*. Durham, NC: Duke University Press.

Kühberger, A. (1998). The influence of framing on risky decisions: A meta-analysis. *Organizational Behavior and Human Decision Processes*, *75*(1), 23–55. doi:10.1006/obhd.1998.2781

Large, A., Beheshti, J., Breuleux, A., & Renaud, A. (1995). Multimedia and comprehension: The relationship among text, animation, and captions. *Journal of the American Society for Information Science*, *46*(5), 340–347.

Lyons, G. (2006). The role of information in decision-making with regard to travel. *Intelligent Transport Systems*, *153*(3), 199–212.

Lyons, G., Avineri, E., Farag, S., & Harman, S. (2007). *Strategic review of travel information*. London: The Department for Transport. Retrieved from http://webarchive.nationalarchives.gov.uk/+/http://www.dft.gov.uk/adobepdf/245385/249577/Strategic_Review_of_Travel_1.pdf

Madrian, B. C., & Shea, D. F. (2001). The power of suggestion: Inertia in 401(k) participation and savings behavior. *Quarterly Journal of Economics*, *116*(4), 1149–1187. doi:10.1162/003355301753265543

Mahnke, F. (1996). *Color, Environment, Human Response*. New York, NY: Van Nostrand Reinhold.

Mandl, H., & Levin, R. (Eds.). (1989). *Knowledge acquisition from text and pictures* (Vol. 58). Advances in Psychology. New York, NY: Elsevier Science Publishers.

Nolan, J. M., Schultz, P. W., Cialdini, R. B., Goldstein, N. J., & Griskevicius, V. (2008). Normative social influence is underdetected. *Personality and Social Psychology Bulletin, 34*, 913. doi:10.1177/0146167208316691

Peirce, S. & Lappin, J. (2005). *Why don't more people use advanced traveler information? Evidence from the Seattle area*. Retrieved from http://ntl.bts.gov/lib/jpodocs/repts_te/14004_files/14004.pdf

Prochaska, J. O., & Norcross, J. C. (2001). Stages of change. *Psychotherapy: Theory, Research, Practice, Training, 38*(4), 443–448.

Prochaska, J. O., Redding, C. A., & Evers, K. E. (2008). The transtheoretical model and stages of change. In K. Glanz, B. K. Rimer & K. Viswanath (Eds.), *Health behavior and health education: Theory, research, and practice* (4th ed., pp. 97–121). San Francisco, CA: Jossey-Bass.

Prochaska, J. O., Velicer, W. F., Rossi, J. S., Goldstein, M. G., Marcus, B. H., Rakowski, W., … Rossi, S. R. (1994). Stages of change and decisional balance for 12 problem behaviors. *Health Psychology, 13*(1), 39–46.

Rose, G., & Ampt, E. (2001). Travel blending: An Australian travel awareness initiative. *Transportation Research Part D: Transport and Environment, 6*(2), 95–110.

Saito, M. (1996). Comparative studies on color preference in Japan and other Asian regions, with special emphasis on the preference for white. *Color Research and Application, 21*(1), 35–49.

Schultz, P. W. (1999). Changing behavior with normative feedback interventions: A field experiment on curbside recycling. *Basic and Applied Social Psychology, 21*(1), 25–36.

Schultz, P. W., Nolan, J. M., Cialdini, R. B., Goldstein, N. J., & Griskevicius, V. (2007). The constructive, destructive, and reconstructive power of social norms. *Psychological Science, 18*(5), 429–434. doi:10.1111/j.1467-9280.2007.01917.x

Slovic, P. (2010). If I look at the mass I will never act: Psychic numbing and genocide. *Emotions and Risky Technologies, 5*(1), 37–59.

SMILE (2006). *CIVITAS SMILE* (Newsletter No. 2). Retrieved from http://www.civitas-initiative.org/docs/CIVITAS%20SMILE%20-%20Newsletter%20No%201.pdf

Spenwyn, J., Barrett, D. J. K., & Griffiths, M. D. (2010). The role of light and music in gambling behaviour: An empirical pilot study. *International Journal of Mental Health Addiction, 8*, 107–118.

Staats, H., Harland, P., & Wilke, H. A. M. (2004). Effecting durable change: A team approach to improve environmental behavior in the household. *Environment and Behavior, 36*(3), 341–367.

Stenning, K., & Oberlander, J. (1995). A cognitive theory of graphical and linguistic reasoning: Logic and implementation. *Cognitive Science, 19*(1), 97–140.

Swim, J., Clayton, S., Doherty, T., Gifford, R., Howard, G., Reser, J., Weber, E. (2009). *Psychology and global climate change: Addressing a multi-faceted phenomenon and set of challenges*. American Psychological Association. Retrieved from http://www.apa.org/science/about/publications/climate-change.aspx

Taniguchi, A., Hara, F., Takano, S., Kagaya, S., & Fujii, S. (2003). Psychological and behavioral effects of travel feedback program for travel behavior modification. *Transportation Research Record: Journal of the Transportation Research Board, 1839*(1), 182–190.

TAPESTRY. (2003). *Travel awareness publicity and education supporting a sustainable transport strategy in Europe: Final publishable report*. Available from http://www.transport-research.info/Upload/Documents/200810/20081002_161028_52677_TAPESTRY.pdf. Accessed on May 7, 2011.

Taylor, M. A. P., & Ampt, E. S. (2003). Travelling smarter down under: Policies for voluntary travel behaviour change in Australia. *Transport Policy*, *10*(3), 165–177.

Thaler, R., & Benartzi, S. (2004). Save more tomorrow™: Using behavioral economics to increase employee saving. *The Journal of Political Economy*, *112*(S1, Papers in Honor of Sherwin Rosen: A Supplement to Volume 112), pp. S164–S187. Retrieved from http://www.jstor.org/stable/10.1086/380085

Thaler, R. H., & Sunstein, C. R. (2008). *Nudge: Improving decisions about health, wealth and happiness*. New Haven, CT: Yale University Press.

Thaler, R. H., Tversky, A., Kahneman, D., & Schwartz, A. (1997). The effect of myopia and loss aversion on risk taking: An experimental test. *Quarterly Journal of Economics*, *112*, 647–661.

Tversky, A., & Kahneman, D. (1974). Judgment under uncertainty: Heuristics and biases. *Science*, *185*(1124), 1131.

Tversky, A., & Kahneman, D. (1991). Loss aversion in riskless choice: A reference-dependent model. *The Quarterly Journal of Economics*, *106*(4), 1039–1061.

van de Kaa, E. J. (2010). Prospect theory and choice behaviour strategies, review and synthesis from social sciences. *European Journal of Transport and Infrastructure Research*, *10*(4), 347–367.

Verplanken, V., Aarts, H., & van Knippenberg, A. (1997). Habit, information acquisition, and the process of making travel mode choices. *European Journal of Social Psychology*, *27*, 539–560.

Waygood, E. O. D., & Avineri, E. (2010). *Research and design report. Carbon Aware Travel Choices (CATCH)*. Retrieved from http://www.carbonaware.eu/fileadmin/user_upload/Deliverables/D1.2_Research_and_design_report.pdf

Waygood, E. O. D., & Avineri, E. (2011). Does '500g of CO_2 for a five mile trip' mean anything? Towards more effective presentation of CO_2 information. *Proceedings of the Transportation Research Board 90th Annual Meeting*, Washington, DC.

CHAPTER 13

THINKING STRATEGICALLY: PATHWAYS TOWARDS LOW-CARBON TRANSPORT

Robin Hickman and David Banister

ABSTRACT

Purpose – *To consider the potential for transport* CO_2 *reductions in Oxfordshire. It examines the* CO_2 *baseline within the county and elsewhere in the United Kingdom, assessing the high and low emitters in terms of relative transport* CO_2 *emissions. A target for the county is developed at 1.2* tCO_2 *per capita by 2030, and ultimately 0.5* tCO_2 *per capita by 2050, and scenario analysis is used to assess the potential interventions, policy packages and different pathways available for* CO_2 *emission reduction.*

Methodology/approach – *The central Oxfordshire transport model (COTOM) is used to quantify the likely impacts of different policy areas over the long term to 2030. It is illustrated how the strategic policy direction can be set within transport, and attempts are made to quantify the scale of likely impacts from the options available.*

Findings – *Current interventions and levels of funding for sustainable transport are not strong enough to significantly reduce transport* CO_2 *emissions in line with the national UK targets and there is a significant underestimation of the scales of change required. There is a major*

Transport and Climate Change
Transport and Sustainability, Volume 2, 341–367

ISSN: 2044-9941/doi:10.1108/S2044-9941(2012)0000002016

discontinuity between the 'greenwash' of the conjecture at the national governmental level and the level of funding necessary to actually implement effective actions at the local level.

Originality/value – *The chapter is original in illustrating how the strategic policy direction can be set within transport at the regional level and also in attempting to quantify the scale of likely impacts.*

Keywords: Transport; climate change; carbon dioxide; motorisation; public transport; walking and cycling

INTRODUCTION

The first modern motor car was designed and built in 1885, by Karl Benz in Germany. Jump forward 125 years, and by 2010, there were over 600 million motor cars worldwide, around one per eleven people. The vast majority are powered by a common technology and power source: the four stroke cycle engine, fuelled by petrol or diesel. The motor car has shaped our cities and urban form, and the use of the car is associated with both instrumental (flexibility and convenience) and affective emotions (a sense of freedom, the thrill of driving, perceived success and enjoyment) (Anable & Gatersleben, 2005). These varied motivations help explain the strength of the desire for motorisation and car ownership, an almost ubiquitous aspiration globally. The development and 'consumerisation' of the product has been one of the technological and business success stories of the 20th century, as illustrated by the Toyota Corolla which is the current best-selling vehicle of all time (32 million vehicles and rising, 1966-), closely followed by the VW Golf (25 million, 1974-), VW Beetle (22 million, 1938-), Ford Model T (16.5 million, 1908–1927) and Chrysler Voyager (11.7 million, 1984-) (Motortrend Forum, 2010). There is significant momentum in the business model, but also some huge problems. In the words of Heathcote Williams (1991, pp. 7–9):

> The first automobile [...] had three wheels, like an invalid car, and ran on alcohol, like many drivers. Since then more than seventeen million people[1] have been killed in an undeclared war. And the rest of the world may be in danger of being run over in a terminal squabble over the oil. Were an alien visitor to hover a few hundred yards above the planet, it could be forgiven for thinking that cars were the dominant life-form, and that human beings were a kind of ambulatory fuel cell, injected when the car wished to move off, and ejected when they were spent.

This chapter examines some of these issues in terms of the possibilities for different types of futures in travel. These are likely to include much less dependence on the conventional engine technologies and fuels and a break in the dominant trend of motorisation. A key argument is that the potential scarcity of oil and the concerns about climate change will mean that our travel in the future (2030) will, by necessity, be very different to the extrapolated 'business as usual' trajectory. Alongside, there will be related benefits in terms of the health dividend (reduced fatalities from reduced motorisation and health benefits associated with more active lifestyles) and the urban fabric dividend (the urban design benefits of reduced car use). Many of these issues have been well known for years, but there has been little real success in terms of moving towards more sustainable means of travel. It is perhaps the finite nature of oil that will make the difference here (Gilbert & Perl, 2010). The US Energy Information Administration (2009), for example, projects that there are just 46 years left of oil consumption assuming proven reserves and current consumption rates. Hence, the time for procrastination has perhaps passed and there is now a much narrowing window of opportunity for change.

The case study for the analysis is Oxfordshire, a county in the southeast of the UK.[2] An initial comparison is made of the CO_2 baseline within the county and elsewhere in the UK; scenario analysis is then used to assess the potential interventions, policy packages and different pathways available for CO_2 emission reductions. A case study from the UK is illuminating in that the UK has been the first country to set international binding carbon targets, together with intervening budgets for five-year periods. The Climate Change Act (2008) requires emissions to be reduced by at least 80% by 2050 (on 1990 levels) and an interim target of at least 34% by 2020, and it also limits the amounts of carbon that can be offset by investing in projects overseas (the 'credit limit'). This means that substantial and real reductions will have to take place across all sectors of the UK economy (including transport). Oxfordshire provides one example of the difficult decisions that have to be taken at the local level to meet these new legal requirements; the particular context means finding solutions for an affluent rural hinterland (the county) as well as a compact urban centre (Oxford), and this may lead to contrary priorities.

The contribution of the chapter is to illustrate the process of quantifying the likely impacts of potential options available and, ultimately, to assess whether substantial transport CO_2 emission reductions can be achieved. There is an emerging literature in this area (see, e.g. Banister et al., 2000; Hickman, Ashiru, Seaborn, & Walters, 2010; Hickman & Banister, 2007;

Schäfer, Heywood, Jacoby, & Waitz, 2009; Sperling & Gordon, 2009; Yang, McCollum, McCarthy, & Leighty, 2009). Our concluding argument is not new but needs to be restated because of its importance in policy terms: the current application and levels of funding in sustainable transport are not enough to significantly reduce transport CO_2 emissions to the levels sought in national targets. There is a major discontinuity between the 'greenwash' of the conjecture at the national governmental level and the level of funding necessary to actual implement effective actions at the local level.

THE OXFORDSHIRE CONTEXT

Oxfordshire provides a useful example, with resonance to many other areas, of differential travel behaviour across the region. There is a car-dependent rural hinterland and a relatively sustainable central urban area in urban mobility terms (Oxford). These travel patterns (Table 1) reflect many factors, including income, car ownership levels, the levels and cost of alternative transport provision, a relatively dispersed urban structure beyond Oxford and also individual aspirations in terms of travel behaviour choices. There is potential for developing a much greater emphasis on walking, cycling and public transport, and also for making significant progress through the introduction of low-emission vehicles. The major difficulty for Oxfordshire is in achieving sustainable behaviour across a region or sub-region, in particular the suburban and rural areas. Even the more carbon-efficient urban centres in European cities – including the classic good practice examples such as Freiburg, Strasbourg, Amsterdam and Zürich – tend to be surrounded by relatively car-dependent suburban and rural areas, and it is these areas that are difficult to change.

Oxfordshire County Council and Oxford City Council are developing planning strategies and transport investment strategies with a forward look to 2026. The Oxford Local Development Framework (LDF), for example, has been prepared by Oxford City Council (2008, core strategy), and there are four other LDFs covering the districts across the county (Cherwell, South Oxfordshire, Vale of White Horse and West Oxfordshire). Each is prepared within the context of the regional South East Plan (Government Office for the South East, 2009) and national planning policy guidance. The current Oxfordshire Local Transport Plan (LTP3) (Oxfordshire County Council, 2011) has a long-term horizon running from 2011 to 2030, with a three- and five-year investment programme that can be updated annually. There are, however, major limitations in terms of funding, with very low levels

Table 1. Oxfordshire – Some Key Statistics.

Issue	Comment			
Population				
Oxfordshire (2001)	605,500			
Oxfordshire (1801)	111,977			
Oxford (2007)	165,000			
Oxford (1801)	12,690			
Area (2007)	2,605 km^2			
Population density	246/km^2			
Commuters	75% of Oxford's population works within Oxford; only 9% travel outside of the county to work; yet around 50% of Oxford's workforce lives beyond its (Oxfordshire County Council, 2011) boundary			
Mode share (2001)	Oxfordshire	Oxford	West Oxfordshire	England
	Main mode for journey to work: 62% car, 10% walk, 7% bus, 7% bicycle, 10% work at home, 2% rail	Journey to work: 42% car, 15% walk, 16% bus, 15% bicycle, 8% work at home, 2% rail	Journey to work: 67% car, 9% walk, 5% bus, 5% bicycle, 10% work at home, 2% rail	Journey to work: 61% car, 10% walk, 8% bus, 3% bicycle, 9% work at home, 7% rail
Cars per household (at least one)	82%			
Transport CO_2 emissions (2008)	3.0 tCO_2 per person			

Source: Oxfordshire County Council (2006, 2007, 2011).

of investment being allowed due to constraints imposed by central government through the allocation of strict cash limits. The current Conservative-Liberal Coalition government is less supportive of regional planning efforts, and the land use planning system as a whole, and this has created even greater uncertainty with further reductions in funding for local authorities. There is little strategic direction being offered, and some confusion as to the most effective process for forward planning at the regional and local levels.

Within Oxfordshire, there are no explicit county-wide or local CO_2 emission reduction targets, or budgets, either across all sectors or for the transport sector. This situation is common across most UK local authorities, and it is clearly at odds with the strong statements from central government on the importance of achieving an 80% CO_2 reduction target by 2050. Although the UK target represents a leading position internationally, within this a target for transport CO_2 emissions has not yet been set. If and when this happens, it is likely to be lower within transport as the costs of reduction are perceived to be considerably higher than those in other sectors of the economy. However, a problem remains: because transport accounts for a high proportion of total emissions, reductions within transport are imperative to meeting national targets. If the total 80% reduction target is to be achieved, even a 90% reduction in the industrial and commercial and domestic sectors means that transport has to achieve a 52% reduction (Hickman, Pharoah, Torrance, & Dyer, 2011).

There is little understanding of the different baselines spatially, how these might evolve over time and the means by which significant alternatives might be conceptualised. The nearest perhaps in Oxfordshire is a traffic growth target which seeks to limit the growth in overall traffic levels on Oxfordshire roads. A baseline is used of 10.9 million vehicle kilometres travelled in 2000, with a target of 'no more than' 12.2 million vehicle kilometres per day to be travelled in 2010. This equates to an approximate 1% per annum maximum traffic growth target, against a projected 2% per annum (Oxfordshire County Council, 2000). Although the development of sustainable transport initiatives has some history over the last decade or so, the development of low-carbon transport strategies that look across a range of technological, network investment and behavioural opportunities are much less well developed. There is a lack of clarity in strategic thinking, insignificant funding and little understanding of the scale of change required in travel behaviour to contribute to national CO_2 reduction targets. Limited progress may have led to a feeling that change is not possible, and this has created frustration at the local level.

BASELINE AND PROJECTIONS

The national UK target of an 80% reduction in transport emissions by 2050 (on 1990 levels) can be applied to Oxfordshire (Table 2).[3] There are some data issues, in that 1990 data is not readily available for the county and an end year of 2030 is used to be consistent with the later analysis in this chapter. A 60% reduction in emissions by 2030 can be derived on a 2005 base, which is broadly equivalent to the national UK target. This would mean a target for Oxfordshire at around 1.2 tCO_2 per capita in 2030. An ultimate target for 2050 would, however, use the UK 1990 baseline and seek an 80% reduction on this, resulting in a target of around 0.5 tCO_2 per capita for transport in Oxfordshire in 2050. Hence, the baselines and timescales used can be very important in target setting.

The recent historic cross-sector CO_2 emissions are given for the districts in Oxfordshire in Table 3, and they cover both overall emissions and the levels of transport-related CO_2 emissions. All areas are reducing in emissions over time, but progress is very slow. Cherwell, South Oxfordshire and Vale of White Horse are associated with the highest emissions in the county, and these represent the affluent, remote rural areas with the highest levels of car use. Cherwell, for example, has a total of 11 tCO_2 per capita, and 5 tCO_2 per capita in transport. The urban area of Oxford is by far the lowest emitter, with a total of 6.9 tCO_2 per capita, and just 0.9 tCO_2 per capita in transport. About 30–45% of emissions are associated with the transport sector in some

Table 2. Potential Transport CO_2 Targets for Oxfordshire.

Area	Aggregate (Per Capita)
Oxfordshire, total CO_2 emissions, 2005	5,880 kt CO_2 (9.4 tCO_2 per capita)
Oxfordshire, transport, CO_2 emissions, 2005	1,964 kt CO_2 (3.1 tCO_2 per capita)
Oxfordshire, target total CO_2 emissions, 2030 (60% reduction on 2005 levels by 2030)	2,352 kt CO_2 (3.8 tCO_2 per capita)
Oxfordshire, target transport CO_2 emissions, 2030 (60% reduction on 2005 levels by 2030)	786 kt CO_2 (1.2 tCO_2 per capita)
UK-based, target transport CO_2 emissions, 2050 (80% reduction on 1990 UK levels by 2050)	1990: 120 Mt CO_2 (2.1 tCO_2 per capita) 2010: 120.6 Mt CO_2 (2.0 tCO_2 per capita) 2050: 24.0 Mt CO_2 (0.42 tCO_2 per capita) *Assumed target: 0.5 tCO_2 per capita*

Table 3. CO_2 Emissions by Local Authority in Oxfordshire.

Local Authority[a]	Year	Total	Road Transport	Population ('000s, Mid-Year Estimate)	Total Per Capita Emissions (t)	Transport Per Capita Emissions (t)	Transport (%)	Reduction (%) to 0.5 tCO_2 Per Capita
Cherwell	2005	1,632	688	137	11.9	5.0	42.2	90.0
Cherwell	2006	1,642	692	138	11.9	5.0	42.1	90.0
Cherwell	2007	1,611	695	138	11.7	5.0	43.1	90.0
Cherwell	2008	1,548	664	139	11.2	4.8	42.9	89.6
Oxford	2005	1,026	147	144	7.1	1.0	14.3	50.0
Oxford	2006	1,026	142	144	7.1	1.0	13.8	50.0
Oxford	2007	965	146	145	6.6	1.0	15.1	50.0
Oxford	2008	1,014	139	147	6.9	0.9	13.7	44.4
South Oxfordshire	2005	1,252	460	129	9.7	3.6	36.7	86.1
South Oxfordshire	2006	1,277	452	129	9.9	3.5	35.4	85.7
South Oxfordshire	2007	1,262	465	129	9.8	3.6	36.8	86.1
South Oxfordshire	2008	1,252	445	130	9.6	3.4	35.5	85.3
Vale of White Horse	2005	1,165	435	118	9.9	3.7	37.4	86.5
Vale of White Horse	2006	1,168	432	118	9.9	3.7	37.0	86.5
Vale of White Horse	2007	1,135	432	118	9.6	3.7	38.0	86.5
Vale of White Horse	2008	1,148	415	118	9.7	3.5	36.1	85.7
West Oxfordshire	2005	805	233	99	8.1	2.3	29.0	78.3
West Oxfordshire	2006	797	229	100	8.0	2.3	28.7	78.3
West Oxfordshire	2007	796	233	102	7.8	2.3	29.3	78.3
West Oxfordshire	2008	779	223	102	7.6	2.2	28.6	77.3
Oxfordshire Total	2005	5,880	1,964	628	9.4	3.1	33.4	83.9
Oxfordshire Total	2006	5,911	1,946	630	9.4	3.1	32.9	83.9
Oxfordshire Total	2007	5,770	1,970	632	9.1	3.1	34.1	83.9
Oxfordshire Total	2008	5,740	1,885	636	9.0	3.0	32.8	83.3
England	*2008*	*506,526*	*131,045*	*61,398*	*8.2*	*2.1*	*25.9*	*76.2*

Source: Data from Department of Energy and Climate Change (2010).
[a]In local authority order.

districts, which is very high in the UK context. A much more 'appropriate' level is below 15% as found in Oxford. The differences in transport emissions spatially are little understood, particularly when related to potential for change. Even in a county as small as Oxfordshire, these variations are large, with income levels and accessibility to public transport and central urban areas all playing large contributory parts. It is the variability in transport emissions that largely account for the increase in emissions beyond the urban area of Oxford.

There are important issues here in terms of target setting, as the different contexts represent very different baselines and opportunities for changing travel behaviour. Hence, national targets will need to be differentially apportioned by area. Urban and rural areas will provide variable opportunities for reductions as well as starting points. Difficult questions hence remain unresolved about whether the high emitting areas should be targeted on the basis of equity, or whether efforts should be concentrated elsewhere, perhaps in areas that may be more receptive to low-carbon travel behaviours, or where actions might be cheaper to introduce. The above approach for target setting in Oxfordshire draws directly on the current baseline, and this means that it would 'penalise' a location with a low CO_2 emissions by suggesting an even lower future emission level, or conversely it could 'favour' a location with high emissions (as in some parts of rural Oxfordshire). There is no analysis or guidance at the regional and local levels as to the most appropriate methodology for setting reduction targets.

Table 4 provides similar data for selected local authorities in England, and again the variation in emissions is very large. Areas are categorised relative to transport CO_2 emissions per capita. The affluent rural areas (e.g. West Berkshire, Chester, Cambridgeshire and Harrogate) are associated with relatively high per capita emissions, both in cross sector and transport terms. Lower incomes areas and larger urban areas (e.g. London, Liverpool and Manchester), historic university cities (e.g. Cambridge, Oxford and York) and very remote urban areas (e.g. Plymouth and Norwich) have lower per capita emissions. Some of the low income northern towns have surprisingly high per capita emissions because of their relatively high levels of car use (e.g. Warrington and Doncaster). The recent low-density developments on the edge of these towns, including out of town retail parks, leisure complexes and residential areas inaccessible to public transport, may have added substantially to the levels of car dependency in these areas. The contributory factors to travel behaviour and resultant CO_2 emissions are thus wide, covering socio-economic, attitudinal, urban planning and

Table 4. CO_2 Emissions by Local Authority in England, 2008.

Local Authority	Total Emissions	Road Transport	Population ('000s, Mid-Year Estimate)	Total Per Capita Emissions (tCO_2)	Transport Per Capita Emissions (tCO_2)[a]	Transport (%)	Reduction (%) to Target (0.5 tCO2 Per Capita)
High emitters (>3 tCO_2 per capita)							
West Berkshire	1,841	809	151	12.2	5.3	43.9	90.6
Richmondshire	537	268	51	10.5	5.2	49.9	90.4
Chester	1,291	461	118	10.9	3.9	35.7	87.2
Harrogate	1,533	587	156	9.8	3.8	38.3	86.8
Warrington	1,934	706	196	9.9	3.6	36.5	86.1
St Albans	1,080	491	135	8	3.6	45.5	86.1
Doncaster	2,546	897	289	8.8	3.1	35.2	83.9
High-medium emitters (2–3 tCO_2 per capita)							
Oxfordshire	5,740	1,885	636	9	3.0	32.8	83.3
Surrey	8,574	3,252	1,101	7.8	3.0	37.9	83.3
Wiltshire	4,694	1,283	454	10.3	2.8	27.3	82.1
Hampshire	10,834	3,515	1,284	8.4	2.7	32.4	81.5
Milton Keynes	1,920	594	233	8.2	2.6	30.9	80.8
Derbyshire	8,945	1,925	758	11.8	2.5	21.5	80.0
Shropshire	2,509	732	291	8.6	2.5	29.2	80.0
Kent	11,879	3,568	1,402	8.5	2.5	30.0	80.0
Devon	6,096	1,834	747	8.2	2.5	30.1	80.0
South East	63,773	20,402	8,369	7.6	2.4	32.0	79.2
Fenland, Cambridgeshire	1,036	189	92	11.3	2.1	18.3	76.2

National (England)	506,526	131,045	61,398	8.2	2.1	25.9	76.2
Leeds	5,465	1,647	779	7	2.1	30.1	76.2
West Sussex	5,368	1,667	789	6.8	2.1	31.0	76.2
Medium-low emitters (1–2 tCO_2 per capita)							
Cornwall	4,131	1,052	529	7.8	2.0	25.5	75.0
Newcastle upon Tyne	1,890	474	278	6.8	1.7	25.1	70.6
Stevenage	560	131	81	7	1.6	23.4	68.8
Manchester	3,215	687	473	6.8	1.5	21.4	66.7
Birmingham	6,534	1,493	1,019	6.4	1.5	22.8	66.7
York	1,182	296	195	6.1	1.5	25.1	66.7
Stoke-on-Trent	1,647	341	239	6.9	1.4	20.7	64.3
Liverpool	2,815	602	441	6.4	1.4	21.4	64.3
Harlow, Essex	645	106	80	8.1	1.3	16.5	61.5
Sheffield	3,666	682	540	6.8	1.3	18.6	61.5
Plymouth	1,414	336	256	5.5	1.3	23.8	61.5
Bristol	2,320	536	426	5.4	1.3	23.1	61.5
Kingston upon Hull	1,738	319	261	6.7	1.2	18.3	58.3
Nottingham	1,806	370	297	6.1	1.2	20.5	58.3
Leicester	1,925	341	304	6.3	1.1	17.7	54.5
Greater London	46,357	8,736	7,668	6	1.1	18.8	54.5
Low emitters (< 1 tCO_2 per capita)							
Oxford	1,014	139	147	6.9	0.9	13.7	44.4
Cambridge	782	109	119	6.6	0.9	13.9	44.4
Norwich	794	128	137	5.8	0.9	16.1	44.4

Source: Data from Department of Energy and Climate Change (2010).
[a]Ranked in descending order by transport per capita emissions (tCO_2).

network investments, including the levels of public transport, walking, cycling and highway infrastructure. In future years, the take-up of low-emission vehicles in different area will also become important. Again, it can be seen that some areas are 'spending' a very high proportion of their emissions on transport, over 40% in some cases. An aspirational target of 0.5 tCO_2 in transport CO_2 emissions by 2050 can be suggested, broadly in line with the national CO_2 reduction target. If this is to be achieved on an equitable basis, then some areas have a very long way to go, with a reduction in transport emissions of up to 90% in some of the high-income rural areas. Other areas need to contribute less, but even those with the lowest levels of CO_2 emissions should be seeking at least a 40% reduction in transport CO_2 emissions. Budget allocations for the intervening years (perhaps on a five-year basis) would also help in moving towards the end state targets, as this encourages early progress to be made in terms of investing in certain transport options, and a more likely achievement of the targets.

SCENARIO ANALYSIS

The scale of change required in reducing transport emissions towards the 0.5 tCO_2 per capita target is therefore formidable. The key issues are to examine what can be undertaken in terms of policy interventions and to assess whether this level of emission reduction is possible. Scenario analysis is used, as described below, to help provide a framework. The first stage is to understand the main trends, in terms of uncertainties, which may be important to transport futures in Oxfordshire (Table 5).

The three most important issues, by level of uncertainty and likely impact, are used to derive the 'possibility space' for future scenario development. Usually, scenarios are derived as a two-by-two matrix (see Van Der Heijden, 1996; Department for Trade and Industry and Office of Science and Technology, 2006), but below the scenarios are developed as a 'trilemma'[4] (Fig. 1).

In Oxfordshire, and indeed much of the UK and beyond, perhaps, London, funding levels are likely to be severely constrained and this will limit the efforts to achieve sustainable travel. The policy intention will still be present, but the implementation programme in terms of schemes and initiatives will be much harder to realise. Funding problems are evident in the short term, but potentially also in the longer term, as local authorities are heavily dependent on central government support. Transport investment is usually

Table 5. Major Uncertainties Affecting Transport.

Uncertainties	Ranking by Importance
Economy and governmental	
Economic growth rate (e.g. GDP or other growth metric)	
Funding for transport investment and urban development	1
Income levels, income inequality	
Tertiary education sector growth	
Socio-demographics	
Population growth	
Age profile (including an ageing population)	
Aspirations and culture – continued 'western consumption' model?	
Technologies	
Clean vehicle technologies	2
Energy and power supply – renewable sources	
Environmental	
Major environmental shocks – changing climate, flooding, water supply	
Quality of environment and ecology	
Urban issues and transport planning	
Environmental issues – stewardship, extent of 'seriousness' given to them in policy making and implementation	3
Urban design and built environment quality	
Extent of development and urban sprawl	
Level of sustainability in travel, investment in public transport, walking and cycling	

relatively low in the priorities for state spending. There is also a wider problem in practice. There is a major aversion to innovation (and necessarily risk) in transport planning, and efforts to adopt the newer public transport options, or public realm (streetscape design), cycling and walking schemes are not often ambitious. This is evident when comparing some of the practice in the UK and that in Copenhagen, Delft or Groningen (cycling and walking), or many of the French cities for tram schemes (Strasbourg, Montpellier, Bordeaux), or even Bogotá and Curitiba (bus rapid transit), San Sebastian, Madrid or Barcelona (public realm), and Freiburg or Malmö (urban extensions). UK practice is failing to deliver to the level and the quality on offer elsewhere. The good practice has been around for years, and there are many advocates pushing for improved decision making and implementation, but internationally leading practice in the UK seems to be not on the same scale as elsewhere.

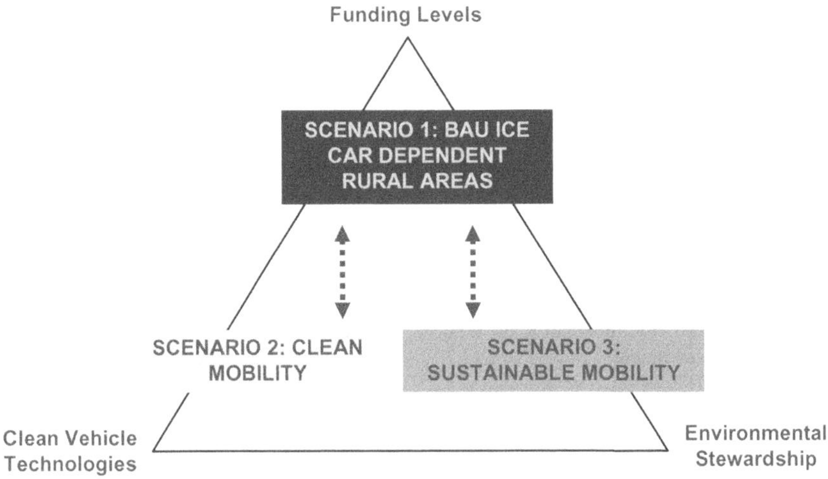

Fig. 1. The Possibility Space for Transport Scenarios in Oxfordshire. BAU ICE, Business as Usual, Internal Combustion Engine Motorisation.

The scenario analysis for Oxfordshire can be used to examine three major potential futures:

- *Scenario 1 – Business as Usual (BAU)*: A continuation of the current policy approach (and levels of implementation), with only ad-hoc progress being made towards sustainable travel behaviour, no real strategic direction, and low funding levels.
- *Scenario 2 – Clean Mobility*: There is a marked change in terms of the increased use of low-emission vehicles, but levels of mobility continue to rise and there are no coordinated efforts to invest in alternatives to the car.
- *Scenario 3 – Sustainable Mobility*: Here the focus is in developing public transport, walking and cycling options, as well as clean vehicle mobility. Urban structure plays an important role in enabling sustainable travel, and there is a reduction in travel by private motor car. Funding priorities change in terms of much higher investment in the non-car modes and the public realm.

Modelling Approach

The approach to understanding the likely pathways and combination of policy alternatives for each of the three scenarios is to develop an inventory of the range of measures likely to reduce CO_2 emissions in Oxfordshire. This inventory covers the network infrastructure, demand management and pricing measures, low-emission vehicle technologies and other policy measures. Individual measures are 'clustered' into policy packages (PPs) as outlined below, and described in more detail later, and the impacts quantified at the policy package level using model runs from the central Oxfordshire multi-modal transport model (COTOM). The results from COTOM are also supplemented with other datasets including vehicle/speed CO_2 emission factors (Department for Transport, 2008), modal CO_2 emission factors (Department for Food and Rural Affairs, 2009) and spatial planning assumptions from the South East Plan (Government Office for the South East, 2009).

- PP1 Rail
- PP2 Bus
- PP3 Walk
- PP4 Cycle
- PP5 Highway Infrastructure
- PP6 Traffic Demand Management
- PP7 Pricing
- PP8 Parking Management
- PP9 Park and Ride
- PP10 Land Use Planning
- PP11 Behavioural Change
- PP12 Low-Emissions Vehicles
- PP13 Alternative Fuels
- PP14 Slower Speeds and Ecological Driving
- PP15 Freight
- PP16 Long Distance Travel Substitution

Each package typically has a level of 'applicability': low, medium and high. This notion of applicability may be viewed in terms of the level of investment, the level of policy support or the scale of the initiative, but it should also be remembered that there are also important issues of feasibility and deliverability. A scenario is then developed using a variety of policy packages and different levels of application within each.

Appraisal Framework

Each scenario is compared relative to its CO_2 reduction potential, but also wider multi-criteria impacts; this is an attempt to show that low-carbon transport pathways may have differential impacts against other policy goals. A multi-criteria appraisal (MCA) framework is thus developed (Table 6).

EXPLORING THE SCENARIOS

One of the key scenarios from the Oxfordshire 'possibility space' is explored in more detail below. Scenario 3 'Sustainable Mobility' represents a balanced approach to reducing transport CO_2 emissions, covering many of the possible policy packages with the exception of highway infrastructure building. The scenario is illustrated in Table 7. The total CO_2 reduction impact is approximately 1.366 $MtCO_2$ per annum in 2030 (a 48.3% reduction relative to the BAU level in 2030). This would mean 2.0 tCO_2 per capita transport emissions in 2030, relative to a BAU of 3.8 tCO_2 per capita (and relative to a 2005 transport baseline of 3.1 tCO_2 per capita and target of 1.2 tCO_2 per capita in 2030 and ultimately 0.5 tCO_2 per capita in 2050). There are, however, some important findings and major problems here. The scenario as developed contains a major level of intervention and it is unaffordable under current funding regimes. Further, although it would represent a major contribution to reducing CO_2 emissions in Oxfordshire, it is still insufficient against required targets. This is a story replayed across many authorities in the UK, and internationally perhaps. In the technical analysis and strategy development, very ambitious futures have been identified, but these are currently not financially (or politically) deliverable. In addition, although they are ambitious, they are not likely to help us meet the extremely demanding legally binding strategic CO_2 reduction targets adopted at the UK level under the Climate Change Act (2008).[5] This discontinuity is a major dilemma for policy makers and not widely appreciated. The adoption of targets is often made in the belief that they are reachable, but the 80% reduction in CO_2 emissions for the UK by 2050 is not likely to be delivered in the UK in the transport sector unless there are immediate and major changes to investment strategies, including the availability of public finance.

The major CO_2 reduction impacts in Oxfordshire arise from the following policy packages:

- PP12 Low-emission vehicles (high): a high level of application is envisaged, the total car fleet averages 95 gCO_2/km and heavy goods vehicles (HGVs) (fully loaded) 800 gCO_2/km by 2030. This level of new

Table 6. MCA Framework.

Environment (Strategic)	Environment (Local)	Accessibility	Safety	Economy
• Total CO_2 emissions by car, LGV, HGV, bus and rail	• Population affected by noise • Number of households experiencing noise levels above 68 db • Population perceiving a noise nuisance • Population experiencing vibration • Number of households experiencing vibration • Air quality index (by nitrogen dioxide, ozone, sulphur dioxide, particulates)	• Hansen measure of accessibility to town centres (by mode) • Hansen measure of accessibility to hospitals • Hansen measure of accessibility to workplaces • 2026 households within 30 minutes of town centers • 2026 households within 30 minutes of hospitals • 2026 households within 30 minutes of workplaces	• Number of personal injury accidents (PIAs), based on Stats 19 data (2007) • Number of accidents resulting in slight injury • Number of accidents resulting in serious injury • Number of accidents involving fatalities	• Journey time reliability • Junction delays • Travel time on links, by mode • Access to jobs

NB. For reasons of space, only partial reporting against CO_2 reduction potentials is given in the following analysis.

Table 7. Sustainable Mobility.

Policy Package	Level of Application	Summary Specification
PP1 Rail	Medium	Interchange improvements, Mainline network frequency upgrades (Reading–Oxford–Midlands) and assumed new networks (Cotswold line upgrade; Evergreen 3, Oxford–Bicester–London Marylebone; and East-West rail, Oxford-Milton Keynes-Cambridge-Felixstowe), new Milton Park station.
PP2 Bus	Medium	Improved Premium Routes to main housing and employment sites; improved bus services between large towns and small towns; electric and hybrid buses in Oxford.
PP3 Walk	High	High quality pedestrian environment and streetscape upgrades in Oxford, larger towns, smaller towns and villages, and connections to Rights of Way network.
PP4 Cycle	High	High quality network improvements, integrated cycle parking in Oxford, larger towns, smaller towns and villages, cycle hire schemes in Oxford and large towns. New links from new developments.
PP5 Highway Infrastructure	BAU	None
PP6 TDM/Active Traffic Management	Medium	Access to Oxford programme, including HOV lane on A34; routeing measures from Transport Networks Review (TNR); Traffic Incident Management (TIM) programme; expansion of real-time monitoring systems across the county.
PP7 Pricing	Medium	Road pricing scheme across Oxfordshire and the UK, charging 20p per kilometre for all vehicles.
PP8 Parking Management	Medium	Reduced parking supply in new developments and maximum standards applied, increased use of controlled parking zones.
PP9 Park and Ride	Medium	Increase capacity at existing sites: Seacourt, Redbridge, Thornhill; new remote park and ride sites for Oxford on A34(s), A40(w) and A34(n)/A41 corridors, used for accessing premium bus routes.
PP10 Land Use Planning	Medium	Development located according to South East Plan but better strategic coordination and sustainability travel aspirations achieve 25% lower car trip rates in new developments.

Table 7. *(Continued)*

Policy Package	Level of Application	Summary Specification
PP11 Smarter Choice Behavioural Measures	Medium	Enhanced travel planning (workplace and schools), widespread travel awareness campaign, including increased availability of pre-trip and en-route information, personalised travel planning in new developments, support for car clubs.
PP12 Low Emissions Vehicles	High (Car/LGV) High (HGV)	A total car/LGV fleet at an average of 95 gCO_2/km and HGVs (fully loaded) at 800 gCO_2/km.
PP13 Alternative Fuels	Medium	Increase in car dieselisation and much greater uptake of biofuels across modes – car 25% diesel (from 15%), 60% petrol, 15% alternative fuels (mix of LPG, electric, biofuel); HGV 75% diesel (from 100%), 25% alternative fuels; bus 60% diesel (from 100%), 40% alternative fuels. No hydrogen in 'medium' level application.
PP14 Slower Speeds and Ecological Driving	Medium	20 mph speed limits in all major towns and 50 mph speed limits on all rural single-carriageway roads; lower speed limits are supported by variable signage and enforcement. There is also a targeted public education campaign concerning ecological driving skills.
PP15 Freight	Medium	Increase in rail freight capacity between the south coast ports at Southampton and the Midlands and North, together with advisory HGV routing.
PP16 Long Distance Travel Substitution	Medium	Improved coach/rail links to a range of major destinations, including London, Oxford and all London airports, and along major corridors – Birmingham/Manchester, Oxford–Southampton (M40/A34 road/rail corridor), Oxford–Bristol (via Swindon).

Scenario Results (relative to BAU 2030)	Absolute	Percentage
Climate: carbon dioxide (CO_2) emission reduction (tCO_2)	−1,365,697	−48.3%
	2.0 tCO_2 per capita (2030) relative to BAU of 3.8 (assumed aggregate BAU in 2030 is 2,664,849 tCO_2)	

vehicle technology penetration will be very difficult to achieve across the whole vehicle fleet. The levels (2006) for the current new car fleet are at around 165 gCO_2/km and HGVs (fully loaded) at 1100 gCO_2/km.

- PP14 Slower speeds and ecological driving (medium): this has a major impact, representing 20 mph speed limits in all major towns and 50 mph speed limits on all rural single-carriageway roads; lower speed limits are supported by variable signage and enforcement. There is also a targeted public education campaign concerning ecological driving skills. This package achieves a −5% reduction in CO_2 emissions.
- PP15 Freight (medium): HGV freight movements represent over 20% of traffic on key routes in Oxfordshire, particularly the A34 and M40, and the large and heavy vehicles are high CO_2 emitters. An assumed increase in rail freight capacity between the south coast ports at Southampton and the Midlands and North, together with advisory HGV routing, would reduce HGV traffic through the county. This package also achieves a −5% reduction in CO_2 emissions.

The other selected policy packages also have a direct impact on CO_2 emissions, but the scales of impact are less than the above. PP1 Rail (medium), PP2 Bus (medium), PP3 Walk (high), PP4 Cycle (high), PP6 TDM/Active Traffic Management (medium), PP7 Pricing (medium), PP8 Parking Management (medium), PP9 Park and Ride (medium), PP10 Land Use Planning (medium), PP11 Behavioural Change (medium), and PP16 Long Distance Travel Substitution (medium), for example, all lead to limited CO_2 emission reductions. Expected impacts are often masked by other traffic filling up the space gained in network improvements. The major gains are made when car and freight emissions are tackled (the vast majority of current emissions). The interpretation of these types of result needs careful consideration. This is not an argument for not investing in the non-car modes, simply a result of the current dominance of car use relative to other travel modes. PPs 1–4, 6–11 and 16 lead to a level of CO_2 reduction and they are also useful for non-carbon reduction issues, such as supporting health, urban fabric and the wider quality-of-life objectives.

MORE RADICAL INTERVENTIONS: TOWARDS A 60% AND DEEPER REDUCTION

Scenario 3 as examined delivers major gains in CO_2 reduction and wider sustainability benefits. However, it appears that reducing transport CO_2

emissions on an area-wide basis by anything more than 20–30% is likely to be very difficult. The scale of intervention being considered, even in the innovative authorities such as Oxfordshire, is nowhere near enough to deliver the 'fare share' of CO_2 reduction required in the transport sector relative to contributing towards the national CO_2 reduction target. Greater innovation is required in considering the range of measures available, in making the most of potential application levels, and in packaging and synergies. But too little analysis is considering these issues. This is where scenario analysis can become very useful: in helping to think through what is currently 'unthinkable' (in technical, funding and political terms) with the aim of changing the parameters of the debate.

Deeper reductions in CO_2 emissions, to move towards national target levels, would require a 'high level' of application of the policy packages, as considered in Oxfordshire, and beyond. This may mean very radical measures and initiatives (for the present-day understanding). This could include the following types of measures:

- New local rail links, a light rapid transit network for Oxford, tram-train services (e.g. serving Reading–Abingdon–Oxford), fare level changes;
- Much more extensive bus services in rural areas, better network planning and integration, bus rapid transit (e.g. serving Oxford and other key centres), taxi-bus type schemes for rural areas;
- A much improved public realm for pedestrians; cycle networks and hire facilities across the county;
- Road pricing (at a high charge per kilometre); car-free developments, higher parking charges and restrictions in parking supply; much extended park and ride with 'remote' parking sites next to major routes;
- Spatial planning shaped more directly to support public transport usage, walking and cycling to a much greater extent than at present;
- Widespread, mass market smarter choice behavioural measures; and slower speed limits on all roads.

International best practice in transport planning, as found in cycling facilities in Groningen or Delft, or public transport in Zürich, Strasbourg or Montpellier, needs to become standard practice. Despite the good intentions of many transport planners over the last 20 years, this has not happened. The effective transferability of good practice and the transitioning processes towards sustainable travel behaviours need a much more coordinated approach, based on greater empirical understanding, perhaps through one or two cities becoming examples of what can actually be achieved in terms of reduced carbon use in transport.

There are, of course, wider issues involved in the transition to different travel behaviour. Improved networks, facilities and urban planning are only part of the rationale for travel, as individual motivations are also a critical component. It is here that some of the emerging literature in transport planning sheds some light, as there are important psychological and social and cultural dimensions to people's choice of travel. For example, Lucas and Jones (2009) and Lucas, Blumenberg, and Weinberger (2011) describe

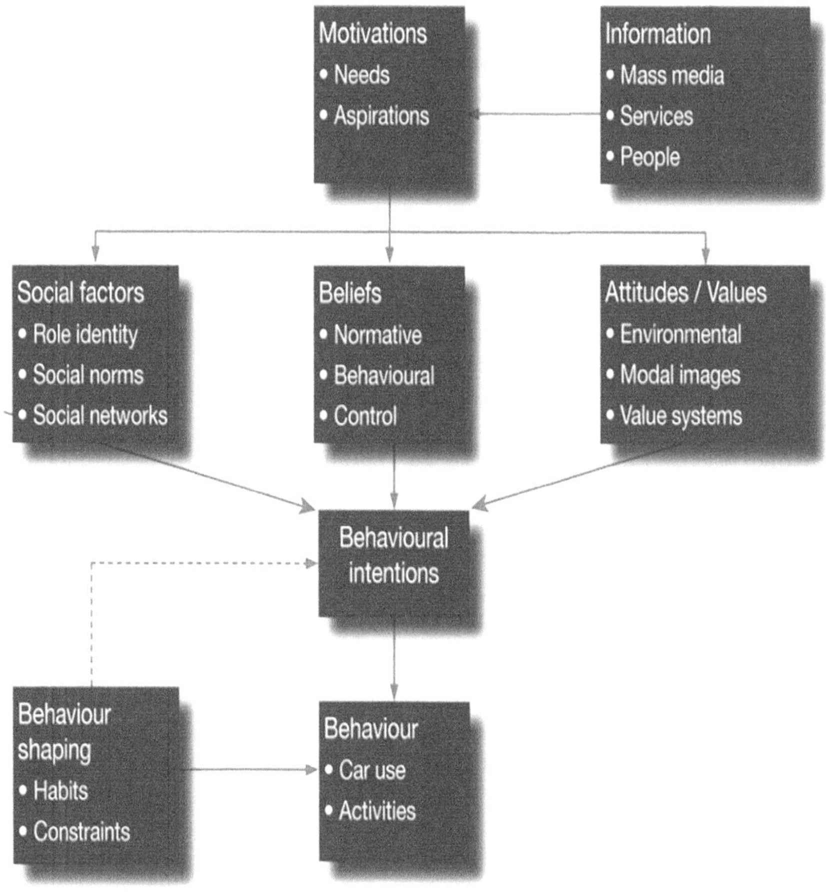

Fig. 2. The Personal Factors Influencing Car Use Behaviours. *Source*: Lucas and Jones (2009).

the motivations that may underlie choice processes. These include needs and aspirations, the social factors (identity and norms), beliefs, attitudes and values and behavioural intentions. These are also influenced by information and marketing, and they interrelate with the networks, opportunities and urban form, as are more conventionally considered in transport planning (Fig. 2). There are also important influences in terms of time–space constraints on travel (Dowling, 2000; Schwanen & Kwan, 2008) and the impact these have on current travel behaviours and future possibilities for changed behaviours. It is these types of issues that tend to be ignored when thinking about and modelling transport behaviours, and also when shaping policies and strategies for future transport investments.

The sociology of mobilities take the debate even further, perceiving transport or mobility as a defining factor within society (Cresswell, 2006; Kaufmann, 2002; Sheller & Urry, 2006; Urry, 2000; Urry, 2007). There is also the existence of social 'practices' (Bourdieu, 1972) within lifestyles and travel. Perhaps surprisingly, the emerging understanding of the rationales behind travel is still embryonic, at all levels from individual motivations to societal representation, including how these might combine, and critically, the potentialities for change. Furthering these understandings seem central to the transition to sustainable travel behaviours.

CONCLUSIONS: INCREASING THE AMBITION

The transport sector therefore needs new thinking and approaches to understanding the scale and immediacy of the CO_2 reductions necessary over the next 20 years. It must also demonstrate and discuss the potential pathways and examine the feasibility for change. In addition, there is an imperative to examine the means by which action can take place, including the responsibilities for both central and local government, so that the necessary powers and funds are available. Any move towards sustainable travel behaviour needs to be area-wide and low-carbon targets should also be consistent with wider sustainability objectives.

This chapter has examined the different baselines for districts in Oxfordshire and also for urban areas more generally in the United Kingdom. It suggests that location and context is of key importance, with the starting points, opportunities and scale of change more varied than imagined. There is uniqueness in all situations. The options for Oxfordshire have been examined, with a core scenario outlined in terms of the likely impacts in CO_2 reduction terms. Major CO_2 reductions are possible, but the reality makes any real progress difficult. Many of the measures that would make significant contributions are deemed impossible, as the responsibilities do not reside at the local level and as funding is not available, and even the policy environment has changed away from needing to take immediate action. The Oxfordshire modelling allows the exploration of the strategic policy choices in a transparent and participatory manner, with a role playing element allowing the different options to be tested by a range of participants. This makes explicit some of the tensions evident for decision makers involved in local transport planning. It is this type of decision making and process that can change the parameters of the debate by making the problems of implementation more explicit, and through allowing a wide range of different interested parties to get involved in examining the alternatives. Such an approach allows the ownership of the decisions to be spread as widely as possible, and this in turn means that public and political acceptability will be enhanced.

Creative combinations of policy measures are required to achieve the target reductions in transport CO_2. As noted earlier, the interventions considered cover the responsibilities of Oxfordshire County Council and other key public and private decision makers, including those at the national and European levels. Reduced national funding levels certainly are causing difficulties in investment programmes. The Sustainable Mobility option (Scenario 3), although representing substantial new investment which is far beyond the current funding levels available, still only delivers an expected level of 2.0 tCO_2 per capita in 2030. This is only a reduction of a third against 2005 levels and quite a way from an assumed target of 1.2 tCO_2 per capita in 2030 and 0.5 tCO_2 per capita in 2050.

Such targets may therefore be impossible to achieve, even if there were funds available and the necessary political and public support for effective implementation. This in turn means that combinations of policies need to provide more than their individual contributions (i.e. synergetic effects), that transport options need to be considered with interventions in other sectors and that reductions in levels of mobility are required. It should also be remembered that carbon reduction is only one policy objective, so impacts across multiple indicators need to be part of any assessment, and this might include multi-criteria or goals achievement assessment (Saxena, 2012; Hickman et al., 2010). All of these are major issues to be resolved and progressed in the next few years. The binding targets set out in the UK Climate Change Act (2008) are extremely difficult to achieve in an affluent, rural county such as Oxfordshire, and it seems that it is problematical even finding the means by which more than a 30% reduction in existing levels can be achieved. It may be easier to achieve the targets set in the more urban areas, such as London (Hickman, Ashiru, & Banister, 2010), but this case study provides evidence that transport may have to take a lower level of reduction on CO_2 emissions than other sectors. Alternatively, the binding targets themselves will have to be reviewed.

The caveat from Thomas More is relevant in this debate (More, 1516, p. 118):

> There are many features of the Utopian commonwealth that I can more easily wish for in our own societies than hope to see realised.

The 'strategic conversation' (Van der Heijden, 1996), discussing the major strategic choices in policy making, becomes central to transport planning: involving ambitious targets, a range of measures and high levels of application, consistent starting points for assessing change and understanding the variations between different locations. But it is also important to realise the need for substantial change in individual lifestyles, and the practices of behaviour and society. Effective realisation means that all of these barriers must be overcome, and in this sense, it is understandable why so little progress has been made to date, either at the national or the local levels, in reducing the levels of CO_2 emissions in transport.

NOTES

1. The current data estimates that there are over 1.2 million deaths each year on the world's roads, and between 20 and 50 million suffer non-fatal injuries (World Health Organization 2009. Global Status Report on Road Safety. Washington DC: WHO.). This is a huge cost that seems to be accepted by society.
2. The analysis draws on work carried out in the Integrated Transport Appraisal and Simulation for Oxfordshire study (INTRA-SIM Oxfordshire). Hickman, R., Ashiru, O., Seaborn, C., & Walters, G. (2010) *INTRA-SIM Oxfordshire*. Final Report. Halcrow, London.
3. Note that the national level has been taken and used locally, and that transport has been allocated its 'fair' share rather than taking a lower level as discussed earlier.
4. A scenario matrix is often used in scenario planning, with two dimensions of uncertainty used as the parameters for the future. In theory, multiple dimensions can be used to explore possible futures, but the most commonly used approaches tend to keep things simple with limited key dimensions and variation developed within the scenario portraits.
5. The UK Climate Change Act (2008) has placed legally binding targets for UK CO_2 reductions of 34% by 2020 and 80% by 2050 on 1990 levels across all sectors of the economy.

ACKNOWLEDGEMENTS

Many thanks to Oxfordshire County Council (Jo Fellows and Roger O'Neill) who provided funding for the Oxfordshire INTRA-SIM study and assisted in the work, and the wider study team who have helped in the modelling and analysis – from the Halcrow Group (including Olu Ashiru, Catherine Seaborn, Wambui Gachaco, Gareth Walters and Will Saltmarsh). The views expressed in this chapter are of course from the authors and do not necessarily reflect those of Oxfordshire County Council or other organisations involved.

REFERENCES

Anable, J., & Gatersleben, B. (2005). All work and no play? The role of instrumental and affective factors in work and leisure journeys by different travel modes. *Transportation Research Part A: Policy and Practice*, *39*, 163–181.

Banister, D., Stead, D., Steen, P., Akerman, J., Dreborg, K., Nijkamp, P., & Schleicher-Tappeser, R. (2000). *European transport policy and sustainable mobility*. London: Spon.

Bourdieu, P. (1972). *Outline of a theory of practice* [Originally published as Esquisse d'une Théorie de la Pratique, by Libraire Droz, Switzerland, and reprinted in English in 1977]. Cambridge: Cambridge University Press.

Cresswell, T. (2006). *On the move: Mobility in the modern Western World*. London: Routledge.

Department for Food and Rural Affairs. (2009). *Modal CO_2 emission factors*. London: DEFRA.

Department for Trade and Industry and Office of Science and Technology. (2006). *Intelligent infrastructure systems. Project overview*. London: DTI and OST.

Department for Transport. (2008). *Vehicle/speed CO_2 emission factors*. London: Department for Transport.

Department of Energy and Climate Change. (2010). *Local and regional CO_2 emissions estimates for 2005–2008 for the UK [Online]*. London: AEA and DECC.

Dowling, R. (2000). Cultures of mothering and car use in suburban Sydney: A preliminary investigation. *Geoforum*, *31*, 345–353.

Gilbert, R., & Perl, A. (2010). *Transport revolutions: Moving people and freight without oil*. Philadelphia, PA: New Society.

Government Office for the South East. (2009). *South east plan*. Guildford, UK: GOSE.

Hickman, R., Ashiru, O., & Banister, D. (2010). Transport and climate change: Simulating the options for carbon reduction in London. *Transport Policy*, *17*(2), 110–125.

Hickman, R., Ashiru, O., Seaborn, C., & Walters, G. (2010). *INTRA-SIM Oxfordshire, final report*. London: Halcrow.

Hickman, R., & Banister, D. (2007). Looking over the horizon: Transport and reduced CO_2 emissions in the UK by 2030. *Transport Policy*, *14*(5), 377–387.

Hickman, R., Pharoah, T., Torrance, J., & Dyer, R. (2011). *Moving towards smarter travel?* London: Halcrow for Sustrans and Friends of the Earth.

Kaufmann, V. (2002). *Re-thinking mobility: Contemporary sociology*. Aldershot, UK: Ashgate.

Lucas, K., Blumenberg, E., & Weinberger, R. (2011). *Auto motives: Understanding car use behaviours*. Bingley, UK: Emerald.

Lucas, K., & Jones, P. (2009). *The car in British society*. London: RAC Foundation.

More, T. (1516). Utopia. Toronto: Broadview Press [published 2010].

Motortrend Forum. (2010). *Top 50 best selling cars*. Retrieved from http://forums.motortrend.com

Oxford City Council. (2008). *Oxford local development framework, core strategy*. Oxford: OCC.

Oxfordshire County Council. (2000). *Road traffic reduction report*. Oxford: OCC.

Oxfordshire County Council. (2006). *Local transport plan 2*. Oxford: OCC.

Oxfordshire County Council. (2007). *Population estimate*. Oxford: OCC.

Oxfordshire County Council. (2011). *Local transport plan 3*. Oxford: OCC.

Saxena, S. (2012). *Low carbon transport pathways in Delhi*. D.Phil. (in development), University of Oxford, UK.

Schäfer, A., Heywood, J. B., Jacoby, H. D., & Waitz, I. A. (2009). *Transportation in a climate-constrained world*. Cambridge, MA: MIT.

Schwanen, T., & Kwan, M.-P. (2008). Space-time constraints, the Internet and mobile phone. *Geoforum*, *39*, 1362–1377.

Sheller, M., & Urry, J. (2006). The new mobilities paradigm. *Environment and Planning A*, *38*, 207–226.

Sperling, D., & Gordon, D. (2009). *Two billion cars. Driving toward sustainability*. New York, NY: Oxford University Press.

Urry, J. (2000). *Sociology beyond societies: Mobilities for the twenty-first century*. London: Routledge.

Urry, J. (2007). *Mobilities*. Cambridge: Polity.

US Energy Information Administration. (2009) *World proved reserves of oil and natural gas, most recent estimates*. Retrieved from http://www.eia.gov/international/reserves.html

Van Der Heijden, K. A. (1996). *Scenarios: The art of strategic conversation*. Chichester, UK: Wiley.

Williams, H. (1991). *Autogeddon*. New York, NY: Arcade Publishing.

World Health Organization. (2009). *Global status report on road safety*. Washington, DC: WHO.

Yang, C., McCollum, D., McCarthy, R., & Leighty, W. (2009). Meeting an 80% reduction in greenhouse gas emissions from transportation by 2050: A case study in California. *Transportation Research D*, *14*(3), 147–156.

CHAPTER 14

CONCLUSIONS

Lee Chapman and Tim Ryley

The transport sector is a major contributor to greenhouse gas emissions, and large sections of this book have discussed methods, tools and instruments, as well as the challenges, that could prove crucial in mitigating the impacts of transport on the environment. However, as Chapter 1 explains, mitigation is only half of the picture. Due to the long residence time of greenhouse gases in the atmosphere, the world is already committed to some degree of climate change and this in turn will begin to impact upon transport systems forcing society to adapt hard infrastructure to ensure that it is resilient for years to come. Consequently, mitigation and adaptation go hand in hand, with successful mitigation reducing the need for adaptation. Hence, a sustainable transport system ultimately needs to consider both mitigation and adaptation in a holistic framework. This book was conceived and developed with this holistic approach in mind, and this final concluding chapter uses these headings to synthesise and discuss the key findings from the various chapters.

MITIGATION

Due to the more pressing need, the majority of material in this book has dealt with mitigation; interventions to reduce emissions of greenhouse gases (IPCC, 2001) away from a scenario of 'business as usual'. An early academic review on climate change mitigation and transport appeared in

Transport and Climate Change
Transport and Sustainability, Volume 2, 369–378

ISSN: 2044-9941/doi:10.1108/S2044-9941(2012)0000002017

2007 (Chapman, 2007), a year before the United Kingdom committed itself to the highly ambitious Climate Change Act 2008. The final act sought an ambitious 80% cut in greenhouse gas emissions by 2050. Although criticised in this book as being unrealistic, this is the level of intervention required if 'dangerous' climate change is to be avoided (defined as greater than 2°C rise in global temperatures). As Chapter 8 (a policy perspective) explains, such targets, however unrealistic, are informed by considerable input from expert opinion (e.g. Delphi studies) and require a detailed knowledge of current emissions as well as accurate predictions/scenarios of future emissions. For this reason, scenarios and the control of uncertainty were discussed in Chapter 2 towards the start of this book. Indeed, the commonly quoted 'business as usual' scenario is in itself too simplistic and highly improbable, not only due to mitigation measures imposed by governments, but also because of the future scarcity of oil which will force change in the medium term regardless. Backcasting is the key tool used to model the continuum of socio-economic scenarios which exist between 'business as usual' and the equally unlikely case of all targets being met. However, the science is inherently difficult and the end result is a wide range of permutations and storylines, largely dependent on mitigation. Early progress towards the 80% target has not been promising, but the decarbonisation of the transport sector is still seen as key in meeting the demands of the Climate Change Act 2008. There is a need to tackle the three primary culprits of greenhouse gas emissions in the transport sector: aviation, freight and car ownership (Chapman, 2007). This book has examined in detail how this could be achieved in all these sectors using a range of aspects relating to technological and behavioural change.

Aviation

One of the biggest concerns in the transport sector regarding emissions is aviation. The scale of growth in the industry (approaching 5% per annum; IATA, 2011) ensures that whilst aviation currently is not the largest contributor to emissions in the transport sector, its relative contribution is increasing rapidly. Aviation is indeed a complicated case to consider. Whilst much of the debate has focused on CO_2 emissions due to long atmospheric residence times, there are many other impacts (often more localised – e.g. contrails) that need to be considered when assessing the total impact of aviation. However, as is the case in much of the climate change debate, it is the international dimension which can cause the most difficulties. Boundaries are a considerable barrier for regulation and multilateral agreement is rarely achieved, as evidenced by the exclusion of international aviation in the Kyoto Protocol.

Whilst there have been a number of visible changes since 2008 (e.g. carbon trading within the EU Emission Trading Scheme) the problems with international regulation initially lead to the conclusion that in the case of aviation at least, technology is probably the primary solution to the problem. Alternative fuels (e.g. biofuels) provide one answer, as are smarter engineering solutions (e.g. blended wings, open rotor engines) or even improved logistics such as those offered by the Single European Skies initiative. However, the advance of technology is not keeping pace with the present growth in aviation and international policy and intervention is now looking increasingly necessary if there is to be a significant change to increasingly air dependant lifestyles. Air Passenger Duty (at a national level) and carbon trading (at an international level) are looking increasingly vital tools to curb the growth in emissions from aviation. However, as discussed subsequently in Chapter 9 (an economic perspective), the most simple way of reducing growth is to limit expansion.

Freight

Aviation also plays a key role in rapidly moving perishable freight across the world. However, shipping remains the fundamental transport component linking production with consumption and accounts for 89.6% of global trade in volume terms. This volume of freight often means that shipping is considered to be a relatively environmentally friendly form of transport, accounting for less than 3% of global CO_2 emissions. However, there is still much that can be undertaken in order to improve the environmental credentials of the international shipping industry – particularly with respect to other (non greenhouse gas) pollutants which are of a particular concern due to the low quality fuel used in the industry. As with aviation, international shipping was not included in the Kyoto Protocol, yet as demonstrated in Chapter 4, and subsequently in the examination of the cost effectiveness of carbon reduction measures in transport (Chapter 9), a number of mitigation measures have been developed to reduce pollution and ultimately greenhouse gas emissions. These sub-divide into technical measures (engine efficiency, improved hull design, low sulphur fuel and alternative fuels) and operational measures (optimising fleet operation, port efficiency savings, reducing ship speeds). The International Maritime Organisation spearheads regulation of the industry to encourage the strategic adoption of such strategies, but the sector appears to be some way off implementing market based measures such as carbon trading under consideration for the aviation sector.

The freight industry also has a significant national dimension. Road transport accounts for 81% of fuel usage in the transport sector and contributes approximately 20% of total transport emissions (Chapman, 2007). Mitigation schemes to reduce the impact of freight transport (as discussed in Chapter 6) focus on the decoupling of freight activity with national economic growth. Modal shift onto the railways is frequently cited, but the focus to date has largely been on ways to increase fuel efficiency in the sector. Simple eco-driving is viewed as an important cost-effective carbon reduction measure (Chapter 9), but dematerialisation of goods, and improved logistics from vehicle routing and utilisation, have an important part to play (Chapter 6).

Car Ownership

Without doubt, the primary challenge the world faces in reducing greenhouse gas emissions from transport is tackling the dependence on the motor car. This is a central theme in this book and such is the complexity of the issue, different aspects have been tackled in many of the chapters. Chapter 10, providing a behavioural perspective, outlines some of the reasons for this: convenience, independence, freedom, comfort, flexibility, security and status. Indeed, Lucas and Jones (2009) found that over 80% of individuals would have difficulties continuing with their present lifestyles without the use of a car.

A mixture of technological and behavioural change is required to reduce the impact of car ownership (Chapman, 2007). For example, Chapter 8, providing a policy perspective, discusses how technological change could be particularly important in the sense that it offers a transitional pathway, thus buying time for behavioural change over the long term. This approach has been backed heavily by policy and regulation, and recent examples include the subsidy of new electric vehicles and hybrids as well as EU emission targets for new vehicles. However, behavioural change is notoriously slow and needs careful planning to be successful. For example, taxation is an approach favoured by governments, being effective as well as quick and easy to implement (Chapter 9), but it is clearly unpopular and does not tackle the root of the problem. Indeed, as shown in Chapter 11, some sectors of society (e.g. rural areas) have a high dependence on the motor car and subsequently spend a higher proportion of their income on travelling. These groups are particularly vulnerable to such measures, which needs to be taken into account by policymakers if mobility gaps are to be avoided. Such problems are often overlooked even though it has long been recognised that investment in transport tends to benefit the rich more than the poor (Banister, 2007). Hence, the more progressive policy measures, including parking and congestion charges, directly target higher income groups. A number of other instruments can also be used, each contributing a gradual reduction in car journeys: travel plans, travel awareness campaigns, public transport marketing, car clubs and car sharing schemes, teleworking, teleconferencing and home shopping (Chapter 10). These so-called 'smart options' can be highly cost effective in realising carbon emission reductions (Chapter 9).

Successful behavioural change is reliant on providing a viable alternative, which in many cases will initiate a 'modal shift' (e.g. changing car journeys to public transport). Policies which enable this are also likely to be progressive in the sense that it will improve the ability to travel of those on low incomes. The nature of the transport mode will vary with the distance travelled and different approaches are needed at national/international (Chapters 8 and 10), regional (Chapter 13) and local (for walking/cycling trips, Chapter 7) levels. At whatever scale, there is considerable difficulty in affecting the switch to sustainable transport solutions. At a national (or indeed continental) level, electric trains (run on renewable or nuclear energy) are a viable carbon reduction measure (Chapter 9). At a local level, Chapter 10 suggests increased bus usage as an example, yet is dismissed due to low public opinion regarding its status, levels of service and cost effectiveness. Chapter 7 highlights the different approach of walking and cycling, discussing ways in which this can be promoted. However, the conclusions are negative in the sense that such moves would have minimal impacts on carbon emissions and could paradoxically actually increase emissions by generating more car journeys via increased leisure-related walking and cycling.

Raising awareness is central to a successful uptake of alternatives, hence the rôle of information to ensure that travel options are effectively communicated. Here, Advanced Traveller Information Systems, which provide integrated multimodal travel information, could potentially be very influential in affecting modal shift (as presented in the review of information systems in Chapter 12). Indeed, the potential exists with such systems to embed carbon calculators to reinforce the environmental cost of the journey as well as price and time considerations (Coulter, Clegg, Lyons, Chatterton, & Musselwhite, 2007). However, as also highlighted in Chapter 12, it has to be realised that suitable alternatives are not currently in place to bring about a change and investment is therefore required.

In summary, it is difficult to perceive a situation where the personal advantages of car travel are sufficiently met by other transport solutions and hence it will be alternative technological advances that will begin to pave the way for significant carbon reductions. Alternative fuels are one such example, with electric cars using energy from sustainable sources perhaps offering the most realistic long-term answer. The middling technology of hybrid vehicles is already proving successful; there are also ways to promote eco-driving (a practice appropriate for any car), as well as providing feedback to the motorist the energy efficiency of their current driving style in a game-like fashion (Chapter 12). However, perhaps the most obvious change to travel behaviour is actually reduced car usage and CO_2 emissions by downsizing to more fuel efficient vehicles combined with eco-driving (Chapter 9) to reduce the cost of motoring. This has certainly been helped by manufacturers producing increasingly efficient vehicles, albeit mostly in response to stringent EU legislation. However, a downside is the reduced cost of motoring that may perhaps promote an increase in travelling (Chapter 9).

ADAPTATION

The majority of the scientific literature on climate change and transport still tends to deal with mitigation and policy. However, as this book demonstrates, there is an important parallel line of debate – the impact of climate change on transport. Already the current climate poses frequent meteorological hazards to transport networks and climate change will ensure that the nature and impact of these hazards will also be subject to alteration. Whilst impacts are not all negative, the consensus is that transport networks will be subject to increased exposure to climate impacts and will therefore require targeted adaptation to continue to function effectively. This has been touched upon in several of the mode specific chapters in this book (Chapters 3–6), which have discussed how adaptive capacity can be increased by preparing the hard infrastructure to cope with future climate stimuli.

Climate Impacts

Even if strong mitigation measures were to be immediately implemented, society would remain committed to some degree of climate change. The potential impacts of even a small amount of climate change on transport are numerous and there is a need to develop effective strategies for society to be able to deal with these impacts. In terms of international transportation, the biggest impact will be from sea-level rise triggered by any increase in global temperatures. The impacts on both international aviation and shipping (Chapters 3 and 4) are similar as both rely on low-lying infrastructure to function which will be at increased risk of inundation. However, a positive impact for shipping and freight will be the new opportunities offered by the increased navigability of high latitude routes. Additional localised impacts on aviation are also highlighted in Chapter 3 and are mostly due to the impact of high temperatures which will degrade take-off performance and possibly damage runways.

Surface transportation is the most exposed element to the localised impacts of climate change (Chapter 5). High summer temperatures will result in road rutting, rail buckling and decreased thermal comfort, whereas more intense winter precipitation will cause flooding, landslips and bridge scour across all modes. For all impacts, it is the extreme events (e.g. heatwaves and storms) that are potentially the most devastating. Storms are particularly problematic, and although there are no clear projections with respect to future frequency and intensity, they already have a significant impact on transport networks (i.e. crosswinds and thunderstorms for aviation, and flooding and high winds on surface transport networks). Thus, extreme events have the potential to instigate multi-mode transport failure which could then cascade onto other critical infrastructure networks. Of course, climate change impacts are not all negative. Indeed, in the case of winter maintenance, all transport networks stand to benefit (see Chapters 3 and 5).

Climate Change Risk Assessments

A challenge exists in how to adapt (or design new) infrastructure to be resilient to future climates. The starting point is a climate change risk assessment. This utilises a climate response model developed for use in the baseline climate, which can then be used to make future forecasts based on climate change scenarios (Chapters 2 and 6). Such information can then be used to engineer a solution which could withstand 100 years+ of climate change. However, in addition to climate scenarios, there is the additional need to include socio-economic scenarios (Chapter 2), underlining the requirement of a holistic approach to climate change risk assessment (Jaroszweski, Chapman, & Petts, 2010). This adds significant complexity to the issue and essentially requires an understanding of *what future society actually needs from a transport network*. This of course will change depending on the timescale. In the short term, investment is essential across all transport modes to improve adaptive capacity. Without this, some modes will remain more resilient than others in particular weather conditions and this in turn will influence passenger behaviour (which may be a good or a bad thing). In the longer term, society may have radically different expectations of the role of transportation should mitigation measures such as home-working and regional production become commonplace. Should this be the universal case, then significant adaptation is not actually required as the actual climate change experienced will not be as severe. However, the pertinent example of the UK road network still persistently following the routes of the Roman Empire used in Chapter 2 highlights the enormity of the challenge in this process. Radical change is very difficult to implement and in the case of climate change, a 'belt and braces' approach is probably essential.

OVERALL CONCLUSIONS

In terms of mitigation, Kyoto targets effectively end in 2012, leaving the international community without a firm focus for future mitigation efforts. In any case, the success of the protocol is questionable. Clearly it offered a useful foundation for a worldwide mitigation effort, but the targets were insufficient to make real impact. The international barriers preventing multilateral agreement on suitable targets (particularly in the aviation and shipping sectors as demonstrated in Chapters 3 and 4) are always going to be very difficult to overcome. Hence, there is a need to promote the importance of each nation reducing greenhouse gas emissions at their own level. Perhaps, given that climate change issues cross the spatial scales, it would be sensible to also think about local mitigation before subsequently scaling up to national and international level. Any solution needs to be considered as the sum of many small actions, perhaps with a mantra of 'think big, act small'. Indeed, Chapter

13 is particularly interesting in this respect as it draws together many of the issues in this book to provide solutions for mitigation at a local level. And yet, whilst this approach has been shown to potentially reduce emissions by 33% by the year 2030, this is still significantly below central government targets. However, since international multilateral agreement has not been successful to date, it could be worth trying a bottom-up approach.

The UK, the focus for much of the case study examples within the book, remains committed to the Climate Change Act 2008 and with such limited early progress towards targets, it seems that there is a general disconnect between policy and goals (particularly demonstrated in Chapters 8 and 9). Policy is not sufficient to deliver the paid changes required. Peak oil will certainly play a role, as will policy and possibly also geo-engineering (Fox & Chapman, 2011), an approach not considered in this book, but there is still a sense that both behavioural and technological responses will be required to at least attempt progression towards the targets. The risky strategy of a sole focus on mitigation underlines the importance of adaptation as a safety net. In effect, both mitigation and adaptation are linked and a holistic framework for research and investment is required to improve the situation with regard to transport and climate change.

The economic case along with the estimated costs of failing to adapt to climate change are set out in the Stern (2006) and Eddington (2006) Reviews (see Chapters 8 and 9). The key finding from the reviews is that early responses (many of which are small scale) are crucial to minimise the costs of climate change. However, at the time of writing, most developed countries are tentatively recovering from a significant economic downturn. This needs to be in the context that developed countries have more resources available to develop a more resilient transport infrastructure and it will be the developing countries that suffer more from the impacts of climate change. The timing of the downturn is a concern, with environmental issues such as climate change taking a lower policy priority and a lack of willingness amongst policymakers and stakeholders to instigate change. Hence, there has been a delay in the action required. It could be argued that as society has already been running out of time to avoid 'dangerous' climate change, this only exacerbates the problem. Although this presents a very pessimistic tone, there is always the faint hope that climate change impacts have been overestimated, that transport policymakers will suddenly change tack on climate change issues, and that there will be further unexpected behavioural and technological changes relating to transport.

REFERENCES

Banister, D. (2007). Cities, mobility and climate change. *Journal of Industrial Ecology*, *11*, 7–10.

Chapman, L. (2007). Transport and climate change: A review. *Journal of Transport Geography*, *15*, 354–367.

Coulter, A., Clegg, S., Lyons, G., Chatterton, T., & Musselwhite, C. B. A. (2007). *Exploring public attitudes to personal carbon dioxide emission information*. London: Department for Transport.

Eddington, R. (2006). *The case for action: Sir Rod Eddington's advice to Government*. HM Treasury. Retrieved from http://www.hm-treasury.gov.uk

Fox, T., & Chapman, L. (2011). Engineering geo-engineering. *Meteorological Applications*, *18*, 1–8.

Intergovernmental Panel on Climate Change (IPCC). (2001). *Climate change 2001: Synthesis report*. A contribution of Working Groups I, II and III to the third assessment report of the IPCC. Cambridge, UK: Cambridge University Press.

International Air Transport Association (IATA). (2011). *Air transport market analysis June 2011*. Retrieved from www.iata.org/economics

Jaroszweski, D., Chapman, L., & Petts, J. (2010). Assessing the potential impact of climate change on transportation: The need for an interdisciplinary approach. *Journal of Transport Geography*, *18*, 331–335.

Lucas, K., & Jones, P. (2009). *The car in British society*. London: RAC Foundation.

Stern, N., Peters, S., Bakhshi, V., Bowen, A., Cameron, C., Catovsky, S., ... Zenghelis, D. (2006). *Stern review: The economics of climate change*. London: HM Treasury.

ABOUT THE AUTHORS

Erel Avineri is associate professor in travel behaviour at the University of the West of England (UWE), Bristol. He has 20 years experience of consultancy, research and teaching in transport, logistics and operations management. Since he joined UWE in 2004 he has lead research in the fields of travel behaviour, travel information systems, road safety and behavioural change. Dr Avineri has gained wide recognition for his research on travel behaviour under uncertainty, the incorporation of socio-psychological aspects into behaviour models and the design of behaviour change policy measures. Applying choice architecture, Dr Avineri studies the effect of 'nudges' on the perception of and attitudes to CO_2 reduction. He holds degrees in industrial engineering and management (BSc) and transportation sciences (MSc, PhD) from Technion-Israel Institute of Technology.

Ian Bache is professor of politics at the University of Sheffield. He has written widely on European and British politics. His publications include: *Cohesion Policy and Multi-level Governance in South East Europe* (co-editor, Routledge, 2011); *Politics in the European Union* (co-author, Oxford University Press, 3rd ed., 2011); *Europeanization and Multi-level Governance: Cohesion Policy in the European Union and Britain* (Rowman & Littlefield, 2008); *The Europeanization of British Politics* (co-editor, Palgrave, 2006; 2008pbk); *Multi-Level Governance* (co-editor, Oxford University Press, 2004; 2005pbk) and *The Politics of European Union Regional Policy* (Sheffield Academic Press, 1998). He is currently conducting research on The Politics of Wellbeing.

David Banister is professor of transport studies at the School of Geography and the Environment (SoGE) in the University of Oxford. Until 2006 he was professor of transport planning at University College London, and he has held research posts at the Warren Centre in the University of Sydney, the Tinbergen Institute in Amsterdam, the University of Bodenkultur in Vienna and most recently as the first BIVEC-GIBET Transport chair, a position set up by the Benelux Interuniversity Association of Transport Researchers. His research expertise is in cities, transport and the environment, and he has published 19 books and over 250 papers on these topics, including several on transport and climate change.

Abigail L. Bristow is professor of transport studies at Loughborough University. She is fellow of the Institute of Acoustics and the Royal Society of the Arts and has served on the board of the Low Carbon Vehicle Partnership. Her main research interests are the externalities generated by transport systems particularly the costs of transportation noise nuisance, the value of tranquillity and carbon mitigation. Research on carbon includes: exploring the potential for significant reductions in passenger transport emissions in the United Kingdom and freight transport emissions in London by 2050; transport in the hydrogen economy; cost effective carbon mitigation; the acceptability and behavioural impact of personal carbon trading and the potential market for gas powered vehicles in the United Kingdom.

Lucy Budd is a lecturer in transport studies in the School of Civil and Building Engineering at Loughborough University where she specialises in commercial aviation operations at a variety of sites and scales. Lucy's work has appeared in leading academic and consumer journals, including *Progress in Human Geography*, *Transactions of the Institute of British Geographers*, *Environment and Planning A* and *Airliner World*, and she has contributed to national television and radio programmes on the subject of flight. Lucy currently sits on the international editorial board of the *Journal of Transport Policy*.

Lee Chapman is a senior lecturer in applied climatology at the University of Birmingham. His research interests lie at the interface of climatology and civil engineering covering the broad theme of the 'impact of weather and climate on the built environment'. This covers a range of topics from the development of 'climate-response' models under the baseline climate to assessing the future impact of climate change on hard infrastructure. In addition to transport and infrastructure, Lee has a keen interest in urban climatology and is the director of the Birmingham Urban Climate Laboratory. Lee now has over 40-refereed journal articles published in international peer-reviewed journals and has been invited to speak and chair sessions at numerous conferences and events across the world.

Alison Chisholm is a research associate in the Department of Planning at Oxford Brookes University and recently worked on the EPSRC-funded *Understanding Walking and Cycling* project. Her research interests are sustainable transport, particularly walking and cycling, social psychology and attitudes/behaviours relating to mobility, children's independent mobility, the effect of the built environment on mobility and the relationship between transport and public health. She holds a masters degree in health psychology and a PhD in sociology.

Kevin Cullinane is director of the Transport Research Institute and professor of international logistics at Edinburgh Napier University. He is fellow of the Chartered Institute of Logistics and Transport, visiting professor at the University of Gothenburg and honorary professor at the University of Hong Kong. He has been a transport adviser to the World Bank and the governments of Scotland, Ireland, Hong Kong, Egypt, Chile, Korea and the United Kingdom. Kevin has published 9 books and over 160-refereed journal and conference papers. He is associate editor of *Transportation Research A* and the *International Journal of Applied Logistics*. He also sits on the editorial boards of seven other journals and is a member of the civil and construction engineering sub-panel for REF 2014.

Robert Dingwall is director of Dingwall Enterprises, offering research and consulting services, and professor of sociology at Nottingham Trent University. He was previously founding director of the Institute for Science and Society at the University of Nottingham, leading a team investigating the social, ethical, legal and cultural implications of developments in science and technology. His interest in transport and climate change developed through work with the EPSRC FUTURENET consortium and with the RCUK Digital Economy Hub in Nottingham. He is currently working on issues in public engagement with strategic planning for technology, including transport.

Murray Goulden is a research fellow at the Horizon Digital Economy Research Institute at Nottingham University. His background is in STS and sociology. He is currently involved in the FUTURENET project, exploring the social dimensions of future UK transport infrastructure and work exploring the impact of smart grid technologies on social practices in the home. His research interests encompass transport and energy in context of climate change and new digital technologies.

Helen Harwatt is a research fellow at the Sustainability Research Institute, University of Leeds. Helen completed a PhD at the Institute for Transport Studies, exploring public response to a proposed Personal Carbon Trading scheme as a measure to significantly reduce carbon emissions from personal transport. Helen has contributed to a number of projects in the area of sustainable transport and is currently working on transition pathways towards a low carbon economy (funded by the Centre for Low Carbon Futures), including the transport sectors contribution to achieving this goal. Her research interests include the potential of sustainable behaviours to transform health and the environment, stimulating the uptake of pro-environmental behaviours and the rebound effects of sustainable behaviour adoption.

Robin Hickman is a senior lecturer at the Bartlett School of Planning, University College London and visiting research associate at the Transport Studies Unit, University of Oxford. He was previously research fellow at the TSU (2009–2011) and associate director and led on transport research at Halcrow (2002–2011). He has wide experience in transport planning, policy research, strategy development, urban planning and the project management of major multi-disciplinary studies. He is a specialist in transport and climate change issues, urban structure and travel, and integrated transport and urban planning strategies.

Elizabeth Hooper is a Doctoral Researcher at the University of Birmingham, researching the impacts of climate change on future transport networks as part of the FUTURENET project funded by the EPSRC, focusing particularly on the impacts of precipitation on the UK road network. Elizabeth studied BSc in Geography at the University of Birmingham and then went on to complete an MSc in Applied Meteorology and Climatology before commencing her PhD studies.

Dave Horton is a sociologist and writer based in Lancaster, UK. His research interests centre around the environment, culture, everyday life, politics, social movements and mobility. His main interest and passion, though, is cycling. Between 2008 and 2011 he contributed to a major UK research project, *Understanding Walking and Cycling*. Dave is a founder member of the Cycling and Society Research Group, and co-edited (with Paul Rosen and Peter Cox) *Cycling and Society* (Ashgate, 2007). He writes about cycling for both the academic and popular press, and blogs at http://thinkingaboutcycling.wordpress.com.

Candice Howarth is a research fellow in the Global Sustainability Institute at Anglia Ruskin University investigating a range of issues relating to climate action. Her PhD research investigated perceived barriers to sustainable travel behaviour and the role of climate change information in overcoming these. She has been involved in numerous projects relating to travel behaviour including the development of a social marketing and travel behaviour strategy for the Highways Agency. She has presented her research at numerous national and international conferences and won the first prize for her work at the Young European Arena of Research conference in 2010. She has a masters in climate change from the University of East Anglia and a bachelors degree in meteorology.

David Jaroszweski is a research fellow at the School of Civil Engineering at the University of Birmingham. David has published in the area of integrated climate change impact assessment for the transport sector. This work involves a multi-disciplinary approach which considers not just how the climate may change but also how the transport network itself may develop and how this will affect its vulnerability to weather. David completed a PhD at the University of Birmingham investigating the potential impacts of climate change on road freight transport in the United Kingdom. He is currently working on the EPSRC-funded Future Resilient Transport Network project (FUTURENET) which is a multimodal climate change impact assessment for UK transportation.

Tim Jones is a research fellow sharing his time between the Faculty of Technology Design and Environment at Oxford Brookes University and the Transport Studies Unit (TSU) at the University of Oxford. His research is primarily focused around walking and cycling for short journeys in urban areas and the application of multiple-methods and theories to understand taken-for-granted travel practices. Recent projects include the EPSRC-funded Understanding Walking and Cycling (UWAC) study (Oxford Brookes) and the EPSRC Impact of Constructing Non-Motorised Networks and Evaluating Changes in Travel (iConnect) project (TSU-Oxford). Tim is currently a member of the National Institute for Clinical Excellence (NICE) Programme Development Group for Walking and Cycling.

Ann Jopson is a research fellow at the Institute for Transport Studies, University of Leeds. Her expertise includes travel behaviour psychology, transport planning and policy (including land use), with a particular emphasis on attitudinal and behavioural measures, and social aspects of transport. More specifically this covers the role of psychology in enhancing analysis to expand our understanding of human reactions to land use and transport policies, with a particular regard to improving the effectiveness of sustainability measures. Ann has recently worked on the EPSRC-funded *Understanding Walking and Cycling* project.

Charlotte Kelly is a senior research fellow in the Institute for Transport Studies and Academic Unit for Health Economics at the University of Leeds. Charlotte has won and led research projects at an EU, National and Local Level and teaches on a range of Leeds University courses including Project Appraisal and Transport Investment Appraisal (which include the valuation of carbon benefits from transport projects). Her current research includes the areas of Multi-level Governance, Transport Policy and Carbon Emissions Management and patient travel for healthcare (carbon footprint and accessibility implications).

Karen Lucas is a senior research fellow with the Transport Studies Unit at Oxford University. For the past 10 years, she has pioneered a programme of research looking at ways to represent and address the local environmental concerns of people living in deprived areas, with a particular focus on transport and social exclusion. She was recently awarded a research fellowship by the Economic and Social Research Council of the United Kingdom to further her studies in this area. A particular current interest is in bringing equities and inclusion perspectives to the development of climate change policies for transport. Karen is also associate editor for the *Journal of Transport Geography* and a member of the editorial board of *Transportation*.

Glenn Lyons is professor of transport and society and founder of the Centre for Transport & Society (CTS) at the University of the West of England, Bristol (UWE). He was CTS director from 2002 to 2010 and is now associate dean (research and knowledge exchange) for the Faculty of Environment and Technology. His research has focused upon the role of new technologies in supporting and influencing travel behaviour and on attitudes towards transport. A former secondee and expert advisor to the Department for Transport's Transport Direct initiative, Glen has led major studies for the DfT and UK research councils into traveller information systems, teleworking, virtual mobility, travel time use, user innovation, road pricing and public and business attitudes to transport.

Greg Marsden is director of the Institute for Transport Studies at the University of Leeds. His research interests relate to the governance of transport and understanding how effective transport strategies are developed, delivered and communicated. He has led research on public attitudes to climate change for the Department for Transport and is currently leading a multi-institution Economic and Social Research Council project looking at the governance of climate change emissions in the transport sector. He joined Leeds after a two-year period as specialist adviser to the UK Parliamentary Transport scrutiny committee, is editor of transport policy and member of the Independent Transport Commission think tank.

Caroline Mullen is a research officer at the Institute for Transport Studies, University of Leeds. She works on ethics and political philosophy, governance and policy applied to transport, environment and health. Her PhD, from the University of Manchester, concerned the moral defensibility of the risk and harms caused by transport.

Kate Pangbourne is a post-doctoral social scientist in the Geography and Environment Department of the University of Aberdeen (Scotland). She gained a PhD in geography (environmental) there in 2010, having also gained a masters in sustainable rural development. Prior to that she spent a decade working in a public sector natural heritage conservation organization. Her current research interests focus on governance theory, collaboration processes, interdisciplinary communication in human-environment interaction research, climate change, peak oil, travel behaviour and the environmental and social sustainability of transport.

Colin G. Pooley is professor of social and historical geography in The Lancaster Environment Centre, Lancaster University, UK. His research focuses on the social geography of Britain and continental Europe since circa 1800, with recent projects focused on residential migration, travel to work and other aspects of everyday mobility. One recent project (*Understanding Walking and Cycling* (UWAC) funded by the EPSRC), which forms the basis for this chapter, examined everyday travel behaviour for short trips in urban areas, using both quantitative and qualitative methods to understand the reasons why people do (and do not) use sustainable means of travel (especially walking and cycling) for short trips in urban areas.

Tim Ryley is a senior lecturer in transport studies at Loughborough University. He has over 16 years transport research experience, with expertise in travel behaviour analysis, and transport and climate change issues, across a range of transport applications. Tim is currently leading a series of travel behaviour research projects, including the EPSRC (Engineering and Physical Sciences Research Council) funded 'The ABC project. Airports and Behavioural Change: towards environmental surface access travel' and 'DRT for DRT: Developing Relevant Tools for Demand Responsive Transport'. Tim is also responsible for a travel behaviour component within the EPSRC-funded FUTURENET (Future Resilient Transport Networks) project, including a major survey of travel under weather uncertainty.

Griet Scheldeman is a research associate in the Lancaster Environment Centre at Lancaster University, UK. Her research interests centre on perceptive and creative processes at play in the relationships between people and their environments. Since completing her PhD in social anthropology (St Andrews, 2006) she has published on adolescents' lives with insulin pumps, and on urban walking and cycling, reflecting on embodiment, everyday experience and sense-making. Most recently Scheldeman conducted the component of ethnographic research in *Understanding Walking and Cycling* (EPSRC, 2008–2011), a multi-disciplinary project investigating people's practices of sustainable transport across Britain. Funded by a postdoctoral scholarship from the Research Council of Norway (2011–2012) she is currently investigating creativity and improvisation in Arctic scientists' field practices.

Miles Tight is professor of transport, energy and environment at the School of Civil Engineering, University of Birmingham and has been researching in the area of sustainable transport for over 25 years. He has lead research funded by the Tyndall Centre for Climate Change Research on reducing transport carbon emissions in London and recently looked at local authority initiatives in this area for DfT. His current work is looking at the potential for large scale change in transport through EPSRC-funded projects Step-Change project (2011–2015) and 'Visions for the Role of Walking and Cycling in 2030' (2008–2012). He has recently completed an involvement in the EPSRC-funded *Understanding Walking and Cycling* project which ended in 2011.

Owen Waygood is assistant professor at the École supérieure d'aménagement du territoire et de développement régional (ÉSAD) at Université Laval, Quebec. At the time of writing, he was a research fellow at the Centre for Transport & Society (CTS) at the University of the West of England, Bristol. Owen's primary research at CTS focused on climate change information and transport behaviour. His work has looked at rational interpretations of carbon dioxide (CO_2), at how presentation can affect interpretation, as well as reported motivation to change behaviour. Owen completed a PhD in travel behaviour from Kyoto University, Japan that focused on children's independent travel, community connections and physical activity.

Alberto M. Zanni is a research associate at the Transport Studies Group, School of Civil and Building Engineering, Loughborough University. Alberto is an applied economist and his research interests broadly cover spatial, transport and environmental economics. His most recent research work has focused on the analysis of the acceptability and behavioural impact of personal carbon trading schemes, freight urban policy, acceptability of low carbon vehicles in the freight sector, social dimensions in individual choices, travel behaviour under uncertainty and ways to reduce carbon emissions from surface access to airport. Alberto previously studied and worked at the Universities of Modena (Italy), Montpellier (France), Madrid (Spain) and Imperial College London.

SUBJECT INDEX